METHODS IN MOLECULAR BIOLOGY

For further volumes:
http://www.springer.com/series/7651

For over 35 years, biological scientists have come to rely on the research protocols and methodologies in the critically acclaimed *Methods in Molecular Biology* series. The series was the first to introduce the step-by-step protocols approach that has become the standard in all biomedical protocol publishing. Each protocol is provided in readily-reproducible step-by-step fashion, opening with an introductory overview, a list of the materials and reagents needed to complete the experiment, and followed by a detailed procedure that is supported with a helpful notes section offering tips and tricks of the trade as well as troubleshooting advice. These hallmark features were introduced by series editor Dr. John Walker and constitute the key ingredient in each and every volume of the *Methods in Molecular Biology* series. Tested and trusted, comprehensive and reliable, all protocols from the series are indexed in PubMed.

Non-Ribosomal Peptide Biosynthesis and Engineering

Methods and Protocols

Edited by

Michael Burkart

Department of Chemistry and Biochemistry, University of California San Diego, La Jolla, CA, USA

Fumihiro Ishikawa

Faculty of Pharmacy, Kindai University, Higashi-Osaka, Osaka, Japan

Humana Press

Editors
Michael Burkart
Department of Chemistry and Biochemistry
University of California San Diego
La Jolla, CA, USA

Fumihiro Ishikawa
Faculty of Pharmacy
Kindai University
Higashi-Osaka, Osaka, Japan

ISSN 1064-3745 ISSN 1940-6029 (electronic)
Methods in Molecular Biology
ISBN 978-1-0716-3213-0 ISBN 978-1-0716-3214-7 (eBook)
https://doi.org/10.1007/978-1-0716-3214-7

This Humana imprint is published by the registered company Springer Science+Business Media, LLC, part of Springer Nature.
The registered company address is: 1 New York Plaza, New York, NY 10004, U.S.A.

Preface

Nonribosomal peptide synthetases (NRPSs) are large multienzyme proteins that generate peptide natural products or their derivatives with large structural and functional diversity. Many of these molecules provide critical therapeutics that are key to modern medicine and agriculture. Common clinical examples include the antibiotic vancomycin, the antitumor agent bleomycin, and the immunosuppressant cyclosporine A, among many others. Not only do these pathways provide these critical existing medicines, but a significant percentage of new drugs continues to be derived from or are analogs of this class of natural products.

Almost all natural product-derived medicines today are manufactured via microbial fermentation, including semi-synthetic analogs that require chemical modifications on top of chemistry installed by a secondary metabolic pathway. This is because nature is still the best synthetic chemist – most of these compounds are so large or complex, total synthesis would not make economic sense for commercial production. These molecules are tailor-made by sophisticated molecular machines like NRPSs, and tailoring enzymes are commonly included in biosynthetic gene clusters (BCGs), offering the molecular complexity we see in these natural products.

Natural product scientists have long wished to harness the biosynthetic potential found in BCGs. The fundamentals of these elegant synthases are deceptively straightforward, and early attempts to control and modify these pathways often led to mixed results. Many technical challenges, including the sheer size of the genes, the codon usage, and specialized translational machinery found in host organisms, among many others, add up to make this research area exceedingly challenging. Nevertheless, hard-won efforts over the last 25 years have led to significant achievements resulting in BCG discovery, pathway elucidation, structural analysis, and pathway reprogramming. Fortunately for these researchers, there have been concomitant advances in tools and equipment that have enabled these achievements and also offered ancillary benefits, such as increased throughput.

Take for instance DNA sequencing and synthesis. In 1997, the prospect of sequencing an entire microbial genome was pure science fiction for most researchers, and DNA synthesis was limited mostly to small primers. Today a genome sequence can be accomplished by mail-in services within one week for a few hundred dollars; and DNA can be synthesized for pennies per base pair, with whole genes available in expression constructs within weeks. (For anyone old enough to have performed their own Sanger sequencing or gene synthesis by primer overlap extension, this is certainly a brave new world!) Similar improvements in structural biology technologies, from protein NMR to cryo-EM, have become cheaper and more accessible, even to non-expert users. All of life science has benefitted from these advancements, and we have seen the impacts in natural product research.

This book is focused on highlighting the practice of new technology as it applies to NRPSs and related carrier protein-dependent synthases, including polyketide synthases (PKS) and fatty acid synthases (FAS). Given rapid advancement of scientific research, there is sometimes the need to delve deeper into recent publications, beyond what is written in Materials and Methods descriptions. Here we have compiled chapters dedicated to the practical application of research, from 18 laboratories at the cutting edge of this new science in natural product research. We have endeavored to highlight the diversity of the field, with examples from multiple sub-disciplines, including enzymology, structural biology, proteomics, chemical biology, natural product chemistry, and bioinformatics. We have also tried to capture some techniques that have not yet been demonstrated on NRPS pathways, but have the potential to be applied in the near future.

We organized this compilation to assist those researchers interested in learning about new techniques and applications, and we hope that these detailed procedures will assist in new technology transfer. While this collection is not exhaustive, and doubt some topics have been missed, we expect that these methods can benefit researchers new to the field, as well as those wishing to adopt a new methodology into their research repertoire. We hope that you will find it useful! We wish to thank each of the chapter authors for their contributions to this book.

Thanks for reading!

La Jolla, CA, USA *Michael Burkart*
Osaka, Japan *Fumihiro Ishikawa*

Contents

Contributors

NADYA ABBOOD • *Max-Planck-Institute for Terrestrial Microbiology, Department of Natural Products in Organismic Interactions, Marburg, Germany; Molecular Biotechnology, Department of Biosciences, Goethe University Frankfurt, Frankfurt am Main, Germany*

DAVID F. ACKERLEY • *School of Biological Sciences, Victoria University of Wellington, Wellington, New Zealand; Centre for Biodiscovery, Victoria University of Wellington, Wellington, New Zealand*

SYED FARDIN AHMED • *Department of Structural Biology, University at Buffalo, SUNY, Buffalo, NY, USA*

NEERU ARYA • *Department of Biophysics and Biophysical Chemistry, Johns Hopkins University School of Medicine, Baltimore, MD, USA*

CHRISTOPHER N. BODDY • *Department of Chemistry and Biomolecular Sciences, University of Ottawa, Ottawa, ON, Canada*

HELGE B. BODE • *Max-Planck-Institute for Terrestrial Microbiology, Department of Natural Products in Organismic Interactions, Marburg, Germany; Molecular Biotechnology, Department of Biosciences, Goethe University Frankfurt, Frankfurt am Main, Germany; Philipps-Universität Marburg, Chemische Biologie, Marburg, Germany; Senckenberg Gesellschaft für Naturforschung, Frankfurt, Germany*

KENAN A. J. BOZHUEYUEK • *Max-Planck-Institute for Terrestrial Microbiology, Department of Natural Products in Organismic Interactions, Marburg, Germany; Molecular Biotechnology, Department of Biosciences, Goethe University Frankfurt, Frankfurt am Main, Germany; Myria Biosciences AG, Basel, Switzerland*

JORDAN T. BRAZEAU-HENRIE • *Department of Chemistry and Biomolecular Sciences, University of Ottawa, Ottawa, ON, Canada*

ALISTAIR S. BROWN • *School of Biological Sciences, Victoria University of Wellington, Wellington, New Zealand; Centre for Biodiscovery, Victoria University of Wellington, Wellington, New Zealand*

MICHAEL BURKART • *Department of Chemistry and Biochemistry, University of California, San Diego, La Jolla, CA, USA*

MAX J. CRYLE • *Monash Biomedicine Discovery Institute, Department of Biochemistry and Molecular Biology, Monash University, Clayton, VIC, Australia*

JULIA DIECKER • *University of Münster, Institute of Biochemistry, Münster, Germany*

WOLFGANG DÖRNER • *University of Münster, Institute of Biochemistry, Münster, Germany*

MATTHIEU DUBAN • *Université de Lille, UMRt BioEcoAgro 1158-INRAe, Métabolites secondaires d'origine microbienne, Charles Viollette Institute, Lille, France*

TADASHI EGUCHI • *Department of Chemistry, Tokyo Institute of Technology, Tokyo, Japan*

ARESKI FLISSI • *Université de Lille, CNRS, Centrale Lille, UMR 9189 CRIStAL, Lille, France*

DOMINIQUE P. FRUEH • *Department of Biophysics and Biophysical Chemistry, Johns Hopkins University School of Medicine, Baltimore, MD, USA*

KEI FUJITA • *Faculty of Pharmaceutical Sciences, Hokkaido University, Sapporo, Japan*

YUKI GOTO • *Department of Chemistry, Graduate School of Science, The University of Tokyo, Tokyo, Japan*

ANDREW M. GULICK • *Department of Structural Biology, University at Buffalo, SUNY, Buffalo, NY, USA; Department of Structural Biology, Jacobs School of Medicine & Biomedical Sciences, Buffalo, NY, USA*

YOSHIMITSU HAMANO • *Graduate School of Bioscience and Biotechnology, Fukui Prefectural University, Fukui, Japan; Fukui Bioincubation Center (FBIC), Fukui Prefectural University, Fukui, Japan*

Y. T. CANDACE HO • *Monash Biomedicine Discovery Institute, Department of Biochemistry and Molecular Biology, Monash University, Clayton, VIC, Australia*

FUMIHIRO ISHIKAWA • *Faculty of Pharmacy, Kindai University, 3-4-1 Kowakae, Higashi-Osaka, Osaka, Japan*

PHILIPPE JACQUES • *Université de Liège, UMRt BioEcoAgro 1158-INRAe, Métabolites secondaires d'origine microbienne, TERRA Teaching and Research Centre, Gembloux Agro-Bio Tech, Gembloux, Belgium*

MASAKAZU KOBAYASHI • *Faculty of Pharmaceutical Sciences, Hokkaido University, Sapporo, Japan*

FUMITAKA KUDO • *Department of Chemistry, Tokyo Institute of Technology, Tokyo, Japan*

VALÉRIE LECLÈRE • *Université de Lille, UMRt BioEcoAgro 1158-INRAe, Métabolites secondaires d'origine microbienne, Charles Viollette Institute, Lille, France*

YUEYING LI • *Department of Chemistry, Texas A&M University, College Station, TX, USA*

MONICA R. MACDONALD • *Department of Structural Biology, University at Buffalo, SUNY, Buffalo, NY, USA*

KENNETH A. MARINCIN • *Department of Biophysics and Biophysical Chemistry, Johns Hopkins University School of Medicine, Baltimore, MD, USA*

CHITOSE MARUYAMA • *Graduate School of Bioscience and Biotechnology, Fukui Prefectural University, Fukui, Japan; Fukui Bioincubation Center (FBIC), Fukui Prefectural University, Fukui, Japan*

KENICHI MATSUDA • *Faculty of Pharmaceutical Sciences, Hokkaido University, Sapporo, Japan; Global Station for Biosurfaces and Drug Discovery, Hokkaido University, Sapporo, Japan*

DESIRAE A. MELLOR • *Department of Chemistry and Biochemistry, University of California, San Diego, La Jolla, CA, USA*

AKIMASA MIYANAGA • *Department of Chemistry, Tokyo Institute of Technology, Tokyo, Japan*

HENNING D. MOOTZ • *University of Münster, Institute of Biochemistry, Münster, Germany*

JEREMY G. OWEN • *School of Biological Sciences, Victoria University of Wellington, Wellington, New Zealand; Centre for Biodiscovery, Victoria University of Wellington, Wellington, New Zealand*

ANDRÉ R. PAQUETTE • *Department of Chemistry and Biomolecular Sciences, University of Ottawa, Ottawa, ON, Canada*

KETAN D. PATEL • *Department of Structural Biology, University at Buffalo, SUNY, Buffalo, NY, USA*

LEONARD PRÄVE • *Max-Planck-Institute for Terrestrial Microbiology, Department of Natural Products in Organismic Interactions, Marburg, Germany; Molecular Biotechnology, Department of Biosciences, Goethe University Frankfurt, Frankfurt am Main, Germany*

MAUDE PUPIN • *Université de Lille, CNRS, Centrale Lille, UMR 9189 CRIStAL, Lille, France*

JENNIFER RÜSCHENBAUM • *University of Münster, Institute of Biochemistry, Münster, Germany*

JAVIER O. SANLLEY • *Department of Chemistry and Biochemistry, University of California, San Diego, La Jolla, CA, USA*
HIROAKI SUGA • *Department of Chemistry, Graduate School of Science, The University of Tokyo, Tokyo, Japan*
JULIEN TAILHADES • *Monash Biomedicine Discovery Institute, Department of Biochemistry and Molecular Biology, Monash University, Clayton, VIC, Australia*
GENZOH TANABE • *Faculty of Pharmacy, Kindai University, 3-4-1 Kowakae, Higashi-Osaka, Osaka, Japan*
TOSHIYUKI WAKIMOTO • *Faculty of Pharmaceutical Sciences, Hokkaido University, Sapporo, Japan; Global Station for Biosurfaces and Drug Discovery, Hokkaido University, Sapporo, Japan*
LAUREN A. WASHBURN • *Department of Chemistry, Texas A&M University, College Station, TX, USA*
CORAN M. H. WATANABE • *Department of Chemistry, Texas A&M University, College Station, TX, USA*
YONGWEI ZHAO • *Monash Biomedicine Discovery Institute, Department of Biochemistry and Molecular Biology, Monash University, Clayton, VIC, Australia*

Part I

Background and Overview

Chapter 1

The Assembly-Line Enzymology of Nonribosomal Peptide Biosynthesis

Chitose Maruyama and Yoshimitsu Hamano

Abstract

Peptide natural products constitute a major class of secondary metabolites produced by microorganisms (mostly bacteria and fungi). In the past several decades, researchers have gained extensive knowledge about nonribosomal peptides (NRPs) generated by ribosome-independent systems, namely, NRP synthetases (NRPSs). NRPSs are multifunctional enzymes consisting of semiautonomous domains that form a peptide backbone. Using a thiotemplate mechanism that employs assembly-line logic with multiple modules, NRPSs activate, tether, and modify amino acid building blocks, sequentially elongating the peptide chain before releasing the complete peptide. Adenylation, thiolation, condensation, and thioesterase domains play central roles in these reactions. This chapter focuses on the current understanding of these central domains in NRPS assembly-line enzymology.

Key words Biosynthesis, NRPS, Peptide, Amide bond, Microorganism

1 Nonribosomal Peptides (NRPs)

Peptide natural products constitute a major class of secondary metabolites produced by microorganisms (mostly bacteria and fungi). Many are used as bioactive compounds with a wide range of applications in medicine, agriculture, and biochemical research. In their biosynthesis, amino acid building blocks are linked via amide bonds (peptide bonds), and this linking process is mediated by ribosomal or ribosome-independent systems. In the past several decades, researchers have gained extensive knowledge about non-ribosomal peptides (NRPs) generated by ribosome-independent systems. Microorganisms employ three types of biosynthetic enzymes to synthesize NRPs: ATP-grasp ligases (also referred to as ATP-dependent carboxylate-amine/thiol ligases) [1], tRNA-dependent amide bond-forming enzymes [2, 3], and NRP synthetases (NRPSs) [4–7]. Unlike in the synthesis of ribosomal peptides, during NRP synthesis, the ATP-grasp ligases and NRPSs can accept

Michael Burkart and Fumihiro Ishikawa (eds.), *Non-Ribosomal Peptide Biosynthesis and Engineering: Methods and Protocols*, Methods in Molecular Biology, vol. 2670, https://doi.org/10.1007/978-1-0716-3214-7_1,

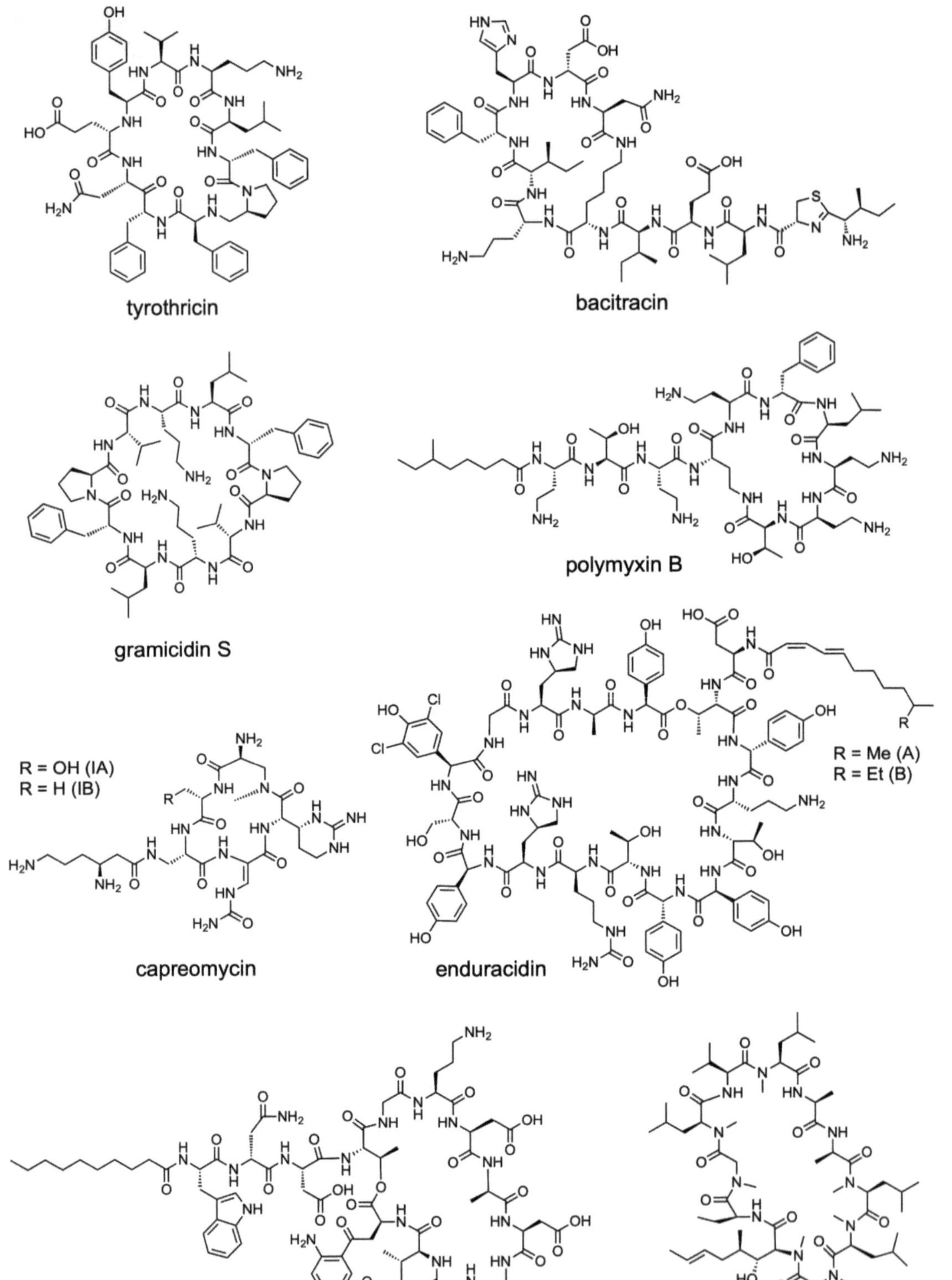

Fig. 1 Chemical structures of marketed NRP drugs. Tyrothricin is an antibacterial (Gram-positive bacteria) drug produced by *Bacillus brevis*. Bacitracin is an antibacterial (Gram-positive bacteria) drug produced by *Bacillus subtilis*. Gramicidin S is an antibacterial (Gram-positive and Gram-negative bacteria and some fungi) drug

unnatural and nonproteinogenic amino acid building blocks as substrates, thereby achieving greater structural diversity. Such structural diversity generates NRPs with various biological actions, and these NRPs have been applied to drug development. For example, tyrothricin (produced by *Bacillus brevis*), bacitracin (*Bacillus subtilis*), gramicidin S (*B. brevis*), polymyxin B (*Bacillus polymyxa*), capreomycin (*Streptomyces capreolus*), enduracidin (*Streptomyces fungicidicus*), and daptomycin (*Streptomyces roseosporus*) are all used as antibacterial drugs (Fig. 1), while cyclosporine A (*Tolypocladium inflatum*) is a potent immunosuppressant (Fig. 1).

2 NRP Biosynthesis

2.1 NRPS Architecture and Its Gene

The producers of NRPS-based compounds are mostly bacteria and fungi. In these microorganisms, the genes required for the biosynthesis of a secondary metabolite are clustered. NRPS genes, such as the biosynthetic gene clusters for polyketides and terpenes, are also often the case. NRPSs have a modular organization (Fig. 2). One module is a section of the NRPS architecture responsible for the elongation of a single peptide. The modules can be further subdivided into domains representing the enzymatic units that catalyze the individual steps of nonribosomal peptide synthesis. Adenylation (A) domains select amino acid building blocks (substrate) and activate them as aminoacyl-*O*-AMPs. The resulting building blocks are subsequently loaded onto the 4′-phosphopantetheine (4′-PP) arm of the adjacent thiolation (T) domain with AMP release, thereby resulting in the formation of an aminoacyl-*S*-enzyme. The T domain is also referred to as a peptidyl carrier protein (PCP) domain. The 4′-PP is post-translationally transferred from Coenzyme A to a conserved serine residue of the PCP by associated phosphopantetheinyl transferases (PPTase), generating the holoform of the T domain. Condensation (C) domains catalyze the formation of a peptide bond between two amino acid building blocks activated as the aminoacyl-*S*-enzymes. These three domains, A, T (PCP), and C, are the basic components of one module. Therefore, an NRPS consisting of, for example, four modules produces a tetrapeptide compound, and its peptide elongation reactions occur on the T domain scaffolds; the peptide synthesis is

Fig. 1 (continued) produced by *Bacillus brevis*. Polymyxin B is an antibacterial (Gram-negative bacteria) drug produced by *Bacillus polymyxa*. Capreomycin is an antituberculous drug produced by *Streptomyces capreolus*. Enduracidin (enramycin) is an antibacterial feed additive produced by *Streptomyces fungicidicus*. Daptomycin is an antibacterial (Gram-positive bacteria) drug produced by *Streptomyces roseosporus*. Cyclosporine A is an immunosuppressive drug produced by *Tolypocladium inflatum*

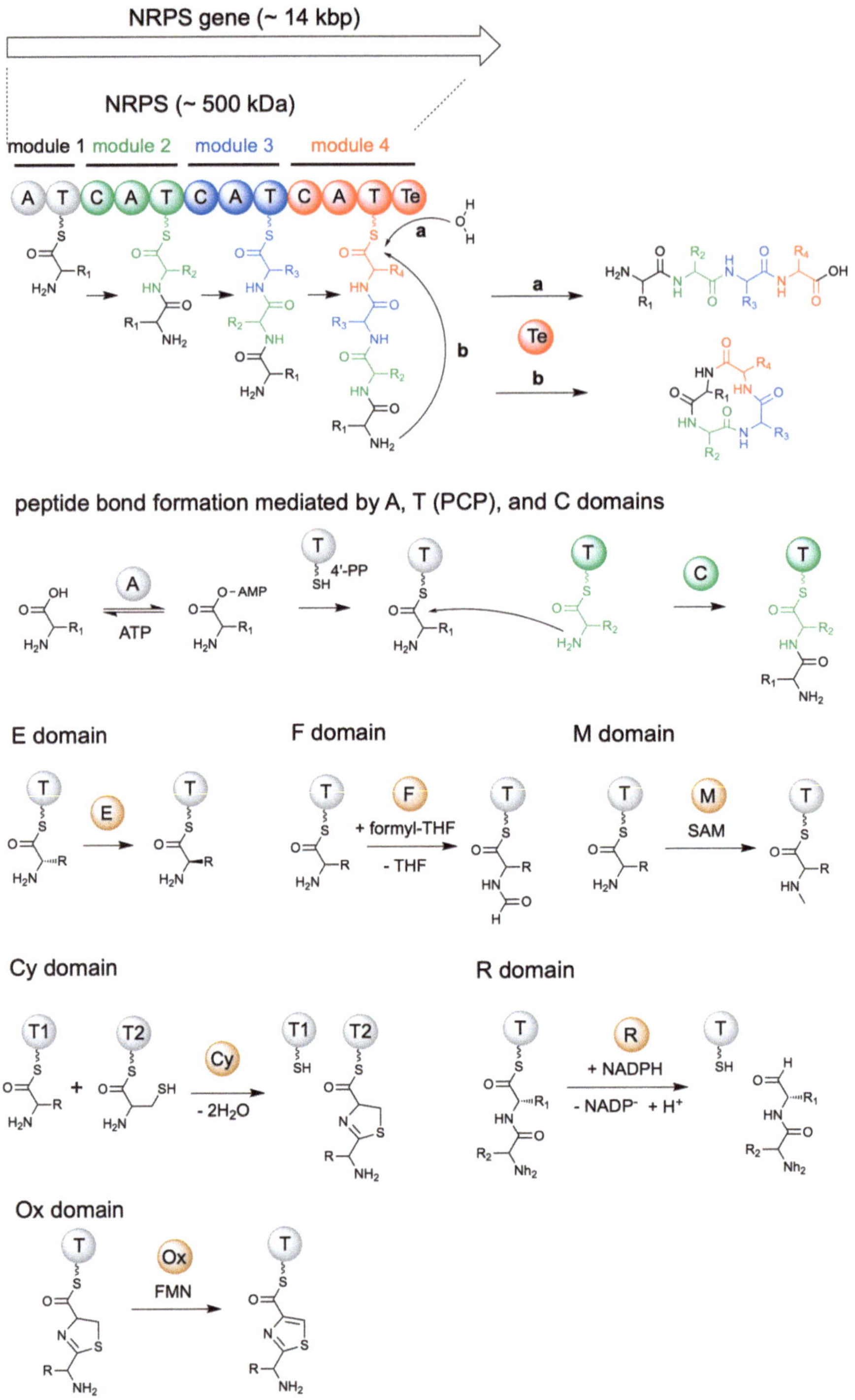

Fig. 2 Domain architecture of a typical linear NRPS and function of the catalytic domains

performed along the NRPS modules as a template (this is the so-called thiotemplate mechanism). The amino acid sequence of a final peptide product is determined by the substrate specificities of the A domains.

In general, an NRPS polypeptide with a size of 500 kDa encoded by an approximately 14-kbp-long gene is needed to produce the tetrapeptide (Fig. 2). Bacitracin (Fig. 1) is the branched cyclic decapeptide. Three NRPS mega-enzymes, BacA (598 kDa), BacB (297 kDa), and BacC (743 kDa), participate in its decapeptide biosynthesis, and twelve modules mediate the peptide chain growth [6]. During the peptide elongation along the NRPS template, the growing chain is often modified by optional domains, including epimerization (E), formylation (F), methylation (M), hetero-cyclization (Cy), reduction (R), and oxidation (Ox) domains (Fig. 2). Finally, a thioesterase (Te) domain cleaves the mature oligopeptide from the NRPS machinery and often mediates macrocyclization during this release step.

2.2 A Domains

A domain in each module selects an amino acid building block (substrate) and activates it as an aminoacyl-*O*-AMP. The aminoacyl-*O*-AMP is subsequently transferred to the 4′-PP arm of the adjacent T domains *via* a thioester linkage; the A domain itself catalyzes these sequential reactions. Based on the primary structure and the catalytic function, the A domain belongs to the ANL (*A*cyl-CoA synthetases, *N*RPS adenylation domains, and *L*uciferase enzymes) superfamily of adenylating enzymes [8]. A detailed comparison of the conserved regions in the NRPS adenylation domains revealed ten conserved regions named A1–A10. These consensus motifs have been rationalized by several X-ray structures and play structural and functional roles [5]. The Asp (A4 motif) and Lys residues (A10 motif) are highly conserved and stabilize the amino and carboxylate groups of an amino acid substrate, respectively. In addition, eight amino acid residues are involved in side-chain recognition, which has led to the establishment of a specificity-conferring code for A domains. Bioinformatic tools have since been developed for the prediction of potential substrates [9], thereby allowing estimation of the product structures of NRPSs in genome mining approaches. Recent studies have suggested that the structural association of the C domain affects the substrate specificity of its neighboring A domain [5]. A more recent investigation presented evidence that any substrate-specifying role for C domains is likely to be the exception rather than the rule and that novel NRPs can be generated by the substitution of A domains alone [10].

In the catalytic process, the conformational changes of the small *C*-terminal subdomain (A_{sub}) are involved in loading and unloading substrates [11]. Initially, the A domain forms an open conformation in which A_{sub} is oriented away from the active site, allowing the binding of an amino acid substrate and ATP in the large *N*-terminal core domain (A_{core}). In the following step, the A domain adopts a catalytically active conformation in which A_{sub} is closed on the active site. After the aminoacyl adenylation, the A domain forms a third conformation in which A_{sub} rotates by 140°, opening space for the 4′-PP arm of the T domain. Then, the A domain loads the amino acid onto the T domain and returns to the initial open state.

In addition to the A domains catalyzing typical enzymatic reactions (adenylation and thiolation), bifunctional A domains have been found and reported. In pacidamycin biosynthesis [12], PacU is a unique stand-alone A domain that can condense diaminobutyric acid-*S*-PacH (T domain) and a freely diffusible alanine molecule after activation as alanyl-*O*-AMP by the PacU adenylating function (Fig. 3a). The stand-alone A domains NovL (Fig. 3b) and CouL (Fig. 3c), which are involved in the biosynthesis of novobiocin and coumermycin, respectively, are known to have such an amide-bond forming activity [13, 14]. The stand-alone A domain ORF19 (Fig. 3d) participates in the biosynthesis of the L-β-lysine oligopeptide chain of streptothricin [15]. ORF19 activates L-β-lysine as L-β-lysyl-*O*-AMP and subsequently mediates an amide-bond formation between L-β-lysyl-*O*-AMP and L-β-lysine-*S*-rORF18 (T domain). More interestingly, using L-β-lysyl-*O*-AMP as extending units, ORF19 can iteratively catalyze the amide-bond formation to grow the L-β-lysine oligopeptide chain on the L-β-lysine-*S*-rORF18 scaffold [15].

Alternatives A domains that require MbtH-like proteins for their adenylation activities have been reported. Ordinarily, the *mbtH*-like protein genes are found in NRPS gene clusters. Gene knock-out experiments have suggested that MbtH-like proteins are often required for the efficient production of NRPs [16, 17]; MbtH-like proteins from other NRPS gene clusters can partially complement each other for the production of the NRPs [17]. Such complementation has even been observed in vitro for MbtH-like proteins of different gene clusters [18]; for example, the activity of the A domain (NovH) for novobiocin biosynthesis was markedly enhanced by the presence of CloY, which is an MbtH-like protein used in clorobiocin biosynthesis.

To investigate the substrate specificity of an A domain, the A domain under consideration is often overexpressed in a heterologous host strain and purified. Traditionally, the enzymatic activity of A domains has been measured by radioactive ATP–[^{32}P]-PPi exchange assays through the detection of ^{32}P-labeled ATP produced by a reversible reaction of the A domain [19, 20]. Following the publication of a nonradioactive high-throughput assay for the screening and characterization of A-domains [21], several research groups reported nonradioactive assay systems using a colored molybdopyrophosphate complex [22, 23].

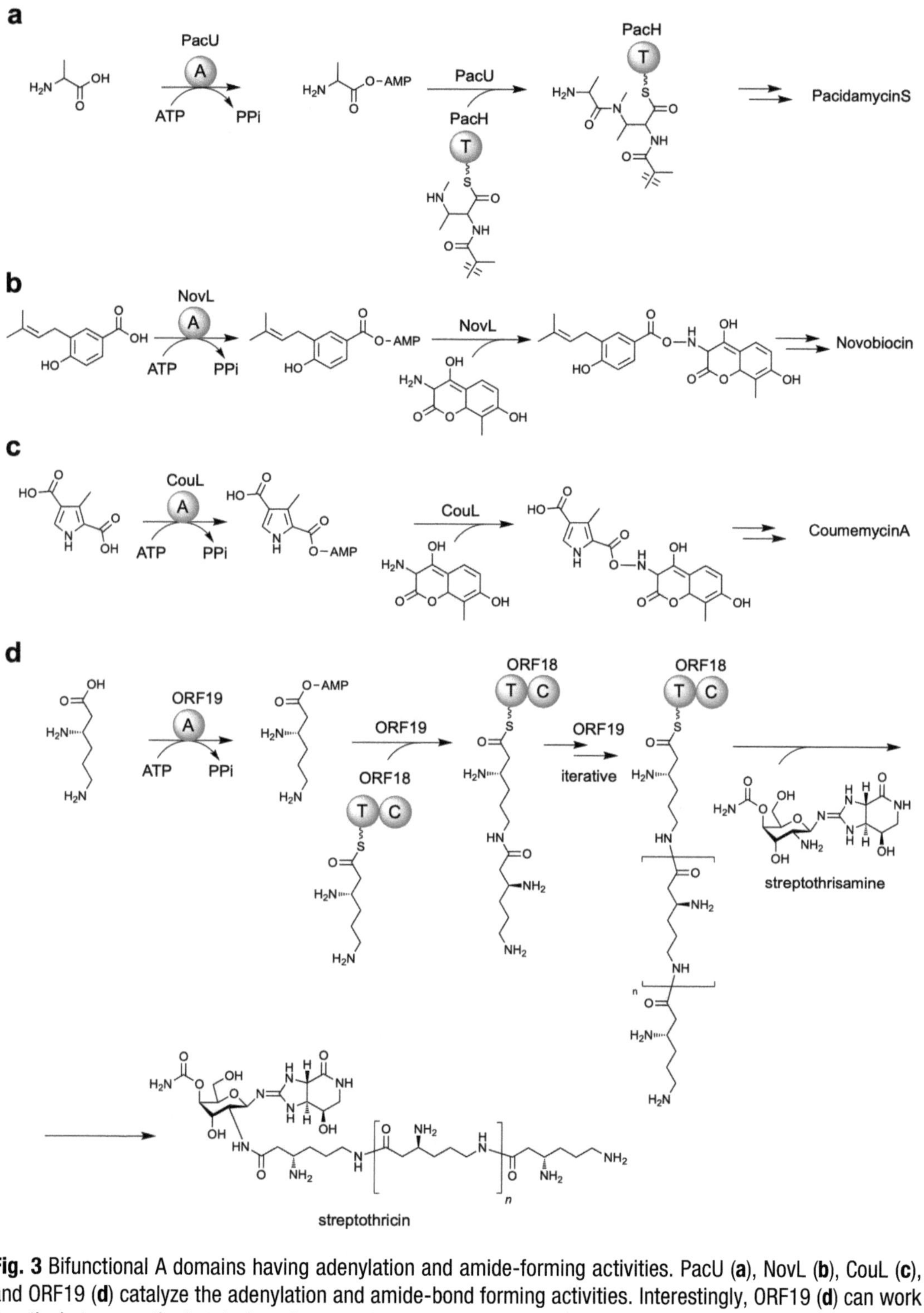

Fig. 3 Bifunctional A domains having adenylation and amide-forming activities. PacU (**a**), NovL (**b**), CouL (**c**), and ORF19 (**d**) catalyze the adenylation and amide-bond forming activities. Interestingly, ORF19 (**d**) can work iteratively to grow the L-β-lysine oligopeptide chain on the L-β-lysine-*S*-rORF18 scaffold

2.3 T Domains

T domains (~80 amino acid proteins) play an important role in delivering an activated amino acid building block, and their function as the flexible robot arm of the NRPS assembly line is an essential requirement for the communication of NRPS domains [5, 24]. The T domain forms a four-helix bundle, with the N terminus of the second helix α2 harboring the highly conserved serine residue (GxxS core motif) that is post-translationally modified with the 4′-PP arm. This modification reaction is mediated by a phosphopantetheine transferase (PPTase); the PPTase converts the apo-T domain to the active holo form by the covalent attachment of a 4′-PP moiety from coenzyme A (CoA) onto the conserved serine residue in the helix α2. Due to their ability to attach CoA substrates directly onto the T domain, PPTases have been utilized in biotechnological applications, including for the attachment of unnatural substrates onto T domains. The PPTase from *Bacillus subtilis*, Sfp, which shows wide substrate specificities with respect to both the T domain and CoA substrates, is a representative example.

Initial NMR studies of T domains have suggested the absence of a distinct substrate-binding pocket and ruled out any substrate specificity of the T domain [5, 25]. However, given the current body of knowledge about T domains, T domains have the potential to exhibit at least some degree of substrate selectivity [5, 24, 26]. To detect the loading of an amino acid substrate onto the 4′-PP arm, autoradiography with ^{3}H- or ^{14}C-labeled amino acid compounds has often been used [27–30]. Peptide mapping of the T domain by high-performance liquid chromatography and mass spectrometry (HPLC-MS) is an alternative method. Finally, the use of high-resolution MS and an HPLC column for protein analysis enables the direct detection of the loading event occurring in T domains [15].

2.4 C Domains

C domains (~50 kDa) catalyze the central coupling reaction to grow the peptide chain on T domain. The C domain is a V-shaped pseudodimer of an N-terminal (C_{NTD}) and a C-terminal subdomain (C_{CTD}) [5]. As members of the chloramphenicol acetyltransferase (CAT) superfamily, the C_{NTD} and C_{CTD} subdomains form a central cleft at their interface, which the donor and acceptor 4′-PP arms have to penetrate from opposite sides to reach the conserved active-site motif HHxxxDG [31]. The second histidine in this motif has been found to act as the general base to promote nucleophilic attack of the α-amino group on the thioester or to stabilize the tetrahedral transition state [5, 32–36]. A comparison of C domain structures suggests that there are opening and closing dynamics between the C_{NTD} and C_{CTD} lobes; the conserved active site is located at the joint of these subdomains. There is debate regarding whether C domains may function as a secondary checkpoint, wherein the C domain must bind the correct substrates in the active site before catalyzing peptide bond formation in addition to creating specific protein–protein interactions with the appropriate donor and acceptor T domains. C domains have three major interaction partners, the intramodule A and T domains as well as the donor T domain of the upstream module. The C-terminally located A domain forms an interface with the C_{CTD} subdomain, which varies in size depending on the catalytic state of the module.

Some C domains have been characterized as bifunctional. In the biosynthesis of arthrofactin produced by a *Pseudomonas* strain, the NRPS modules activate L amino acids and epimerize them as covalently tethered pantetheinyl thioesters. The condensation/epimerization (C/E) domains have both peptide bond-forming condensation and upstream amino-acyl-*S*-pantetheinyl donor epimerization activities [37]. The presence of C/E domains was further generalized to all NRPSs assembling lipopeptides for bacteria belonging to the genera *Pseudomonas*, *Burkholderia* [38], and *Xanthomonas* [39]. Interestingly, NRPSs for lipopeptide seem to have dual C/E domains, although other NRPSs have E domains. C, C/E, and E domains have the conserved active-site motif HHxxxDG.

C domains may be replaced by heterocyclization (Cy) domains, which catalyze both the peptide bond formation and subsequent cyclization of Cys, Ser, and Thr to form a thiazoline, oxazoline, or methyloxazoline ring, respectively (Fig. 4) [40]. These heterocyclic rings are important for the bioactivity of NRPs such as bacitracin A [41], bleomycin [42], argyrin [43], yersiniabactin [44], and colibactin [45]. Cy domains were first observed in bacitracin synthetase [41] and catalyze two separate reactions. First, the Cy domain catalyzes condensation between the aminoacyl/peptidyl-PCP donor and the serinyl/threoninyl/cysteinyl-T domain acceptor. Then, it catalyzes a two-step cyclodehydration between the thiol or hydroxyl of the side chain and the carbonyl of the newly formed amide bond (Fig. 4) [40]. Cy domains have an extremely conserved

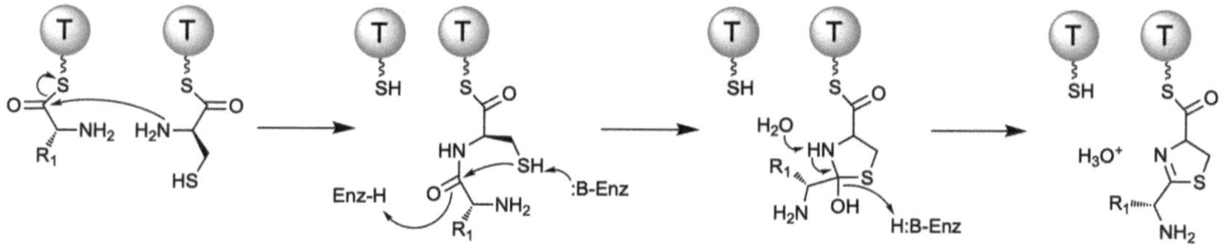

Fig. 4 Cyclization reaction (thiazoline) catalyzed by the Cy domain. The initial condensation reaction is performed by the nucleophilic attack of the donor substrate carbonyl by the α-amino group of the acceptor substrate. The thiol of the acceptor cysteine sidechain (or the hydroxyl of the serine and threonine sidechains) is deprotonated by the enzyme to allow an attack on the carbonyl of the newly formed amide bond. This is followed by a dehydration reaction, resulting in a thiazoline (or (methyl) oxazoline ring) formation

DxxxxD motif in the place of the HHxxxDG [41]. Mutation of the aspartate residues diminishes or abolishes catalytic activity [40], suggesting that the aspartate residues play an important role in catalysis or the structure of the active site. In addition, a triple mutation that replaces the DxxxxD motif with HHxxxDG also abolishes both condensation and heterocyclization [40]. This result showed that Cy domains cannot be transformed into C domains simply by interchanging these active site motifs, although they are evolutionarily related.

Generally, C domains catalyze the condensation between substrates tethered to the T domains, and then the peptide product is released from the NRPS scaffold by the Te domain catalysis. However, some C domains were found to form an amide bond between an amino acid substrate on the T domain and a freely diffusible acceptor substrate, thereby releasing the peptide product. For example, the C domain of ORF 18 catalyzes the condensation of L-β-lysine oligopeptides covalently bound to ORF 18 with a freely diffusible intermediate (streptothrisamine) to release the ST products (Fig. 3d). Interestingly, ORF18 has an HQxxxDM sequence instead of HHxxxDG.

2.5 Te Domains

Once NRP synthesis is completed, the T domain of a termination module transfers the mature peptide to the Ser or Cys residue (GxS/CxG core motif) of the Te domain (~30 kDa), which is located at the C-terminal and catalyzes peptide release from the NRPS scaffold (Fig. 2) [5]. Peptide release from the Te domain occurs either by hydrolysis (water as a nucleophile) or aminolysis (amine as a nucleophile), thereby liberating linear products [5]. Furthermore, Te domains often function as cyclases by constraining the peptide's conformation such that it undergoes intramolecular formation of a lactone or lactam [5]. TE domains belong to the α/β hydrolase family of enzymes and have a Ser-His-Asp catalytic triad. They also have a 40-residue lid region that lines the substrate and alternates between open and closed states [46]. TE domains are further classified into type I, which hydrolyze a mature peptide from the T domain using diverse catalytic strategies, and type II TE domains that recognize and hydrolyze PCPs with incorrectly loaded cargo that can stall the biosynthetic machinery [46].

3 Outlook

Over the last two decades, researchers have gained a much better understanding of how NRPSs work. However, success stories in NRPS engineering are still rare. Domain- or module-swapping approaches have been used to generate and modify peptide structures. A domains were targeted first due to their role in substrate recognition. A landmark success was the substitution of the A domain in the surfactin NRPS in *Bacillus subtilis* [47]. The targeted exchange of the leucine-activating module in SrfA-A resulted in the production of five new surfactin derivatives. This result indicated that domain swapping is possible in principle, but unexpected obstacles, such as intermodular communications and downstream specificity filters, can compromise engineering efficiency. Recent studies, and especially structural biological studies employing high-resolution structural data, have provided comprehensive insight into the various protein–protein interactions between the T domain and its partner enzymes, including the PPTase, A domain, C domain, TE domain, and other tailoring enzymes within the NRPS [4, 5, 24].

Following the first X-ray crystal structure analysis of the T-A domain complex (EntB-EntE) in enterobactin NRPS [48], the T-A domain interactions were investigated in EntF A-T domains (enterobactin) [49], LgrA A-T domains (gramicidin) [11], PltF A domain-PltL T domain (pyoluteorin) [50, 51], and HitB A domain-HitD T domain (hitachimycin) [52]. In these experiments, the adenosine vinylsulfonamide (AVS) inhibitor was used as a chemical probe that mimicked the substrate of the aminoacyl-adenylate intermediate. Although these significant data are helpful in understanding the A-T domain interaction, the mechanism of T-A domain binding remains unclear. Further analysis to elucidate the dynamic status of the PCP-A domain interface will also assist in the efforts to perform A domain substitutions.

The earliest study that structurally analyzed the protein–protein interactions of C domains was performed using SrfA-C of the surfactin NRPS [53]. The SrfA-C acceptor T-C domain structure revealed a protein–protein interface formed by the C domain *N*-terminal lobe and *C*-terminal lobe with the T domain helix $\alpha2$ and helix $\alpha3$, respectively. Recently, x-ray crystallography with a series of constructs of the dimodular NRPS protein linear gramicidin synthetase subunit A (LgrA) revealed the presence of a protein–protein interface that is mainly dependent on hydrophobic interactions [54]. These interactions are located at the T domain loop 1, helix $\alpha2$, and helix $\alpha3$ regions and the C domain C-terminal lobe. Furthermore, this study revealed the first instance of both acceptor and donor T domains occupying their respective sites on the C domain. Structural information on the interactions between the C domain and its donor and acceptor T domains is important for NRPS engineering, but collecting such information remains a challenge due to multiple factors. In particular, two substrate-loaded T domains must be bound at the donor and acceptor sites in order to evaluate the active site interactions that affect substrate selectivity. Emerging techniques in structural biology, such as cryo-electron microscopy (EM), could capture the C domain with a combination of donor and acceptor T domains and unveil the effect of T domain-bound substrates in forming a protein–protein interface.

Funding

This work was supported in part by a JSPS KAKENHI Grant for Scientific Research on Innovative Areas 16H06445 (Y.H.), by a JSPS KAKENHI Grant-in-Aid for Transformative Research Areas (A) 22H05119 (C.M.), by the JSPS A3 Foresight Program (Y.H.), and by JSPS KAKENHI grants 22 K05412 (C.Y.) and 20H02918 (Y.H.).

References

1. Winn M, Richardson SM, Campopiano DJ, Micklefield J (2020) Harnessing and engineering amide bond forming ligases for the synthesis of amides. Curr Opin Chem Biol 55:77–85. https://doi.org/10.1016/j.cbpa.2019.12.004
2. Maruyama C, Hamano Y (2020) tRNA-dependent amide bond-forming enzymes in peptide natural product biosynthesis. Curr Opin Chem Biol 59:164–171. https://doi.org/10.1016/j.cbpa.2020.08.002
3. Moutiez M, Belin P, Gondry M (2017) Aminoacyl-tRNA-utilizing enzymes in natural product biosynthesis. Chem Rev 117(8):5578–5618. https://doi.org/10.1021/acs.chemrev.6b00523
4. Stanisic A, Kries H (2019) Adenylation domains in nonribosomal peptide engineering. Chembiochem 20(11):1347–1356. https://doi.org/10.1002/cbic.201800750
5. Sussmuth RD, Mainz A (2017) Nonribosomal peptide synthesis-principles and prospects. Angew Chem Int Ed Engl 56(14):3770–3821. https://doi.org/10.1002/anie.201609079
6. Schwarzer D, Finking R, Marahiel MA (2003) Nonribosomal peptides: from genes to products. Nat Prod Rep 20(3):275–287. https://doi.org/10.1039/B111145K
7. Mootz HD, Schwarzer D, Marahiel MA (2002) Ways of assembling complex natural products on modular nonribosomal peptide synthetases. Chembiochem 3(6):490–504. https://doi.org/10.1002/1439-7633(20020603)3:6<490::AID-CBIC490>3.0.CO;2-N
8. Gulick AM (2009) Conformational dynamics in the Acyl-CoA synthetases, adenylation domains of non-ribosomal peptide synthetases, and firefly luciferase. ACS Chem Biol 4(10):811–827. https://doi.org/10.1021/cb900156h
9. Rottig M, Medema MH, Blin K, Weber T, Rausch C, Kohlbacher O (2011) NRPSpredictor2—a web server for predicting NRPS adenylation domain specificity. Nucleic Acids Res 39(Web Server issue):W362–W367. https://doi.org/10.1093/nar/gkr323
10. Calcott MJ, Owen JG, Ackerley DF (2020) Efficient rational modification of non-ribosomal peptides by adenylation domain substitution. Nat Commun 11(1):4554. https://doi.org/10.1038/s41467-020-18365-0
11. Reimer JM, Aloise MN, Harrison PM, Schmeing TM (2016) Synthetic cycle of the initiation module of a formylating nonribosomal peptide synthetase. Nature 529(7585):239–242. https://doi.org/10.1038/nature16503
12. Zhang W, Ntai I, Bolla ML, Malcolmson SJ, Kahne D, Kelleher NL, Walsh CT (2011) Nine enzymes are required for assembly of the pacidamycin group of peptidyl nucleoside antibiotics. J Am Chem Soc 133(14):5240–5243. https://doi.org/10.1021/ja2011109
13. Steffensky M, Li SM, Heide L (2000) Cloning, overexpression, and purification of novobiocic acid synthetase from streptomyces spheroides NCIMB 11891. J Biol Chem 275(28):21754–21760. https://doi.org/10.1074/jbc.M003066200
14. Schmutz E, Steffensky M, Schmidt J, Porzel A, Li SM, Heide L (2003) An unusual amide synthetase (CouL) from the coumermycin A1 biosynthetic gene cluster from Streptomyces rishiriensis DSM 40489. Eur J Biochem 270(22):4413–4419. https://doi.org/10.1046/j.1432-1033.2003.03830.x
15. Maruyama C, Toyoda J, Kato Y, Izumikawa M, Takagi M, Shin-ya K, Katano H, Utagawa T, Hamano Y (2012) A stand-alone adenylation domain forms amide bonds in streptothricin biosynthesis. Nat Chem Biol 8(9):791–797. https://doi.org/10.1038/nchembio.1040
16. Wolpert M, Gust B, Kammerer B, Heide L (2007) Effects of deletions of mbtH-like genes on clorobiocin biosynthesis in streptomyces coelicolor. Microbiology 153(Pt 5):

1413–1423. https://doi.org/10.1099/mic.0.2006/002998-0
17. Lautru S, Oves-Costales D, Pernodet JL, Challis GL (2007) MbtH-like protein-mediated cross-talk between non-ribosomal peptide antibiotic and siderophore biosynthetic pathways in Streptomyces coelicolor M145. Microbiology 153(Pt 5):1405–1412. https://doi.org/10.1099/mic.0.2006/003145-0
18. Boll B, Taubitz T, Heide L (2011) Role of MbtH-like proteins in the adenylation of tyrosine during aminocoumarin and vancomycin biosynthesis. J Biol Chem 286(42):36281–36290. https://doi.org/10.1074/jbc.M111.288092
19. Bryce GF, Brot N (1972) Studies on the enzymatic synthesis of the cyclic trimer of 2,3-dihydroxy-N-benzoyl-L-serine in Escherichia coli. Biochemistry 11(9):1708–1715. https://doi.org/10.1021/bi00759a028
20. Rusnak F, Faraci WS, Walsh CT (1989) Subcloning, expression, and purification of the enterobactin biosynthetic enzyme 2,3-dihydroxybenzoate-AMP ligase: demonstration of enzyme-bound (2,3-dihydroxybenzoyl)adenylate product. Biochemistry 28(17):6827–6835
21. McQuade TJ, Shallop AD, Sheoran A, Delproposto JE, Tsodikov OV, Garneau-Tsodikova S (2009) A nonradioactive high-throughput assay for screening and characterization of adenylation domains for nonribosomal peptide combinatorial biosynthesis. Anal Biochem 386(2):244–250. https://doi.org/10.1016/j.ab.2008.12.014
22. Katano H, Tanaka R, Maruyama C, Hamano Y (2012) Assay of enzymes forming AMP+PPi by the pyrophosphate determination based on the formation of 18-molybdopyrophosphate. Anal Biochem 421(1):308–312. https://doi.org/10.1016/j.ab.2011.10.031
23. Katano H, Watanabe H, Takakuwa M, Maruyama C, Hamano Y (2013) Colorimetric determination of pyrophosphate anion and its application to adenylation enzyme assay. Anal Sci 29(11):1095–1098
24. Corpuz JC, Sanlley JO, Burkart MD (2022) Protein-protein interface analysis of the non-ribosomal peptide synthetase peptidyl carrier protein and enzymatic domains. Synth Syst Biotechnol 7(2):677–688. https://doi.org/10.1016/j.synbio.2022.02.006
25. Haslinger K, Redfield C, Cryle MJ (2015) Structure of the terminal PCP domain of the non-ribosomal peptide synthetase in teicoplanin biosynthesis. Proteins 83(4):711–721. https://doi.org/10.1002/prot.24758
26. Jaremko MJ, Lee DJ, Opella SJ, Burkart MD (2015) Structure and substrate sequestration in the pyoluteorin type II peptidyl carrier protein PltL. J Am Chem Soc 137(36):11546–11549. https://doi.org/10.1021/jacs.5b04525
27. Du L, Shen B (1999) Identification and characterization of a type II peptidyl carrier protein from the bleomycin producer Streptomyces verticillus ATCC 15003. Chem Biol 6(8):507–517. https://doi.org/10.1016/S1074-5521(99)80083-0
28. Yamanaka K, Maruyama C, Takagi H, Hamano Y (2008) Epsilon-poly-L-lysine dispersity is controlled by a highly unusual nonribosomal peptide synthetase. Nat Chem Biol 4(12):766–772. https://doi.org/10.1038/nchembio.125
29. Yamanaka K, Kito N, Kita A, Imokawa Y, Maruyama C, Utagawa T, Hamano Y (2011) Development of a recombinant epsilon-poly-L-lysine synthetase expression system to perform mutational analysis. J Biosci Bioeng 111(6):646–649. https://doi.org/10.1016/j.jbiosc.2011.01.020
30. Chen H, Walsh CT (2001) Coumarin formation in novobiocin biosynthesis: beta-hydroxylation of the aminoacyl enzyme tyrosyl-S-NovH by a cytochrome P450 NovI. Chem Biol 8(4):301–312. https://doi.org/10.1016/s1074-5521(01)00009-6
31. Marahiel MA, Stachelhaus T, Mootz HD (1997) Modular peptide synthetases involved in nonribosomal peptide synthesis. Chem Rev 97(7):2651–2674. https://doi.org/10.1021/cr960029e
32. Stachelhaus T, Mootz HD, Bergendahl V, Marahiel MA (1998) Peptide bond formation in nonribosomal peptide biosynthesis. Catalytic role of the condensation domain. J Biol Chem 273(35):22773–22781. https://doi.org/10.1074/jbc.273.35.22773
33. Bergendahl V, Linne U, Marahiel MA (2002) Mutational analysis of the C-domain in nonribosomal peptide synthesis. Eur J Biochem 269(2):620–629. https://doi.org/10.1046/j.0014-2956.2001.02691.x
34. Roche ED, Walsh CT (2003) Dissection of the EntF condensation domain boundary and active site residues in nonribosomal peptide synthesis. Biochemistry 42(5):1334–1344. https://doi.org/10.1021/bi026867m
35. Samel SA, Schoenafinger G, Knappe TA, Marahiel MA, Essen LO (2007) Structural and functional insights into a peptide bond-forming bidomain from a nonribosomal peptide synthetase. Structure 15(7):781–792. https://doi.org/10.1016/j.str.2007.05.008
36. Samel SA, Czodrowski P, Essen LO (2014) Structure of the epimerization domain of tyrocidine synthetase A. Acta Crystallogr D Biol Crystallogr 70(Pt 5):1442–1452. https://doi.org/10.1107/S1399004714004398

37. Balibar CJ, Vaillancourt FH, Walsh CT (2005) Generation of D amino acid residues in assembly of arthrofactin by dual condensation/epimerization domains. Chem Biol 12(11): 1189–1200. https://doi.org/10.1016/j.chembiol.2005.08.010
38. Dashti Y, Nakou IT, Mullins AJ, Webster G, Jian X, Mahenthiralingam E, Challis GL (2020) Discovery and biosynthesis of bolagladins: unusual lipodepsipeptides from burkholderia gladioli clinical isolates*. Angew Chem Int Ed Engl 59(48):21553–21561. https://doi.org/10.1002/anie.202009110
39. Royer M, Koebnik R, Marguerettaz M, Barbe V, Robin GP, Brin C, Carrere S, Gomez C, Hugelland M, Voller GH, Noell J, Pieretti I, Rausch S, Verdier V, Poussier S, Rott P, Sussmuth RD, Cociancich S (2013) Genome mining reveals the genus Xanthomonas to be a promising reservoir for new bioactive non-ribosomally synthesized peptides. BMC Genomics 14:658. https://doi.org/10.1186/1471-2164-14-658
40. Bloudoff K, Schmeing TM (2017) Structural and functional aspects of the nonribosomal peptide synthetase condensation domain superfamily: discovery, dissection and diversity. Biochim Biophys Acta Proteins Proteom 1865(11 Pt B):1587–1604. https://doi.org/10.1016/j.bbapap.2017.05.010
41. Konz D, Klens A, Schorgendorfer K, Marahiel MA (1997) The bacitracin biosynthesis operon of Bacillus licheniformis ATCC 10716: molecular characterization of three multi-modular peptide synthetases. Chem Biol 4(12): 927–937. https://doi.org/10.1016/s1074-5521(97)90301-x
42. Shen B, Du L, Sanchez C, Edwards DJ, Chen M, Murrell JM (2002) Cloning and characterization of the bleomycin biosynthetic gene cluster from streptomyces verticillus ATCC15003. J Nat Prod 65(3):422–431. https://doi.org/10.1021/np010550q
43. Vollbrecht L, Steinmetz H, Hofle G, Oberer L, Rihs G, Bovermann G, von Matt P (2002) Argyrins, immunosuppressive cyclic peptides from myxobacteria. II. Structure elucidation and stereochemistry. J Antibiot (Tokyo) 55(8):715–721. https://doi.org/10.7164/antibiotics.55.715
44. Gehring AM, Mori I, Perry RD, Walsh CT (1998) The nonribosomal peptide synthetase HMWP2 forms a thiazoline ring during biogenesis of yersiniabactin, an iron-chelating virulence factor of Yersinia pestis. Biochemistry 37(33):11637–11650. https://doi.org/10.1021/bi9812571
45. Vizcaino MI, Crawford JM (2015) The colibactin warhead crosslinks DNA. Nat Chem 7(5):411–417. https://doi.org/10.1038/nchem.2221
46. Little RF, Hertweck C (2022) Chain release mechanisms in polyketide and non-ribosomal peptide biosynthesis. Nat Prod Rep 39(1): 163–205. https://doi.org/10.1039/d1np00035g
47. Schneider A, Stachelhaus T, Marahiel MA (1998) Targeted alteration of the substrate specificity of peptide synthetases by rational module swapping. Mol Gen Genet 257(3): 308–318. https://doi.org/10.1007/s004380050652
48. Sundlov JA, Shi C, Wilson DJ, Aldrich CC, Gulick AM (2012) Structural and functional investigation of the intermolecular interaction between NRPS adenylation and carrier protein domains. Chem Biol 19(2):188–198. https://doi.org/10.1016/j.chembiol.2011.11.013
49. Drake EJ, Miller BR, Shi C, Tarrasch JT, Sundlov JA, Allen CL, Skiniotis G, Aldrich CC, Gulick AM (2016) Structures of two distinct conformations of holo-non-ribosomal peptide synthetases. Nature 529(7585):235–238. https://doi.org/10.1038/nature16163
50. Jaremko MJ, Lee DJ, Patel A, Winslow V, Opella SJ, McCammon JA, Burkart MD (2017) Manipulating protein-protein interactions in nonribosomal peptide synthetase type II peptidyl carrier proteins. Biochemistry 56(40):5269–5273. https://doi.org/10.1021/acs.biochem.7b00884
51. Corpuz JC, Podust LM, Davis TD, Jaremko MJ, Burkart MD (2020) Dynamic visualization of type II peptidyl carrier protein recognition in pyoluteorin biosynthesis. RSC Chem Biol 1(1): 8–12. https://doi.org/10.1039/c9cb00015a
52. Miyanaga A, Kurihara S, Chisuga T, Kudo F, Eguchi T (2020) Structural characterization of complex of adenylation domain and carrier protein by using pantetheine cross-linking probe. ACS Chem Biol 15(7):1808–1812. https://doi.org/10.1021/acschembio.0c00403
53. Tanovic A, Samel SA, Essen LO, Marahiel MA (2008) Crystal structure of the termination module of a nonribosomal peptide synthetase. Science 321(5889):659–663. https://doi.org/10.1126/science.1159850
54. Reimer JM, Eivaskhani M, Harb I, Guarne A, Weigt M, Schmeing TM (2019) Structures of a dimodular nonribosomal peptide synthetase reveal conformational flexibility. Science 366(6466). https://doi.org/10.1126/science.aaw4388

Chapter 2

Structural Studies of Modular Nonribosomal Peptide Synthetases

Ketan D. Patel, Syed Fardin Ahmed, Monica R. MacDonald, and Andrew M. Gulick

Abstract

The non-ribosomal peptide synthetases (NRPSs) are a family of modular enzymes involved in the production of peptide natural products. Not restricted by the constraints of ribosomal peptide and protein production, the NRPSs are able to incorporate unusual amino acids and other suitable building blocks into the final product. The NRPSs operate with an assembly line strategy in which peptide intermediates are covalently tethered to a peptidyl carrier protein and transported to different catalytic domains for the multiple steps in the biosynthesis. Often the carrier and catalytic domains are joined into a single large multidomain protein. This chapter serves to introduce the NRPS enzymes, using the nocardicin NRPS system as an example that highlights many common features to NRPS biochemistry. We then describe recent advances in the structural biology of NRPSs focusing on large multidomain structures that have been determined.

Key words Non-ribosomal Peptide Synthetase, NRPS, Natural products, Secondary metabolites, Structural biology, Modular enzymes, Conformational changes, Protein–protein interactions

1 Introduction to Modular NRPS Enzymes

To adapt to the diverse range of environments in which they are found, microbes produce a variety of chemically diverse natural products. These small molecules are secreted into the environment where they perform a variety of functions including providing communication between other members of the same species, eliminating competing species, and promoting growth through the acquisition of necessary nutrients. Because these molecules exert a biological effect in natural environments, including in the context of the host-pathogen interaction, many natural products have found their way into the pharmaceutical industry, with as many as two-thirds of small molecule FDA-approved drugs being natural products or molecules derived or inspired by compounds found in

Michael Burkart and Fumihiro Ishikawa (eds.), *Non-Ribosomal Peptide Biosynthesis and Engineering: Methods and Protocols*, Methods in Molecular Biology, vol. 2670, https://doi.org/10.1007/978-1-0716-3214-7_2,

nature [1]. The discovery of new natural products and the understanding of the fascinating enzymes involved in their biosynthesis has therefore been a rich field of study for many investigators in the field of microbiology, biochemistry, chemical biology, and medicinal chemistry.

Some peptide natural products are produced by a family of large, modular enzymes known as the non-ribosomal peptide synthetases (NRPSs). The NRPS peptide products have many different biological functions, including acting as cytotoxins, antibiotics, quorum signals, and peptide siderophores that are involved in the acquisition of iron and other metal ions. As these are produced without the assistance of mRNAs, tRNAs, or ribosomes, the NRPS products are much more diverse than conventional biological peptides and proteins. As many as several hundred compounds serve as building blocks for peptide biosynthesis, including standard, proteinogenic amino acids as well as other amino acids, aromatic acids, and fatty acids [2].

The structural diversity of NRPS products is further enhanced by other features. Notably, the peptides are often found as cyclic products that form when the entire peptide cyclizes through the N- and C-terminal groups. Also quite common are peptides where an internal side chain, such as the amine of a lysine or ornithine residue or the hydroxyl of a serine or threonine, cyclizes through attack on the C-terminal carboxylate to form a macrocyclic amide or ester [3, 4]. Finally, many NRPS-derived products are further chemically modified, during or following the synthesis of the peptide, with additional changes including methylations, hydroxylations, halogenations, or epimerizations.

What is perhaps most remarkable about the NRPS enzymes is the modular basis for biosynthesis that uses large enzymes with multiple catalytic domains that combine to produce the peptide product with an assembly line architecture [5–9]. While many variations of this approach have been identified, a consensus NRPS system contains a single module for the incorporation of each amino acid building block into the final product. Each module contains the appropriate catalytic domains needed for that residue incorporation. Additionally, each module contains a thiolation or peptidyl carrier protein domain (PCP) that covalently binds the incorporated amino acid and the growing peptide during transit through that module. Large conformational changes in the multidomain protein deliver the amino acid substrates and peptide intermediates to the neighboring catalytic domains. To allow this covalent attachment, the PCP domains are post-translationally modified to attach a phosphopantetheine cofactor to a conserved serine residue converting the *apo* carrier domain to a *holo* domain [10, 11]. The pantetheine, derived from coenzyme A, contains a terminal thiol group to which the amino acid and peptide substrates are bound as a thioester. The pantetheine thus serves as a ~ 15 Å arm that can present the substrate into the active site of the neighboring catalytic domains. However, it is noteworthy that the active sites of the catalytic domains can be positioned more than 50 Å from each other, illustrating that rotations of the pantetheine alone are not sufficient to present the substrate to different active sites. Thus, the entire PCP must move to properly guide the peptide to different catalytic domains in an organized fashion.

We present here the NRPS enzymes and biosynthetic pathway that are responsible for the production of the antibiotic nocardicin (*see* Fig. 1). This system has been characterized functionally by Townsend and colleagues and illustrates both the conventional and more rarely found aspects of NRPS biosynthetic clusters [12–14]. This system serves to present the shared features of NRPS that exist in nearly all systems as well as other, less common NRPS functions that should be considered when examining a novel pathway. The nocardicin NRPS system therefore can provide a background into NRPSs needed for the readers of additional chapters in this volume.

The final product of the nocardicin NRPS system is the β-lactam antibiotic nocardicin A, which is derived from two molecules of L-hydroxyphenylglycine (L-HPG) and a single serine that is cyclized to the lactam ring. The nocardicin tripeptide is first produced as a pentapeptide that is cleaved to release the three C-terminal residues as nocardicin G. The aminoacid building blocks of nocardicin are therefore three molecules of HPG as well as the standard amino acids arginine and serine [12, 13].

NRPS enzymes can exist as a single multidomain protein or can be split into multiple multidomain enzymes; additionally, free-standing domains are not uncommon and can be used in conjunction with multidomain NRPSs. The nocardicin NRPS enzymes of *Nocardia uniformis* contain five modules that are used to produce the initial pentapeptide. The five modules are spread across two polypeptides, NocA and NocB, requiring both intramolecular and intermolecular domain interactions [12]. Notably, the fourth module is split, with the initial domain located on NocA and the subsequent domains located on NocB. Each module contains the PCP domain harboring the pantetheine cofactor. Upstream of the PCP is an adenylation domain that catalyzes a two-step reaction to activate the amino acid and attach it to the pantetheine. The activation occurs through an initial adenylation reaction in which the carboxylate of the substrate attacks ATP to produce the aminoacyl adenylate and pyrophosphate. The pantetheine thiol then attacks the amino acid to release AMP and form the covalently bound amino acyl thioester (*see* Fig. 1a). Once two amino acids are loaded onto adjacent PCP domains, they meet at a condensation domain that serves to catalyze peptide bond formation, transferring the amino acid from the upstream PCP to the amino moiety of the downstream loaded substrate (*see* Fig. 1b). This dipeptide

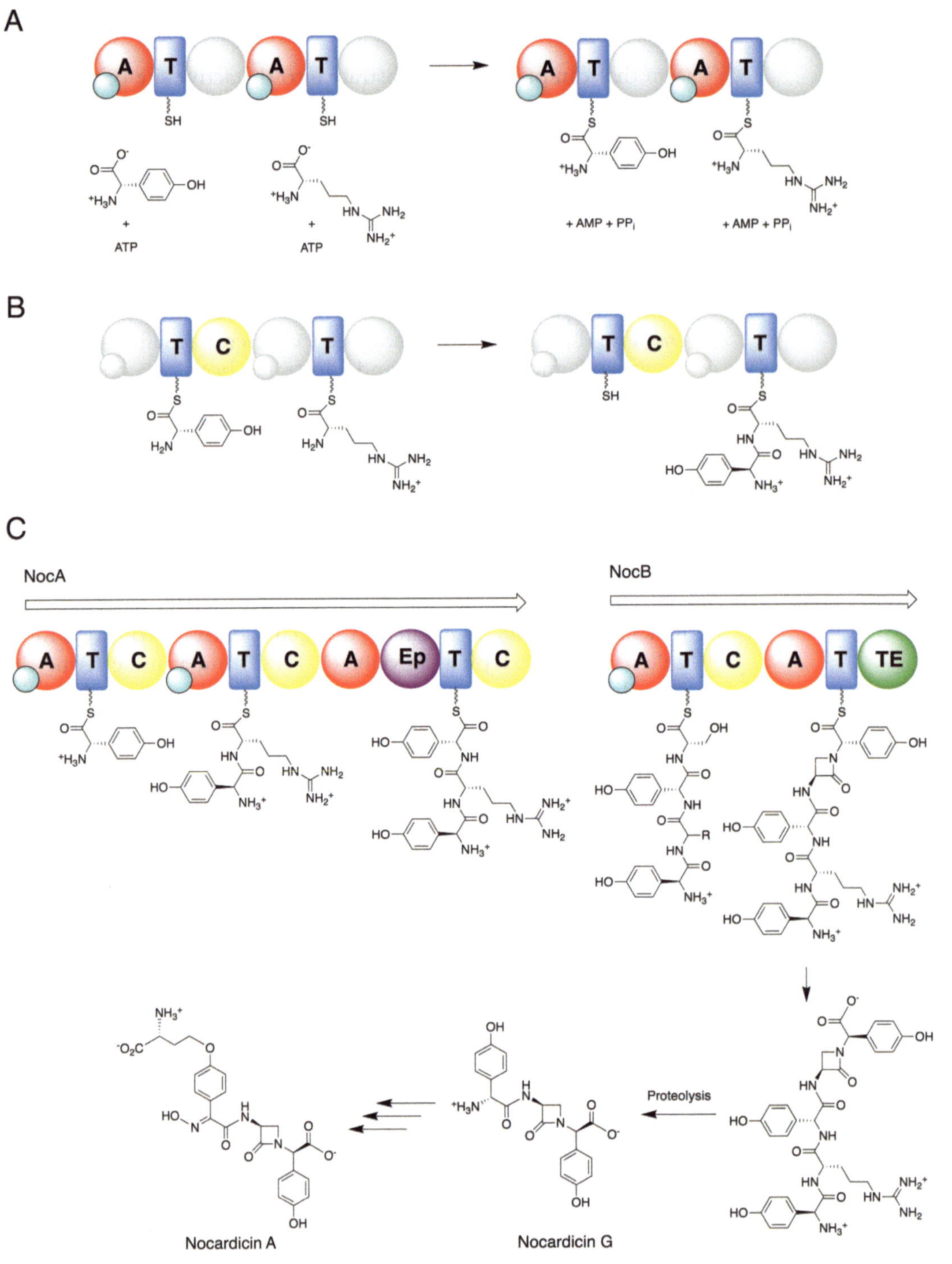

Fig. 1 The NRPS biosynthesis pathway for the β-lactam antibiotic nocardicin A. (**a**). Initial reaction catalyzed by the first two modules of NocA. The hydroxyphenylglycine and arginine residues are loaded onto the carrier proteins (blue rectangle, T) as dictated by the adenylation (red sphere, A) domains. (**b**). The condensation (yellow sphere, C) domain catalyzes peptide bond formation transferring the HPG residue from the upstream to the downstream module. (**c**). Assembly line biosynthesis showing the extending peptide as it progresses. The terminal thioesterase (green sphere, TE) domain catalyzes epimerization and hydrolysis of the pentapeptide. This NRPS product is then cleaved to nocardicin G, which is subsequently processed to nocardicin A. MbtH-like proteins that bind to some adenylation domains are shown in cyan

can then be transferred at the condensation domain of the third module to the downstream PCP forming the tripeptide on the PCP of the third module. Continued elongation of the peptide results in a pentapeptide covalently attached to the fifth PCP, located on NocB (*see* Fig. 1c). To release the peptide, the thioesterase domain that resides at the C-terminus of NocB catalyzes hydrolysis of the thioester.

In addition to this conventional NRPS peptide assembly line strategy, the nocardicin system highlights several more unusual features that are present in many NRPS biosynthetic clusters. The third module of NocA contains an additional domain that catalyzes epimerization of the L-HPG to D-HPG [12]. These extra domains may be inserted between domains in the NRPS assembly line as seen here in NocA or may be inserted within the boundaries of another domain. Additionally, the nocardicin biosynthetic gene cluster encodes NocI, a member of the family of MbtH-like proteins (MLPs). Originally identified in the mycobactin NRPS system [15], these ~8 kDa proteins contain a central three-stranded β-sheet that is stacked against a single α-helix [16]; an additional helix often exists at the C-terminus. These proteins bind to some NRPS adenylation domains and play roles in protein solubility and activity [17, 18]. The NocI MLP protein interacts with three of the adenylation domains from NocA and NocB, specifically the first, second, and fourth modules [13]. Structures of MLP protein complexes with adenylation domains illustrate a common binding motif that exists on the MLPs and the adenylation domain to support their interaction [19].

Some NRPS domains catalyze reactions beyond their expected roles of amino acid activation, peptide bond formation, or thioester release. The NocB protein harbors two such domains. The condensation domain of the fifth module converts the serine residue into the β-lactam ring [20, 21], while the thioesterase domain catalyzes an initial epimerization reaction prior to release of the peptide [22, 23]. Finally, nocardicin biosynthesis illustrates that, upon release of the peptide, additional enzymatic steps can be catalyzed by other, free-standing enzymes that result in the final active product. For nocardicin, this includes the initial proteolytic step to release the tripeptide nocardicin G, followed by maturation that includes the addition of the homoserine moiety to the first HPG hydroxyl and the formation of the oxime at the N-terminus [12]. While other NRPS systems include other notable features, including additional cyclization, halogenation, glycosylation, and methylation steps, the nocardicin NRPS demonstrates how a fundamental catalytic strategy can be further elaborated by the addition of other catalytic domains that reside within the NRPS multidomain proteins as well as on free-standing, more traditional domains.

2 Catalytic Domains of Non-ribosomal Peptide Synthetases

Since the discovery of their broader catalytic mechanism in the 1990s [24–26], the NRPS enzymes have been a target for structural interrogation for decades. Efforts to unravel the structural features that guide the delivery of the carrier proteins and bound substrates to active sites in a methodical fashion have used a variety of biochemical, biophysical, and structural techniques. While there remains much to be learned, a comprehensive picture is emerging of the interactions and conformational changes that govern transit of the nascent peptide through the NRPS assembly line. Structural and functional studies of NRPS proteins initially examined free-standing NRPS domains and genetically truncated single domains. While biochemical studies of multidomain enzymes were within reach from the earliest days of study of the NRPS enzymes, structural studies lagged behind; however, a number of didomain and multidomain structures have now been solved, providing views of the enzymes at different stages of the catalytic cycle. In this chapter, we briefly remind the reader of the structures of individual domains and then focus our attention on complete modular structures of multidomain NRPSs. Additional, more extensive reviews present structural aspects of NRPS enzymology [2, 27–36].

The adenylation domains of NRPS enzymes have been most extensively studied. The enzymes catalyze a two-step reaction in which the substrate first reacts in an adenylation reaction. In this reaction, the carboxylate of the amino acid attacks the α-phosphate of ATP to form an aminoacyl adenylate intermediate. In a second step, the pantetheine cofactor from the PCP attacks the adenylate to displace AMP and form the amino acyl thioester. The adenylation domains are approximately 550 residues in length and consist of two subdomains. A large N-terminal subdomain (also called A_{core}) consists of the first ~450 residues with the remainder of the domain forming a ~ 100 residue C-terminal subdomain (A_{sub}). The active site is located at the interface of these two subdomains. Critically, adenylation domains adopt two distinct catalytic conformations to catalyze the adenylation and thioester-forming partial reactions. This conformational change occurs when the C-terminal subdomain rotates by ~140° [37, 38], presenting opposite faces to a single active site used for both reactions. As described below, this conformational change appears to play a major role in the proper migration of the PCP from the adenylation and condensation domains.

The condensation domains of NRPS enzymes bind to two PCP domains. An upstream amino acid- or peptide-bound PCP serves as the donor, while the amino group on an amino acid bound to the downstream PCP serves as the acceptor. Approximately 440 residues in length, the condensation domain contains two structurally homologous subdomains that have been referred to as the N-terminal and C-terminal lobes [28]. Structures of different condensation domains show that the two lobes can open and close to adopt different conformations. Two distinct PCP binding interfaces are used for the donor and acceptor PCP. Interestingly, the condensation domains appear to be structurally related to other NRPS auxiliary domains that catalyze epimerization and cyclization reactions [28, 39–41].

The ~260 residue NRPS thioesterase domains are members of the α /β hydrolase family of enzymes involved in many hydrolytic reactions [4]. The domain contains a central β-sheet that is surrounded on both sides by α-helices. The domains contain a conserved catalytic triad composed of an anionic aspartic (or rarely glutamic) acid residue, a histidine, and a serine or cysteine nucleophile. The nucleophile attacks the peptide bound to the pantetheine to form an enzyme-bound intermediate as a covalent acyl ester or thioester. This intermediate is then released through attack by a water molecule, resulting in the hydrolytic release of the linear peptide, or by attack from a nucleophilic group from the peptide at either the N-terminus or from an internal side chain. The thioesterase domain contains several helices that cap the active site and adopt alternate conformations in the presence or absence of ligands.

While the adenylation, condensation, and thioesterase form the core catalytic domains that function together with the PCP domain in the fundamental NRPS pathway, additional domains are found within NRPS enzymes that contribute chemical diversity to the peptide products. These domains include SAM-dependent N-methyltransferases [42], formyltransferases [43], and cyclization and epimerization domains, which share structural and sequence homology with condensation domains. Additionally, some NRPS enzymes terminate not in a thioesterase domain but in a reductase domain that releases the peptide aldehyde in an NAD(P)H-dependent manner [44].

3 Structures of Modules of NRPS Enzymes

Here we present the structures of modular NRPS proteins that have been determined through early 2022. This chapter updates our prior review of the structural biology of NRPS enzymes in an earlier volume of *Methods in Molecular Biology* [45]. As the didomain complexes have recently been reviewed by Burkart [36], we focus on enzymes that contain a PCP domain along with more than two catalytic domains. We describe for each the product, if known, the broad overview of the biosynthetic pathway and NRPS domain architecture. Finally, for each protein, we describe the structural organization observed in the structure(s), with a particular focus on the catalytic states observed and the implications of the structure for the understanding of the broader NRPS structural cycle.

3.1 SrfA-C

After many years that revealed the structural biology of individual domains, the first insights into the structural cycle of NRPS modular activity came from the 2008 structure determination of SrfA-C from the labs of Mohamed Marahiel and Lars-Oliver Essen [46]. SrfA-C forms the seventh module of the NRPS system from *Bacillus subtilis* that produces the lipopeptide surfactant known as surfactin. While the first six modules of the surfactin NRPS are expressed on two separate ~400 kDa proteins, SrfA-A and SrfA-B that each contains three modules, the final termination module is expressed on its own as a 1274 residue protein. Catalyzing the incorporation of the terminal leucine residue as well as the cyclization and release of surfactin, SrfA-C has a domain organization of condensation–adenylation–PCP–thioesterase domains.

This landmark structure (*see* Fig. 2, PDB **2VSQ**) demonstrated a number of key features that provided the basis for understanding NRPS structural biology and moved the field beyond the two dimensional assembly line representations such as those shown in Fig. 1. The SrfA-C structure showed the condensation domain formed a large interface with the N-terminal subdomain of the

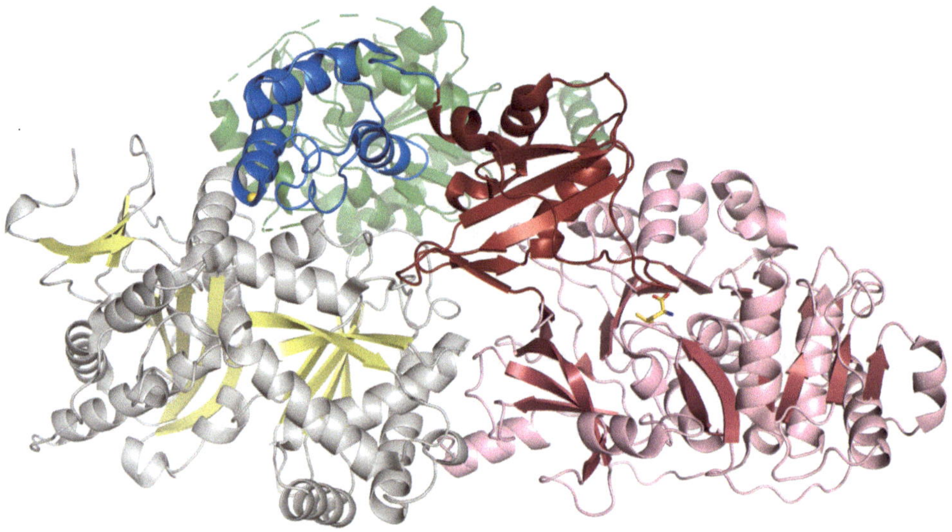

Fig. 2 Structure of the SrfA-C termination module (PDB: **2VSQ**). The structure of SrfA-C is shown with condensation domain (grey/yellow), adenylation domain (N-terminal subdomain in pink, C-terminal subdomain in dark red), PCP in blue, and thioesterase domain in green. Ser1003, the site of pantetheinylation, which was mutated to an alanine residue, is highlighted in gold. A molecule of the substrate leucine is shown within the adenylation domain active site. The color scheme will be used for domains throughout the remaining structural figures, except where highlighted for additional domains

adenylation domain. These two domains, which together account for two-thirds of the ~146 kDa protein, were proposed to form a stable catalytic platform that presented the openings to their active sites on one face. The *apo* PCP, although lacking the pantetheine cofactor, was positioned close to the condensation domain in a position that was reasonably interpreted to represent the functional binding for the acceptor PCP that is awaiting the delivery of the upstream PCP and peptide. The adenylation domain contained a molecule of the substrate leucine in the active site. Importantly, the adenylation domain did not adopt one of the known catalytic conformations for either the adenylation or thioester-forming states. Finally, the C-terminal thioesterase domain of SrfA-C made limited interactions with the remainder of the protein, with the active site directing towards the back face of the PCP and condensation domains.

Critically, the SrfA-C structure allowed for the measurements of the distance from the pantetheine binding site to the three different active sites. While the pantetheine arm was long enough to bridge the 16 Å gap to the condensation domain active site, the distances to the adenylation and thioesterase domain active sites at 57 Å and 43 Å, respectively, were longer than an extended cofactor and implied that large conformational rearrangements would be required for the PCP to reach the other catalytic centers.

3.2 AB3403

Two proteins that contain the same catalytic architecture as SrfA-C were solved in 2016 that built on the results of the early structure and illustrated distinct catalytic conformations of the termination modules [37]. The structure the AB3403 illustrated the functional interaction of the *holo*-PCP domain with the condensation domain, while the associated structure of EntF (*see* Subheading 3.3) described the structure of *holo*-PCP bound to the adenylation domain in the thioester forming conformation.

The AB3403 protein is part of an uncharacterized NRPS biosynthetic pathway of *Acinetobacter baumannii*. The pathway contains a free-standing adenylating enzyme, a free-standing carrier protein [47], and a modular NRPS with a condensation–adenylation–PCP–thioesterase domain organization. The unwieldy name of AB3403 comes from the gene annotation of the *A. baumannii* strain from which it was cloned [48] that has unfortunately been renamed within the NCBI database; the gene annotation in the more common laboratory strain *A. baumannii* ATCC 17978 is A1S_0115. In addition to the NRPS proteins, the biosynthetic gene cluster contains several other enzymes that may play a role in the enzymatic transformations of acyl-CoA or acyl-acyl carrier protein thioesters. Although the product of this biosynthetic pathway is unknown, this operon is among the most highly upregulated in *A. baumannii* during growth as a biofilm [49, 50] and has further been implicated in bacterial motility and quorum sensing [51].

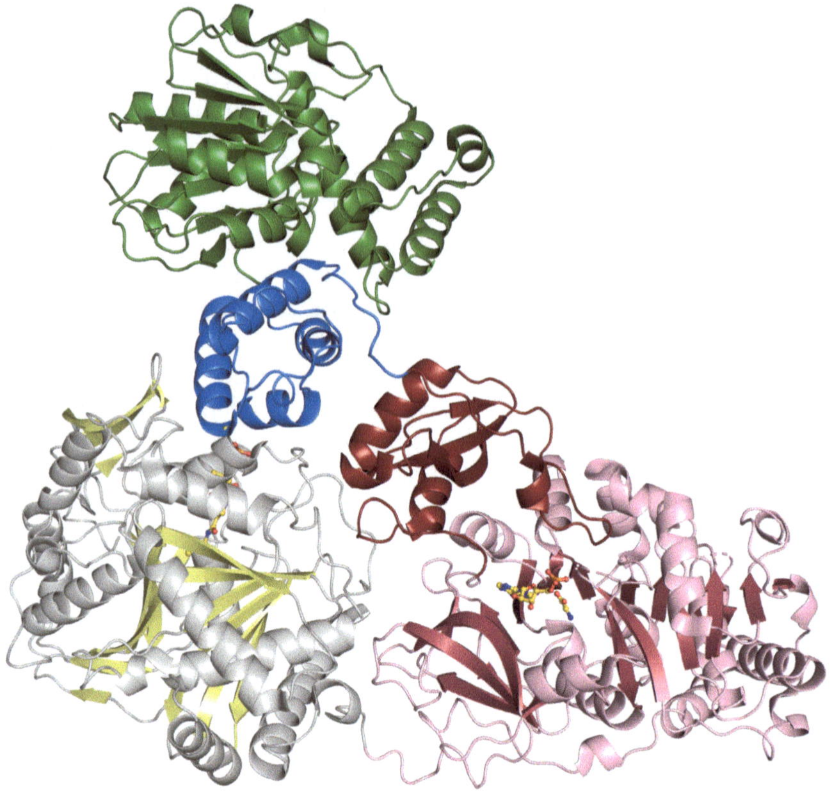

Fig. 3 Structure of the AB3403 termination module (PDB: **4ZXI**). The structure of AB3403 is shown, highlighting the interaction of the *holo*-PCP with the condensation domain. The adenylation domain contains a molecule of AMP and a molecule of glycine, adopting the adenylate-forming conformation

Overall, the structure of AB3403 (*see* Fig. 3, PDB **4ZXI**) was similar to SrfA-C. However, several key differences suggested the enzyme was adopting catalytically relevant conformations. The *holo*-PCP domain extended the pantetheine into the condensation domain active site. The thiol group of the cofactor is positioned 3.7 Å from a histidine residue that orients the substrate. The adenylation domain in one liganded structure contained Mg•AMP and glycine, which was included as a potential substrate molecule. As with SrfA-C, the thioesterase domain interacted mostly with the PCP domain, cradling the back face of the domain that was interacting functionally with the condensation.

Comparison of the adenylation domain to previously characterized adenylation domains in the catalytic conformations [38, 52, 53] showed that this domain adopted the adenylate-forming conformation. The structure therefore demonstrated that the condensation and adenylation domains of AB3403 could adopt catalytic conformations simultaneously. This implied that while the carrier protein was presenting the substrate to the condensation domain to await the delivery of the upstream peptide (or potentially the fatty acyl-ACP from the free-standing acyl carrier protein [47]), the adenylation domain could adenylate another substrate molecule that subsequently awaits off-loading of the product and the arrival of the newly freed *holo*-carrier protein. This simultaneous activity of the condensation and adenylation domains could enable a more efficient structural cycle [37].

3.3 EntF

A third termination module was also solved that provided a different catalytic state from that observed in AB3403 and approximated in SrfA-C [37]. Many NRPS pathways are involved in the production of peptide siderophores, small molecules involved in iron uptake [54, 55]. The enterobactin pathway of *Escherichia coli* and many Gram-negative Enterobacter species contains three NRPS proteins with a similar architecture to the *A. baumannii* pathway just described. The NRPS proteins of the enterobactin biosynthetic pathway are EntE, a free-standing adenylation domain, EntB, which provides the carrier protein, and EntF, a four domain termination module containing the condensation–adenylation–PCP–thioesterase domains [56, 57]. The enterobactin pathway begins with EntE activating a molecule of 2,3-dihydroxybenzoic acid (DHB) and installing that on the carrier domain of EntB. Interestingly, EntB contains a catalytic isochorismatase domain that acts together with EntC and EntA to produce DHB [58]. The EntF module loads a molecule of serine, which gets acylated by the DHB molecule within the condensation domain. This intermediate is transferred to the catalytic serine of the thioesterase domain while two additional DHB-serine amides are produced and sequentially installed on the first. The final cyclization within the thioesterase domain allows for release of the macrocyclic trilactone, joined through ester linkages between the serine side chain and carboxylates [57, 59–62].

The *holo*-EntF crystal structure (*see* Fig. 4, PDB **5JA1**) was determined in the presence of a mechanism-based inhibitor designed to bind in the adenylation pocket and covalently interact with the pantetheine thiol of the carrier protein [37, 63]. This serine adenosine vinylsulfonamide inhibitor was used to mimic the seryladenylate intermediate. The vinyl group serves as a Michael acceptor for attack of the thiol, leading to a covalent analog of the transition state for the thioester-forming reaction [31, 64]. Inhibitors of this approach have been used in a number of adenylation-PCP didomain protein structure studies as well as several larger modular constructs.

The EntF structure maintained the core condensation–adenylation platform, with the C-terminal subdomain of the adenylation domain adopting the thioester-forming conformation. The PCP domain was bound to the adenylation pocket in the same conformation as had been observed previously in didomain structures that similarly trapped this functional interaction [65–70]. The

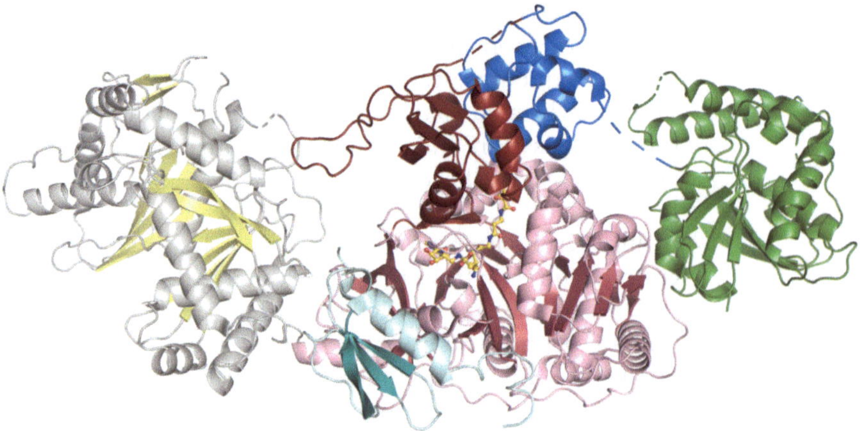

Fig. 4 Structure of the EntF termination module (PDB: **5JA1**). The structure of EntF is shown, highlighting the interaction of the *holo*-PCP with the adenylation domain. Shown in cyan is the YbdZ MLP domain interacting with the face of the adenylation domain

thioesterase domain of EntF was highly mobile. In one structure, the domain was disordered in the crystal lattice, while in another the thioesterase domain was pulled completely onto the other side of the module in comparison to where it was observed in the SrfA-C and AB3403 structures. Notably, the relative orientation of the PCP and C-terminal subdomain to each other was not conserved in EntF in comparison to these earlier structures. This suggests that the subdomain and PCP do not move as a rigid body. Rather, positioning of the adenylation domain in the thioester-forming conformation brings the PCP into close proximity to allow it to form a competent binding partnership.

The EntF structure was solved in the absence and presence of two different MLPs, its natural partner YbdZ protein and a homolog PA2412 from *P. aeruginosa* [37, 63]. The structures illustrate the conventional view [19] of MLP-adenylation domain interactions, with Ala826 from the EntF adenylation domain inserting into a conserved tryptophan pocket formed by Trp27 and Trp37 on the YbdZ MLP domain. The EntF adenylation domain was similar in the presence and absence of its partner MLP, suggesting that binding of an MLP does not result in any major structural changes.

3.4 LgrA

LgrA, the first protein of the biosynthetic NRPS cluster for linear gramicidin composed of 16 modules, has been structurally characterized, provided views of a number of modular and dimodular structures in important conformational states. The LgrA protein contains two modules that serve to incorporate valine and glycine residues at the first two positions of the gramicidin pentapeptide [71]. LgrA contains seven domains, starting with a formyltransferase domain that adds an N-terminal formate to the peptide. The remaining six domains are a standard adenylation-PCP for the initial valine module, followed by condensation–adenylation–PCP–epimerization domains for the incorporation of the glycine residue; interestingly, the epimerization domain is believed to be inactive [71]. Several constructs of LgrA have been crystallized with ligands, providing views of the auxiliary formyltransferase domain and also of multiple stages during the catalytic cycle [43, 72].

An initial study solved five structures of the first module of LgrA, containing the formyltransferase-adenylation-PCP tridomain construct [43]. The formyltransferase domain interacted with the N-terminal subdomain of the adenylation domain, using an interface that was near—though distinct from—the standard condensation domain binding surface. Multiple states of the small C-terminal subdomain of the adenylation domain were observed in different structures. Importantly, a structure for the thioester-forming conformation was observed upon loading the PCP with a valine amino CoA analog providing a non-hydrolyzable mimic of the loaded thioester (*see* Fig. 5a, PDB **5ES8**). A final structure showed the conformation of the PCP positioned to donate the loaded pantetheine to the formyltransferase domain.

A second series of structures exploited the LgrA protein lacking only the inactive epimerization domain, providing views of the dimodular protein [72]. The structures include the first four, five, or six domains of LgrA. Although several of the structures were determined at 6 Å resolution, the availability of the high-resolution structures to initially solve the structures and conservative modeling approaches enabled the determination of reliable structures and important insights into the nature of the cross-module interactions. As expected, the core didomain interactions, either the formyltransferase-adenylation of the first module or the condensation–adenylation of the second module, remain constant. The conformation of the adenylation domain is dictated by the catalytic adenylate- and thioester-forming reactions, in one case trapped with the vinylsulfonamide analog. However, the relative conformations of the two modules appear quite dynamic, supporting much less well-defined interactions across the NRPS module boundaries.

A structure of the six domain structure (FATCAT) containing all but the final epimerization domain was determined (*see* Fig. 5b, PDB **6MFZ**). The *holo* protein contained phosphopantetheine groups on the two carrier domains, which were partly able to be modeled in the low-resolution structure. The PCP domains were positioned near the donor and acceptor sites of the condensation domain, allowing the structure to illustrate the conformation of the peptide bond-forming step. In this conformation, the condensation domain made the conventional interface with the downstream glycine activating adenylation domain. However, the condensation domain contacted only the dynamic C-terminal subdomain of the upstream adenylation domain, suggesting limited interactions

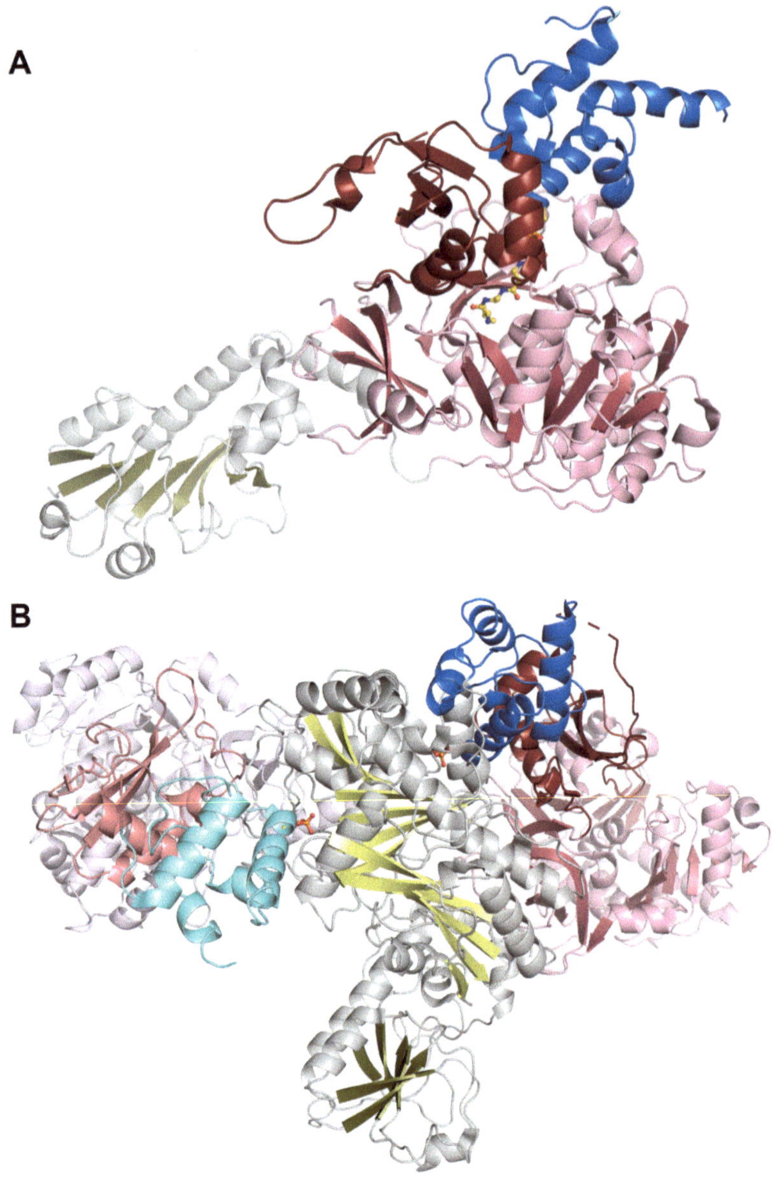

Fig. 5 Structure of the LgrA initiation module A. (PDB: **5ES8**) The structure of the first module of LgrA shows the formyltransferase domain (white, with olive strands), along with the adenylation and carrier domain. The PCP is loaded with a nonhydrolyzable thioester analog. B. (PDB: **6MFZ**) A six-domain construct of LgrA shows the interactions of the two modules via the condensation domain. The adenylation and carrier domains of module 2 are shown in light pink and cyan on the left side of the molecule. The two PCP domains meet at the donor (PCP_1) and acceptor (PCP_2) faces of the condensation domain. The cofactors beyond the phosphate groups are disordered

across modules. This was further supported by small angle X-ray scattering data that failed to match theoretical data generated from any of the crystallographic conformations. Combined, these structures support a dynamic dimodular conformation with limited interactions between modules.

3.5 DhbF

The dimodular structure of DhbF was solved in 2017 using X-ray crystallography techniques as well as negative stain EM to provide a more conclusive picture on the overall protein architecture [73]. DhbF is composed of two modules that incorporate the substrates glycine and threonine. In the biosynthetic pathway, DhbF is preceded by the stand-alone adenylation domain, DhbE, and isochorismatase-aryl carrier protein DhbB that load a molecule of 2,3-dihydroxybenzoate. Together, these proteins synthesize bacillibactin, a trilactone cyclic siderophore.

The two modules of DhbF contain a domain configuration of condensation–adenylation–PCP–condensation–adenylation–PCP–thioesterase. A protein construct containing the adenylation and peptidyl carrier protein of the first module and the condensation domain of the second module was crystallized, providing a 3.1 Å structure (*see* Fig. 6, PDB **5 U89**). Also present in the structure bound to the adenylation domain was an MLP domain, which was shown to be important for the function and crystallization of DhbF [73]. The adenylation domain of this structure was crystallized with the inhibitor glycine-AVS, to trap the complex in a thioester-forming conformation. The structure showed that the adenylation domain and the condensation of the downstream module have no direct interactions; rather, their interactions are mediated solely by the PCP domain. A novel conformation between the adenylation domain and PCP in the first module was also observed in which the PCP domain is rotated almost 90° compared to the previous conformations observed in other structures of adenylation–PCP complexes. This interaction may represent a novel interface between the adenylation and PCP domain or may illustrate a conformation that is used for release of the loaded PCP from the adenylation domain active site [73].

To obtain three-dimensional information on the organization of the full-length DhbF module, negative stain EM was utilized. To improve the homogeneity of protein sample, the terminal thioesterase domain was truncated and both glycine-AVS and threonine-AVS inhibitors were used to trap the protein in the thioester-forming conformation in both modules. Four distinct lobes, two "top" and two "bottom," were identified and corresponded with the first and second module of the protein. An N-terminal histidine-tagged $DhbF_{\Delta Te}$-MLP sample was prepared and treated it with nickel-nitrilotriacetic acid (Ni-NTA)-Nanogold to locate the termini and define the directionality of the envelopes. This technique determined the two lobes oriented towards the bottom of

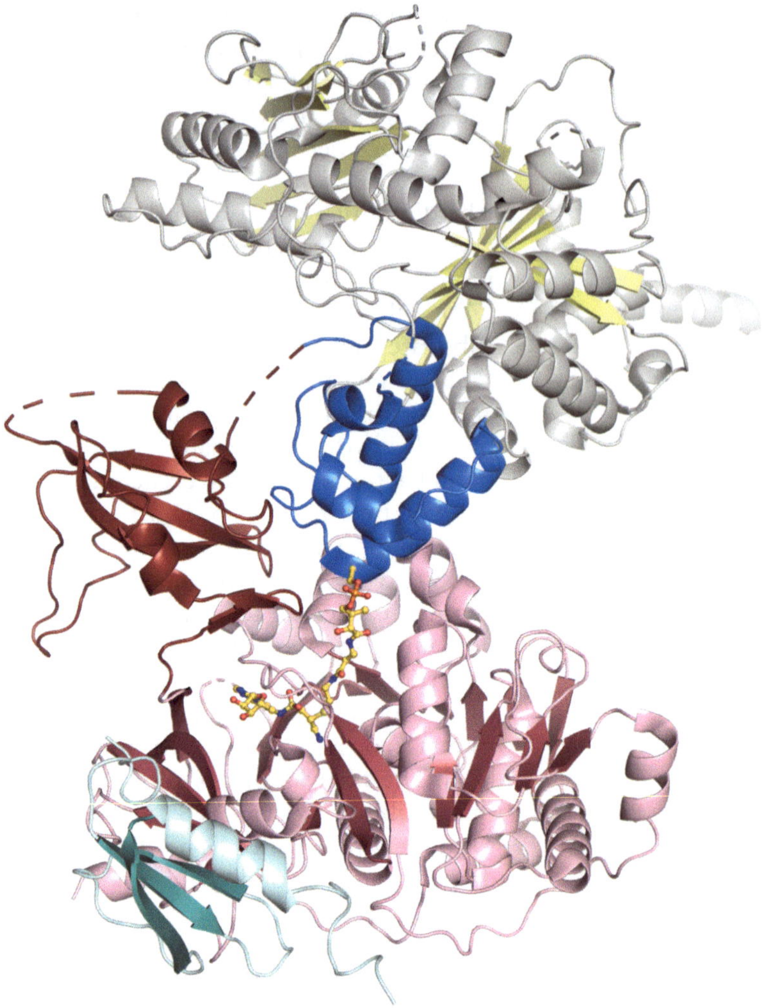

Fig. 6 Structure of DhbF (PDB: **5 U89**). The structure of the cross-module three domain fragment of DhbF highlights the adenylation domain interaction with an MLP and a trapped PCP domain interacting through the AVS ligand. The downstream condensation domain makes no interactions with the adenylation domain and interacts through the PCP

the structure to be the condensation and adenylation module 1 and the two at the top to be the condensation and adenylation domain of module 2. This analysis provides key insights into the organization of a multimodular NRPS.

3.6 FmoA3

The NRPS protein FmoA3, involved in the production of the chloroindole capped peptides JBIR-34 and JBIR-35, was structurally characterized providing a view of a three-domain module with a cyclization-adenylation-PCP architecture [74]. In *Streptomyces* sp. Sp080513GE-23, FmoA3 works with other NRPS proteins FmoA2, FmoA4, and FmoA5, to produce an indole peptide containing a methyloxazoline ring with weak radical scavenging activity [74]. The biosynthesis starts with FmoA2, comprising adenylation-PCP-cyclization-methyl transferase domains, activating 6-chloro-4-hydroxy-indole-3-carboxylic acid, and the adenylation domain of FmoA3, activating α-methyl-L-serine [75]. The cyclization domain of FmoA2 first forms a peptide bond between the indole derivative and α-methyl-L-serine forming the dipeptide on the FmoA3 PCP domain. The dipeptide then migrates to the FmoA3 cyclization domain, which catalyzes cyclization of Ser to make the oxazoline ring. The dipeptide is then transferred to FmoA4 for further elongation by FmoA4 and FmoA5 which add L-Ala and L-Ser, respectively. Cyclization domains are known to perform both peptide bond formation and heterocyclization. However, FmoA3 cyclization domain only performs cyclization after FmoA2 performs peptide bond formation. The separated enzyme activity of cyclization domain between two modules is similar to tandom cyclization domains, where each domain performs one of the peptide bond formation and heterocyclization steps, observed previously in vibriobactin biosynthesis [76].

Three crystal structures of the FmoA3 module were solved, providing structures of unliganded state, as well complexed with AMP-PNP or α-methyl-L-Seryl-AMP (*see* Fig. 7, PDB **6LTB**). All three structures used an *apo* protein as the pantetheine binding serine residue was mutated to alanine. The crystal structure with α-methyl-L-seryl-AMP shows a dimeric structure as head-to-tail homodimer, while the other crystal structures showed similar dimeric interface with protomer from another asymmetric unit. The dimeric organization of FmoA3 in solution was confirmed by size exclusion chromatography-multiangle light scattering (SEC-MALS) and further supported by PISA analysis of the crystal structures. Finally, a cryo-EM structure of *apo* form also showed similar dimeric arrangement of two FmoA3 proteins. While most NRPS proteins were thought to function as monomers, several recent studies (see also PchE, Subheading 3.8) that employ cryo-EM have identified modular NRPSs to adopt a dimeric form.

FmoA3 revealed that the interface between the cyclization domain, which is structurally similar to condensation domains, and the adenylation domain interface is similar to known module structures with condensation and adenylation domains. The C-terminal lobe of cyclization/condensation domain interacts with core N-terminal domain of the adenylation domain. This interaction is governed primarily by the loop between α1 and β1 strand, α3 and α7 of the C-terminal lobe of cyclization/condensation domain, which interacts with a loop between two strands of core adenylation domain. Orientations of the two domains in FmoA3 compared to known module structures are also similar, indicating various modules utilize condensation–adenylation domain interface as anchor to perform various reactions of NRPS

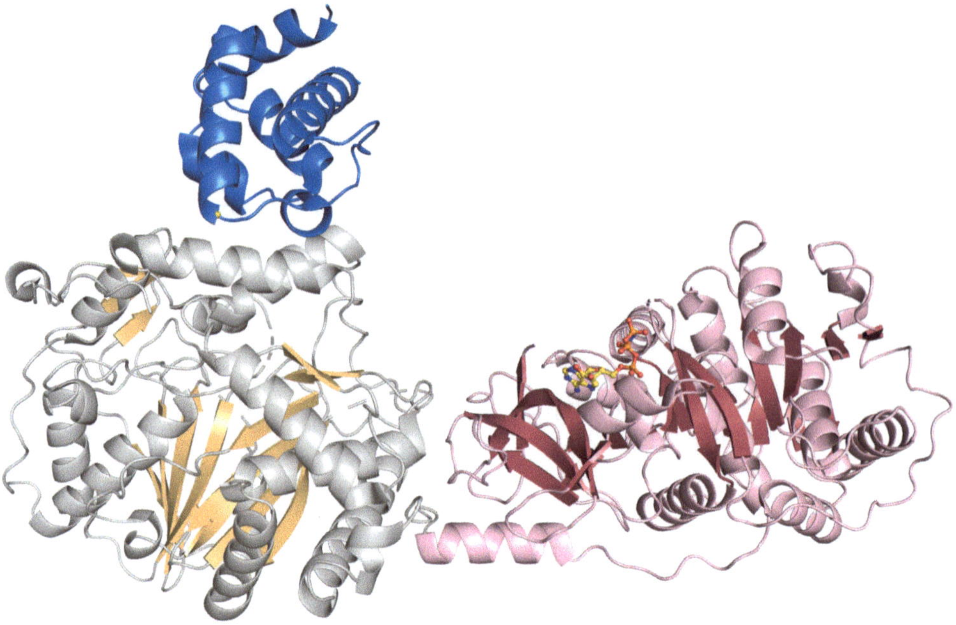

Fig. 7 Structure of FmoA3 (PDB: **6LTB**). The crystal structure of FmoA3 contains the cyclization domain (pale orange sheets) and N-terminal subdomain of the adenylation domain, which interact in a manner that is similar to the condensation–adenylation domain interface. The *apo*-PCP domain is positioned near the condensation domain, with the site of pantetheinylation highlighted with a gold sphere. A molecule of AMPPNP in the adenylation domain active site is also shown

cycle. One crystal structure revealed adenylation C-terminal subdomain in adenylation conformation when bound to AMP and α-methyl-L-serine, while PCP domain was observed at an unusual position near acceptor site in two crystal structures. However, the phosphopantetheine-binding serine residue in PCP was observed facing away from the active site of cyclization domain indicating the observed position of PCP, illustrating that this may be an effect of crystal packing. The structures also provided the foundation for analysis of the role of several catalytic residues in the cyclization domains of both the FmoA2 and FmoA3 proteins confirming the role of the two domains in the peptide bond formation and cyclization steps, respectively.

3.7 ObiF1

Another NRPS enzyme that possesses a similar domain architecture to AB3403 and SrfA-C is the N-acyl-α-amino-β-lactone-producing enzyme ObiF1. ObiF1 forms the second module of the NRPS system from *Burkholderia diffusa* that produces the antimicrobial β-lactone obafluorin [77, 78]. ObiF1 has a domain organization of condensation–adenylation–PCP–thioesterase–MLP, allowing the structure to highlight the β-lactone-producing thioesterase domain, as well as an unusual C-terminal MLP [69]. The study also showed that an interaction between this MbtH-like protein

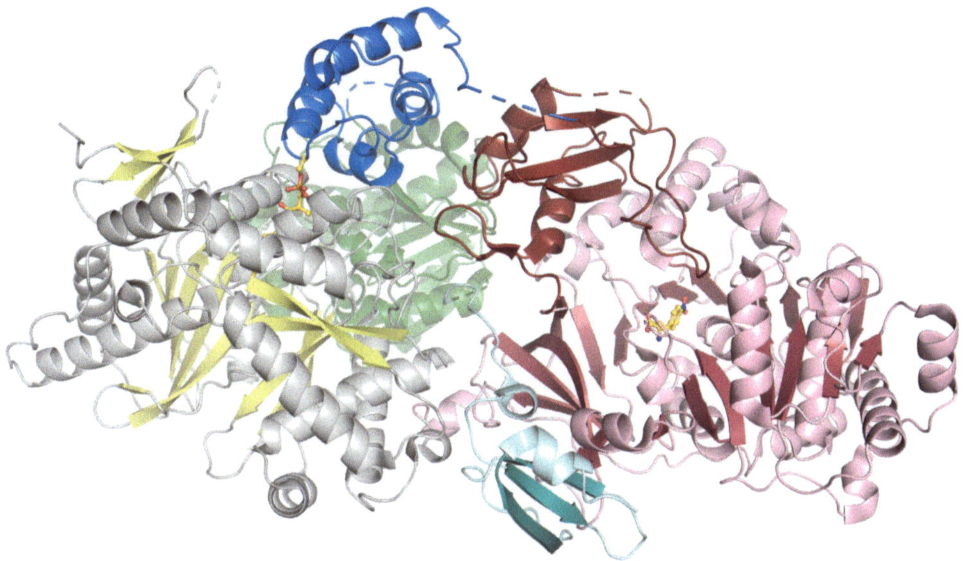

Fig. 8 Structure of ObiF1 (PDB: **6N8E**). The structure of the ObiF1 termination module illustrates a catalytic conformation similar to AB3403. The *holo*-PCP interacts with the condensation domain, while the adenylation domain is bound to a molecule of β-hydroxy-*p*-nitro-homophenylalanine. A linker joins the thioesterase domain and the C-terminal MLP domain that binds to the adenylation domain

(MLP) with an upstream adenylation domain is necessary for the proper functioning of ObiF1.

Similar to the structure of AB3403, the phosphopantetheine arm of the PCP domain is docked into the condensation domain active site cleft (*see* Fig. 8, PDB **6N8E**). Comparison of ObiF1 with homologous multidomain NRPS enzymes showed the condensation domain adopted a more closed conformation, similar to that observed in the calcium-dependent antibiotic (CDA) synthetase [79]. The adenylation domain adopted the catalytic adenylate-forming conformation, and the thioesterase domain illustrated the catalytic triad configuration consistent with other serine and cysteine hydrolases.

The most unusual feature of the ObiF1 structure was guided by the presence of the C-terminal MLP domain. The ObiF1 MLP adopted the common interaction with the upstream adenylation domain, first observed in the structure of SlgN1 [19]. To accommodate the upstream interaction, the linker that connects the thioesterase to the MLP passes through a channel formed between the adenylation and condensation domains, interacting with several residues in the adenylation domain.

Upon observing this unexpected interaction between the C-terminal MLP domain with the upstream adenylation domain, functional assays were conducted to probe the MLP dependence of ObiF1 and ensure that the observed structural interaction was indeed important for protein activity or stability [69]. A truncated

ObiF1 lacking the MLP as well as mutated ObiF1 proteins with changes in the adenylation domain that disrupt the MLP interaction was unstable and aggregated during purification. Furthermore, in vitro reconstitution assays showed that disrupting MLP interactions significantly affects ObiF1 activity, demonstrating that the terminal MLP interactions with the upstream adenylation domain were required for both the stability and activity of ObiF1. Activity could be restored by providing the MLP *in trans* as a free-standing protein, illustrating that the fusion of the MLP to the ObiF1 module was not critical for activity. Whereas most MLPs are free-standing, a few have been observed fused to the N-terminus of an adenylation domain with which they interact. The ObiF1 MLP showed that internal MLPs can also interact with non-adjacent adenylation domains to influence the overall structure of an NRPS enzyme.

3.8 PchE

Pyochelin is a non-ribosomal peptide siderophore synthesized by a cluster containing PchD, PchE, PchF, and PchG [80, 81]. PchD is a salicyl-AMP ligase, which activates salicylate and transfers it to N-terminal aryl carrier protein (ArCP) domain of PchE. PchE is an NRPS protein with an aryl carrier protein, cyclization, adenylation [epimerization], and PCP domain architecture. Interestingly, the epimerization domain is inserted into a loop on the small C-terminal subdomain of the adenylation domain [82]. The interrupted adenylation domain of PchE activates a cysteine and transfers it to cyclization domain which performs condensation and cyclodehydration to introduce the first thiazoline ring. The E_1 domain converts L-cysteine to D- isomer, and the salicyl-thiazoline intermediate is transferred to PchF for addition of a second cysteine that also gets cyclized as well as methylated. Meanwhile, the intermediate bound to the PchF PCP domain is delivered to the free-standing PchG reductase to reduce the second thiazoline to thiazolidine ring as well as to the methyltransferase domain that is embedded within the PchF adenylation domain [83, 84]. Finally, the product is transferred to TE domain which performs a hydrolytic release.

Cryo-EM structures of PchE were solved in three different conformations that represented various stages of the catalytic cycle [85]. Surprisingly, PchE in all three conformations was observed in the dimeric state where two monomers interact in head-to-tail orientation as also reported for FmoA3 module structure [74]. The cyclization domain of PchE has a similar fold as in condensation domain. In conformation 1, the loaded ArCP domain is observed interacting cyclization domain. Interestingly, the presence of interrupted epimerization domain did not affect the domain rotation required for two reactions of adenylation domain. The adenylation domain C-terminal subdomain in conformation 1 was observed in thioester-forming conformation with Cys-AMP

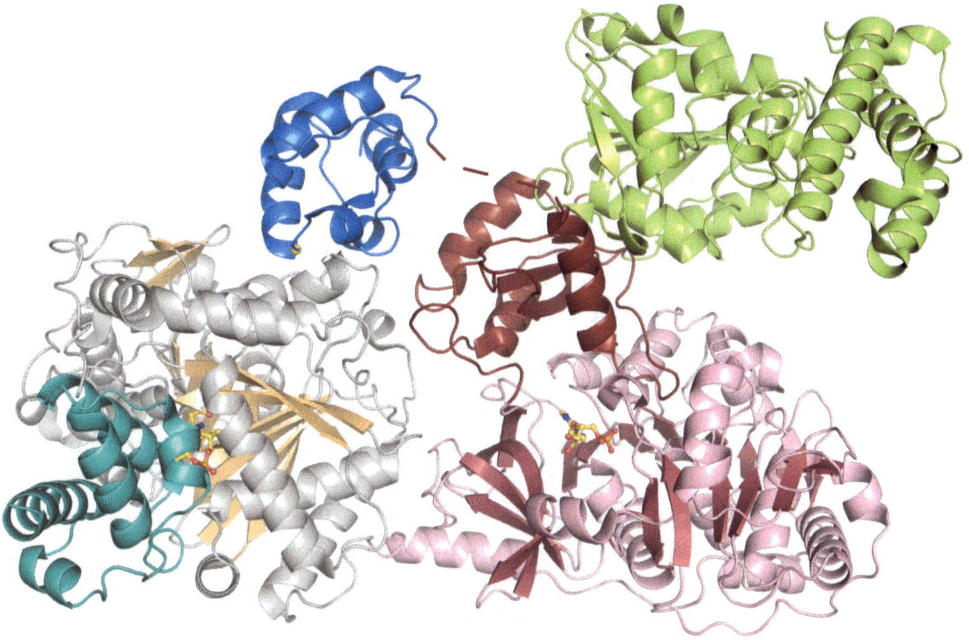

Fig. 9 Structure of PchE (PDB: **7EN2**). Conformation 2 of PchE shows the aryl carrier protein domain (teal) and downstream PCP (blue) meeting at the cyclization domain (pale orange sheets). The epimerization domain (green) interrupts the C-terminal subdomain of the adenylation domain. The panetheine from the aryl carrier domain and a molecule of AMP are also present. The Cα of the pantetheinylation site of the PCP is shown with a gold sphere

ligand and Mg^{2+} ion. However, a low-resolution map of PCP was observed bound to E_1 domain, indicating an epimerization reaction conformation.

In conformation 2 (*see* Fig. 9, PDB **7EN2**), the adenylation domain is in the adenylate-forming conformation with A10 motif lysine interacting with AMP. The interrupted C-terminal subdomain has close contacts with embedded epimerization domain. As a result, the epimerization and C-terminal adenylation domains rotate as a rigid body adopting the adenylation and thioester-forming conformations. More importantly, the two carrier domains were observed to interact at the cyclization domain indicating the condensation state. In conformation 3, the adenylation domain shows both adenylation and thiolation-forming conformations in the two chains of the dimer. The downstream PCP domain that is loaded with the aryl-thiazoline condensation and cyclization product remains within the cyclization domain in a post-condensation conformation.

Overall, three conformations of PchE observed in cryo-EM maps provide insight into various catalytic steps of the PchE reaction cycle. Starting with ArCP interacting with the cyclization domain at donor site in conformation 1 represents delivery of ArCP and salicylic acid to PchE. The adenylation domain adopts the thiolate-forming conformation. Conformation 2 shows both carrier domains interacting at the cyclization domain, representing the condensation and heterocyclization catalytic steps.

The orientation and interface of the cyclization and N-terminal adenylation domains are similar to known NRPS module structures and in all three conformations of PchE. In contrast, the dynamic carrier domains, the adenylation C-terminal subdomain, and the epimerization domains showed various positions around the platform provided by cyclization and core adenylation domain during the catalytic cycle. The didomain interface was also found important for dimerization. In conformation 1, PchE form a "H" shaped head-to-tail dimer where core didomain from each monomer interacts and makes the bridge. Conformation 2 was the most compact and complicated dimer where didomain platform along with the adenylation C-terminal subdomain and the PCP domains from one monomer interacts with second monomer in head-to-tail orientations. Finally, conformation 3 has a similar interface of the dimer as in conformation 1. However, unlike FmoA3, where dimers are quite planar to each other, PchE, by virtue of the dynamic domains that surround the cyclization-adenylation platform, displayed a slightly curved arch-shaped dimeric architecture. Intriguingly, though PchE was observed as dimer in all three conformations, the carrier domains were not observed to cross-react to any domain of the other monomer. Instead, the ArCP and PCP within the same monomer interacted with the cyclization domain. Thus, despite the dimeric structure, PchE is proposed to be monofunctional where each monomer functions individually.

3.9 BmdBC

The biosynthetic gene cluster for the tryptamide algicide bacillamide D has three proteins, namely, BmdA, BmdB, and BmdC. BmdA is a tryptophan decarboxylase that converts L-Trp to tryptamine [86]. BmdB is a two-module NRPS protein with an adenylation-PCP-cyclization-adenylation-PCP-condensation domain architecture. The two BmdB adenylation domains activate L-Ala and L-Cys, respectively, allowing the cyclization domain to perform peptide bond formation followed by heterocyclization, forming a thiazoline ring in the dipeptide intermediate. The BmdB condensation domain condenses the dipeptide with tryptamine yielding pro-bacillamide. BmdC is an oxidase which performs dehydrogenation of the thiazoline ring of pro-bacillamide to the final thiazole in the product bacillamide D.

This cluster served as a target for structural studies, providing the first structure of a NRPS module complexed with an oxidase protein [87]. The BmdB module 2 ($BmdB_{M2}$), containing cyclization, adenylation, and PCP domains, complexed with BmdC ($BmdB_{M2}$–BmdC complex) was examined using both X-ray crystallography (*see* Fig. 10, PDB **7LY2**) and cryo-EM. A third structure was solved using in situ proteolyzed complex of only $BmdB_{M2}$ adenylation domain with the oxidase domain of BmdC by X-ray

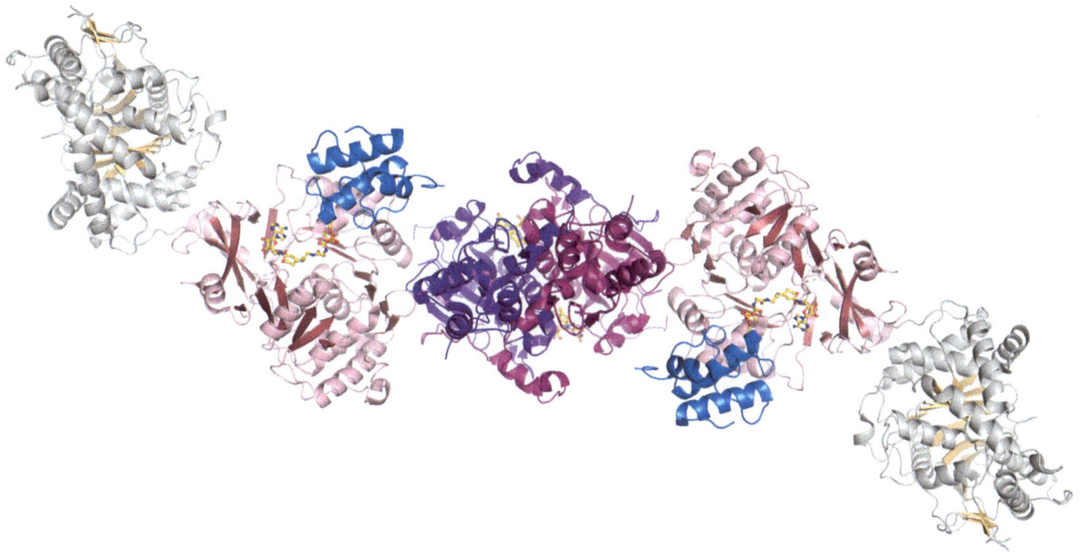

Fig. 10 Structure of BmdBC dimer (PDB: **7LY2**). The structure of complex between BmdB and BmdC shows a dimeric complex. Each monomer of the BmdC dimer (different shades of purple) interacts with a module of BmdB containing the cyclization domain (orange sheets) plus the adenylation domain. The C-terminal subdomain is disordered in the structure. The PCP is shown along with the pantetheine cofactor bound to an AVS inhibitor in the adenylation domain active site. Each subunit of the oxidase dimer contains a flavin cofactor, also highlighted in orange

crystallography. The $BmdB_{M2}$–BmdC complex crystal structure showed an elongated dimer with protomer from another asymmetric unit. Two $BmdB_{M2}$ protomers are at the two ends of the structure, while two BmdC chains form the oxidase dimer at the center. The three active sites for adenylation, cyclization, and oxidation in their respective domains form an in-plane catalytic arc. Each active site opens towards the concave surface which decreases the travel distance of the PCP required for transport of the intermediates across Bmd_{M2} module and BmdC. The crystal structure of in situ proteolyzed complex $BmdB_{M2}$ adenylation:BmdC revealed a network of hydrogen bonding and salt bridges at the interface. Compared to other known structures of NRPS modules, the cyclization-adenylation didomain makes a similar stable interface, while the distal N-lobe of Cy2 shows some flexibility in both EM and crystal structures.

The Cryo-EM map showed a similar dimer with some variability in position of the $BmdB_{M2}$ module compared to the crystal structure. The structure of the oxidase BmdC was also solved identifying a similar dimer to that seen in the $BmdB_{M2}$:BmdC complex. Size exclusion chromatography experiments of mixing BmdB with increasing amounts of BmdC led to shift the peak earlier than expected $BmdB_{M2}$:BmdC 1:2 complex. The molecular weight of the complex in solution corresponded to dimer of BmdC and two molecules of BmdB as observed in crystal structure and

Cryo-EM map. Hence, each chain of the BmdC dimer is able to interact with one molecule of BmdB with similar interactions.

Apart from dimerization, the interaction of BmdC to BmdB was required for oxidation of pro-bacillamide. Free pro-bacillamide in solution is not oxidized by BmdC or when BmdB and BmdC are separated by a 10 kDa MW membrane. However, only when both BmdB and BmdC are in same solution with FMN, L-alanine, L-cysteine, tryptamine, and ATP, oxidation of pro-bacillamide to bacillamide was observed. On other hand, mutation of BmdC to disrupt the interaction with BmdB reduced the oxidation by 50%, indicating that BmdC acts optimally when in complex with BmdB during the assembly-line synthesis. However, formation of the larger complex does not appear to be provide a catalytic advantage for assembly-line synthesis of pro-bacillamide. Assays with individual Ser mutants of BmdB PCP_1 and PCP_2 that abolish pantetheinylation and force the handoff of peptide intermediates from one NRPS chain to another could not complement each other for pro-bacillamide synthesis in the presence of BmdC. This indicates that the conventional assembly line biosynthesis down a single protein chain is employed by the Bmd NRPS system.

4 Conclusions and Future Directions

Recent determination of many multidomain structures that highlight the conformations of complete NRPS modules and dimodular structures advances our understanding of the structural biology of NRPS proteins. For many years, the SrfA-C [46] provided the only view of the organization of NRPS domains. Recent technical advances ranging from NRPS-specific approaches that derived from the earlier studies of smaller protein constructs such as the use of chemical probes [31, 36], as well as advances in the miniaturization of crystallization techniques and higher powered synchrotron X-ray sources, have resulted in a number of new multidomain structures. These new structures provide valuable insight into the conformational dynamics that NRPSs employ in their assembly line enzymology. The future for NRPS structural biology remains bright, with recent examples of the use of single particle analysis by cryo-EM into the field [74, 85, 87].

Some features of NRPS catalysis remain consistent between the different protein systems under study. First, as originally suggested by the SrfA-C structure [46], the condensation (or cyclization) domain and the N-terminal subdomain of the adenylation domain form a relatively stable interaction upon which the mobile domains migrate to allow delivery of the PCP to different active sites. Second, the adoption of two specific catalytic conformations of the adenylation domain, defined originally by us as domain alternation [38], enables the small C-terminal subdomain to adopt two conformations. Although the PCP and the C-terminal subdomain do not move as a rigid body, the two catalytic conformations of the adenylation domain appear to promote delivery of the PCP between the adenylation and condensation domains. Importantly, the ability of an NRPS module to adopt a conformation first approximated by SrfA-C [46] and later observed precisely with AB3403 and ObiF1 [37, 69] in which the PCP is delivered to the condensation domain and the adenylation domain adopts the adenylate-forming reaction demonstrates that the NRPS module can operate in an efficient three-state structural cycle. While the loaded PCP is awaiting delivery of the upstream peptide, the adenylation domain is able to activate a new amino acid residue to "prime the pump" for another round once the extended peptide is delivered downstream to the next module or termination domain [30, 37].

Additionally, though different states have been captured in the NRPS modular structures and some domains such as the thioesterase domain have proven to be highly mobile and potentially adopting a position in the crystal lattice based primarily on where they can fit, the didomain interactions [68] between the PCP and different catalytic domains have been remarkably consistent. Even with limited sequence homology at these protein interfaces, there do appear to be favored interactions that are adopted as the PCP approaches a catalytic domain. These latest structures are providing more insights into the ability of modular NRPS enzymes to accommodate additional domains that catalyze critical steps to diversify the NRPS products.

It remains an exciting time for NRPS structural biology and the field is poised for additional insights through the use of cryo-EM. The ability of artificial intelligence approaches to protein structure [88, 89], while still challenged by the difficulties of multidomain proteins (at least currently), also provides great promise toward the development of great insights into the unique activities of different NRPS biosynthetic clusters. This volume of *Methods in Molecular Biology* should provide additional insights into tools and techniques that will allow for continued advances in our understanding of these fascinating multidomain enzymes.

Acknowledgement

Work in our lab is supported with funding from the National Institutes of General Medical Sciences, NIH (GM-136235).

Conflict of Interest Statement The authors declare no conflicts of interest with the contents of this article.

References

1. Newman DJ, Cragg GM (2020) Natural products as sources of new drugs over the nearly four decades from 01/1981 to 09/2019. J Nat Prod 83(3):770–803. https://doi.org/10.1021/acs.jnatprod.9b01285
2. Süssmuth RD, Mainz A (2017) Nonribosomal peptide synthesis-principles and prospects. Angew Chem Int Ed Engl 56(14): 3770–3821. https://doi.org/10.1002/anie.201609079
3. Keating TA, Ehmann DE, Kohli RM, Marshall CG, Trauger JW, Walsh CT (2001) Chain termination steps in nonribosomal peptide synthetase assembly lines: directed acyl-S-enzyme breakdown in antibiotic and siderophore biosynthesis. Chembiochem 2(2):99–107
4. Horsman ME, Hari TP, Boddy CN (2016) Polyketide synthase and non-ribosomal peptide synthetase thioesterase selectivity: logic gate or a victim of fate? Nat Prod Rep 33(2): 183–202. https://doi.org/10.1039/c4np00148f
5. Walsh CT (2016) Insights into the chemical logic and enzymatic machinery of NRPS assembly lines. Nat Prod Rep 33(2):127–135. https://doi.org/10.1039/c5np00035a
6. Walsh CT (2008) The chemical versatility of natural-product assembly lines. Acc Chem Res 41(1):4–10
7. Sieber SA, Marahiel MA (2005) Molecular mechanisms underlying nonribosomal peptide synthesis: approaches to new antibiotics. Chem Rev 105(2):715–738
8. Marahiel MA (1997) Protein templates for the biosynthesis of peptide antibiotics. Chem Biol 4(8):561–567
9. Finking R, Marahiel MA (2004) Biosynthesis of nonribosomal peptides1. Annu Rev Microbiol 58:453–488. https://doi.org/10.1146/annurev.micro.58.030603.123615
10. Lambalot RH, Gehring AM, Flugel RS, Zuber P, LaCelle M, Marahiel MA, Reid R, Khosla C, Walsh CT (1996) A new enzyme superfamily - the phosphopantetheinyl transferases. Chem Biol 3(11):923–936
11. Beld J, Sonnenschein EC, Vickery CR, Noel JP, Burkart MD (2014) The phosphopantetheinyl transferases: catalysis of a post-translational modification crucial for life. Nat Prod Rep 31(1):61–108. https://doi.org/10.1039/c3np70054b
12. Davidsen JM, Townsend CA (2012) In vivo characterization of nonribosomal peptide synthetases NocA and NocB in the biosynthesis of nocardicin a. Chem Biol 19(2):297–306. https://doi.org/10.1016/j.chembiol.2011.10.020
13. Davidsen JM, Bartley DM, Townsend CA (2013) Non-ribosomal propeptide precursor in nocardicin a biosynthesis predicted from adenylation domain specificity dependent on the MbtH family protein NocI. J Am Chem Soc 135(5):1749–1759. https://doi.org/10.1021/ja307710d
14. Townsend CA (2016) Convergent biosynthetic pathways to β-lactam antibiotics. Curr Opin Chem Biol 35:97–108. https://doi.org/10.1016/j.cbpa.2016.09.013
15. Quadri LE, Sello J, Keating TA, Weinreb PH, Walsh CT (1998) Identification of a mycobacterium tuberculosis gene cluster encoding the biosynthetic enzymes for assembly of the virulence-conferring siderophore mycobactin. Chem Biol 5(11):631–645
16. Drake EJ, Cao J, Qu J, Shah MB, Straubinger RM, Gulick AM (2007) The 1.8 a crystal structure of PA2412, an MbtH-like protein from the pyoverdine cluster of Pseudomonas aeruginosa. J Biol Chem 282(28):20425–20434
17. Felnagle EA, Barkei JJ, Park H, Podevels AM, McMahon MD, Drott DW, Thomas MG (2010) MbtH-like proteins as integral components of bacterial nonribosomal peptide synthetases. Biochemistry 49(41):8815–8817. https://doi.org/10.1021/bi1012854
18. Zhang W, Heemstra JR Jr, Walsh CT, Imker HJ (2010) Activation of the pacidamycin PacL adenylation domain by MbtH-like proteins. Biochemistry 49(46):9946–9947. https://doi.org/10.1021/bi101539b
19. Herbst DA, Boll B, Zocher G, Stehle T, Heide L (2013) Structural basis of the interaction of MbtH-like proteins, putative regulators of nonribosomal peptide biosynthesis, with adenylating enzymes. J Biol Chem 288(3): 1991–2003. https://doi.org/10.1074/jbc.M112.420182
20. Gaudelli NM, Long DH, Townsend CA (2015) β-Lactam formation by a non-ribosomal peptide synthetase during antibiotic biosynthesis. Nature 520:383–387. https://doi.org/10.1038/nature14100
21. Long DH, Townsend CA (2021) Acyl donor stringency and Dehydroaminoacyl intermediates in beta-lactam formation by a non-ribosomal peptide Synthetase. ACS Chem Biol 16(5):806–812. https://doi.org/10.1021/acschembio.1c00117
22. Gaudelli NM, Townsend CA (2014) Epimerization and substrate gating by a TE domain in

β-lactam antibiotic biosynthesis. Nat Chem Biol 10(4):251–258. https://doi.org/10.1038/nchembio.1456

23. Patel KD, d'Andrea FB, Gaudelli NM, Buller AR, Townsend CA, Gulick AM (2019) Structure of a bound peptide phosphonate reveals the mechanism of nocardicin bifunctional thioesterase epimerase-hydrolase half-reactions. Nat Commun 10(1):3868. https://doi.org/10.1038/s41467-019-11740-6
24. Cane DE, Walsh CT, Khosla C (1998) Harnessing the biosynthetic code: combinations, permutations, and mutations. Science 282(5386):63–68
25. Keating TA, Walsh CT (1999) Initiation, elongation, and termination strategies in polyketide and polypeptide antibiotic biosynthesis. Curr Opin Chem Biol 3(5):598–606
26. Marahiel MA, Stachelhaus T, Mootz HD (1997) Modular peptide Synthetases involved in nonribosomal peptide synthesis. Chem Rev 97(7):2651–2674
27. Ackerley DF, Challis GL, Cryle MJ (2018) Understanding biosynthetic protein-protein interactions. Nat Prod Rep 35(11):1118–1119. https://doi.org/10.1039/c8np90037j
28. Bloudoff K, Schmeing TM (2017) Structural and functional aspects of the nonribosomal peptide synthetase condensation domain superfamily: discovery, dissection and diversity. Biochim Biophys Acta 1865:1587. https://doi.org/10.1016/j.bbapap.2017.05.010
29. Bonhomme S, Dessen A, Macheboeuf P (2021) The inherent flexibility of type I non-ribosomal peptide synthetase multienzymes drives their catalytic activities. Open Biol 11(5):200386. https://doi.org/10.1098/rsob.200386
30. Gulick AM (2016) Structural insight into the necessary conformational changes of modular nonribosomal peptide synthetases. Curr Opin Chem Biol 35:89–96. https://doi.org/10.1016/j.cbpa.2016.09.005
31. Gulick AM, Aldrich CC (2018) Trapping interactions between catalytic domains and carrier proteins of modular biosynthetic enzymes with chemical probes. Nat Prod Rep 35(11):1156–1184. https://doi.org/10.1039/c8np00044a
32. Izore T, Cryle MJ (2018) The many faces and important roles of protein-protein interactions during non-ribosomal peptide synthesis. Nat Prod Rep 35(11):1120–1139. https://doi.org/10.1039/c8np00038g
33. Jaremko MJ, Davis TD, Corpuz JC, Burkart MD (2020) Type II non-ribosomal peptide synthetase proteins: structure, mechanism, and protein-protein interactions. Nat Prod Rep 37(3):355–379. https://doi.org/10.1039/c9np00047j
34. Little RF, Hertweck C (2022) Chain release mechanisms in polyketide and non-ribosomal peptide biosynthesis. Nat Prod Rep 39(1):163–205. https://doi.org/10.1039/d1np00035g
35. Reimer JM, Haque AS, Tarry MJ, Schmeing TM (2018) Piecing together nonribosomal peptide synthesis. Curr Opin Struct Biol 49:104–113. https://doi.org/10.1016/j.sbi.2018.01.011
36. Corpuz JC, Sanlley JO, Burkart MD (2022) Protein-protein interface analysis of the non-ribosomal peptide synthetase peptidyl carrier protein and enzymatic domains. Synth Syst Biotechnol 7(2):677–688. https://doi.org/10.1016/j.synbio.2022.02.006
37. Drake EJ, Miller BR, Shi C, Tarrasch JT, Sundlov JA, Allen CL, Skiniotis G, Aldrich CC, Gulick AM (2016) Structures of two distinct conformations of holo-non-ribosomal peptide synthetases. Nature 529(7585):235–238. https://doi.org/10.1038/nature16163
38. Gulick AM (2009) Conformational dynamics in the acyl-CoA synthetases, adenylation domains of non-ribosomal peptide synthetases, and firefly luciferase. ACS Chem Biol 4:811–827. https://doi.org/10.1021/cb900156h
39. Dekimpe S, Masschelein J (2021) Beyond peptide bond formation: the versatile role of condensation domains in natural product biosynthesis. Nat Prod Rep 38(10):1910–1937. https://doi.org/10.1039/d0np00098a
40. Samel SA, Czodrowski P, Essen LO (2014) Structure of the epimerization domain of tyrocidine synthetase a. Acta Crystallogr D Biol Crystallogr 70(Pt 5):1442–1452. https://doi.org/10.1107/S1399004714004398
41. Chen WH, Li K, Guntaka NS, Bruner SD (2016) Interdomain and Intermodule organization in epimerization domain containing nonribosomal peptide Synthetases. ACS Chem Biol 11(8):2293–2303. https://doi.org/10.1021/acschembio.6b00332
42. Mori S, Pang AH, Lundy TA, Garzan A, Tsodikov OV, Garneau-Tsodikova S (2018) Structural basis for backbone N-methylation by an interrupted adenylation domain. Nat Chem Biol 14(5):428–430. https://doi.org/10.1038/s41589-018-0014-7
43. Reimer JM, Aloise MN, Harrison PM, Schmeing TM (2016) Synthetic cycle of the initiation module of a formylating nonribosomal peptide

synthetase. Nature 529(7585):239–242. https://doi.org/10.1038/nature16503

44. Kinatukara P, Patel KD, Haque AS, Singh R, Gokhale RS, Sankaranarayananan R (2016) Structural insights into the regulation of NADPH binding to reductase domains of nonribosomal peptide synthetases: a concerted loop movement model. J Struct Biol 194(3): 368–374. https://doi.org/10.1016/j.jsb.2016.03.014
45. Miller BR, Gulick AM (2016) Structural biology of nonribosomal peptide synthetases. Methods Mol Biol 1401:3–29. https://doi.org/10.1007/978-1-4939-3375-4_1
46. Tanovic A, Samel SA, Essen LO, Marahiel MA (2008) Crystal structure of the termination module of a nonribosomal peptide synthetase. Science 321(5889):659–663
47. Allen CL, Gulick AM (2014) Structural and bioinformatic characterization of an Acinetobacter baumannii type II carrier protein. Acta Crystallogr D Biol Crystallogr 70(Pt 6): 1718–1725. https://doi.org/10.1107/S1399004714008311
48. Adams MD, Goglin K, Molyneaux N, Hujer KM, Lavender H, Jamison JJ, MacDonald IJ, Martin KM, Russo T, Campagnari AA, Hujer AM, Bonomo RA, Gill SR (2008) Comparative genome sequence analysis of multidrug-resistant Acinetobacter baumannii. J Bacteriol 190(24):8053–8064. https://doi.org/10.1128/JB.00834-08
49. Rumbo-Feal S, Gomez MJ, Gayoso C, Alvarez-Fraga L, Cabral MP, Aransay AM, Rodriguez-Ezpeleta N, Fullaondo A, Valle J, Tomas M, Bou G, Poza M (2013) Whole transcriptome analysis of *Acinetobacter baumannii* assessed by RNA-sequencing reveals different mRNA expression profiles in biofilm compared to planktonic cells. PLoS One 8(8):e72968. https://doi.org/10.1371/journal.pone.0072968
50. Rumbo-Feal S, Perez A, Ramelot TA, Alvarez-Fraga L, Vallejo JA, Beceiro A, Ohneck EJ, Arivett BA, Merino M, Fiester SE, Kennedy MA, Actis LA, Bou G, Poza M (2017) Contribution of the *a. baumannii* A1S_0114 gene to the interaction with eukaryotic cells and virulence. Front Cell Infect Microbiol 7:108. https://doi.org/10.3389/fcimb.2017.00108
51. Clemmer KM, Bonomo RA, Rather PN (2011) Genetic analysis of surface motility in Acinetobacter baumannii. Microbiology 157(Pt 9): 2534–2544. https://doi.org/10.1099/mic.0.049791-0
52. Conti E, Stachelhaus T, Marahiel MA, Brick P (1997) Structural basis for the activation of phenylalanine in the non- ribosomal biosynthesis of gramicidin S. EMBO J 16(14): 4174–4183
53. May JJ, Kessler N, Marahiel MA, Stubbs MT (2002) Crystal structure of DhbE, an archetype for aryl acid activating domains of modular nonribosomal peptide synthetases. Proc Natl Acad Sci U S A 99:12120–12125
54. Miethke M, Marahiel MA (2007) Siderophore-based iron acquisition and pathogen control. Microbiol Mol Biol Rev 71(3):413–451
55. Lamb AL (2015) Breaking a pathogen's iron will: inhibiting siderophore production as an antimicrobial strategy. Biochim Biophys Acta 1854(8):1054–1070. https://doi.org/10.1016/j.bbapap.2015.05.001
56. Ehmann DE, Shaw-Reid CA, Losey HC, Walsh CT (2000) The EntF and EntE adenylation domains of Escherichia coli enterobactin synthetase: sequestration and selectivity in acyl-AMP transfers to thiolation domain cosubstrates. Proc Natl Acad Sci U S A 97(6): 2509–2514
57. Gehring AM, Mori I, Walsh CT (1998) Reconstitution and characterization of the Escherichia coli enterobactin synthetase from EntB, EntE, and EntF. Biochemistry 37(8): 2648–2659
58. Drake EJ, Nicolai DA, Gulick AM (2006) Structure of the EntB multidomain nonribosomal peptide synthetase and functional analysis of its interaction with the EntE adenylation domain. Chem Biol 13(4):409–419
59. Liu J, Duncan K, Walsh CT (1989) Nucleotide sequence of a cluster of Escherichia coli enterobactin biosynthesis genes: identification of entA and purification of its product 2,3-dihydro-2,3-dihydroxybenzoate dehydrogenase. J Bacteriol 171(2):791–798
60. Nahlik MS, Brickman TJ, Ozenberger BA, McIntosh MA (1989) Nucleotide sequence and transcriptional organization of the Escherichia coli enterobactin biosynthesis cistrons entB and entA. J Bacteriol 171(2):784–790
61. Rusnak F, Sakaitani M, Drueckhammer D, Reichert J, Walsh CT (1991) Biosynthesis of the Escherichia coli siderophore enterobactin: sequence of the entF gene, expression and purification of EntF, and analysis of covalent phosphopantetheine. Biochemistry 30(11): 2916–2927
62. Raymond KN, Dertz EA, Kim SS (2003) Enterobactin: an archetype for microbial iron transport. Proc Natl Acad Sci U S A 100(7): 3584–3588
63. Miller BR, Drake EJ, Shi C, Aldrich CC, Gulick AM (2016) Structures of a nonribosomal peptide synthetase module bound to MbtH-like

proteins support a highly dynamic domain architecture. J Biol Chem 291(43): 22559–22571. https://doi.org/10.1074/jbc.M116.746297

64. Qiao C, Wilson DJ, Bennett EM, Aldrich CC (2007) A mechanism-based aryl carrier protein/thiolation domain affinity probe. J Am Chem Soc 129(20):6350–6351

65. Mitchell CA, Shi C, Aldrich CC, Gulick AM (2012) Structure of PA1221, a nonribosomal peptide synthetase containing adenylation and peptidyl carrier protein domains. Biochemistry 51(15):3252–3263. https://doi.org/10.1021/bi300112e

66. Sundlov JA, Gulick AM (2013) Structure determination of the functional domain interaction of a chimeric nonribosomal peptide synthetase from a challenging crystal with noncrystallographic translational symmetry. Acta Crystallogr D Biol Crystallogr 69(Pt 8): 1482–1492. https://doi.org/10.1107/S0907444913009372

67. Sundlov JA, Shi C, Wilson DJ, Aldrich CC, Gulick AM (2012) Structural and functional investigation of the intermolecular interaction between NRPS adenylation and carrier protein domains. Chem Biol 19(2):188–198. https://doi.org/10.1016/j.chembiol.2011.11.013

68. Corpuz JC, Podust LM, Davis TD, Jaremko MJ, Burkart MD (2020) Dynamic visualization of type II peptidyl carrier protein recognition in pyoluteorin biosynthesis. RSC Chem Biol 1(1): 8–12. https://doi.org/10.1039/c9cb00015a

69. Kreitler DF, Gemmell EM, Schaffer JE, Wencewicz TA, Gulick AM (2019) The structural basis of N-acyl-alpha-amino-beta-lactone formation catalyzed by a nonribosomal peptide synthetase. Nat Commun 10(1):3432. https://doi.org/10.1038/s41467-019-11383-7

70. Gahloth D, Dunstan MS, Quaglia D, Klumbys E, Lockhart-Cairns MP, Hill AM, Derrington SR, Scrutton NS, Turner NJ, Leys D (2017) Structures of carboxylic acid reductase reveal domain dynamics underlying catalysis. Nat Chem Biol 13(9):975–981. https://doi.org/10.1038/nchembio.2434

71. Kessler N, Schuhmann H, Morneweg S, Linne U, Marahiel MA (2004) The linear pentadecapeptide gramicidin is assembled by four multimodular nonribosomal peptide synthetases that comprise 16 modules with 56 catalytic domains. J Biol Chem 279(9): 7413–7419. https://doi.org/10.1074/jbc.M309658200

72. Reimer JM, Eivaskhani M, Harb I, Guarne A, Weigt M, Schmeing TM (2019) Structures of a dimodular nonribosomal peptide synthetase reveal conformational flexibility. Science 366(6466). https://doi.org/10.1126/science.aaw4388

73. Tarry MJ, Haque AS, Bui KH, Schmeing TM (2017) X-ray crystallography and electron microscopy of cross- and multi-module nonribosomal peptide synthetase proteins reveal a flexible architecture. Structure 25:783–793. https://doi.org/10.1016/j.str.2017.03.014

74. Katsuyama Y, Sone K, Harada A, Kawai S, Urano N, Adachi N, Moriya T, Kawasaki M, Shin-Ya K, Senda T, Ohnishi Y (2021) Structural and functional analyses of the Tridomain-nonribosomal peptide Synthetase FmoA3 for 4-Methyloxazoline ring formation. Angew Chem Int Ed Engl 60(26):14554–14562. https://doi.org/10.1002/anie.202102760

75. Muliandi A, Katsuyama Y, Sone K, Izumikawa M, Moriya T, Hashimoto J, Kozone I, Takagi M, Shin-ya K, Ohnishi Y (2014) Biosynthesis of the 4-methyloxazoline-containing nonribosomal peptides, JBIR-34 and -35, in Streptomyces sp. Sp080513GE-23. Chem Biol 21 (8):923-934:923. https://doi.org/10.1016/j.chembiol.2014.06.004

76. Marshall CG, Burkart MD, Keating TA, Walsh CT (2001) Heterocycle formation in vibriobactin biosynthesis: alternative substrate utilization and identification of a condensed intermediate. Biochemistry 40(35): 10655–10663

77. Schaffer JE, Reck MR, Prasad NK, Wencewicz TA (2017) β-Lactone formation during product release from a nonribosomal peptide synthetase. Nat Chem Biol 13(7):737–744. https://doi.org/10.1038/nchembio.2374

78. Scott TA, Heine D, Qin Z, Wilkinson B (2017) An L-threonine transaldolase is required for L-threo-β-hydroxy-α-amino acid assembly during obafluorin biosynthesis. Nat Commun 8: 15935. https://doi.org/10.1038/ncomms15935

79. Bloudoff K, Rodionov D, Schmeing TM (2013) Crystal structures of the first condensation domain of CDA Synthetase suggest conformational changes during the synthetic cycle of nonribosomal peptide Synthetases. J Mol Biol 425(17):3137–3150. https://doi.org/10.1016/j.jmb.2013.06.003

80. Quadri LE, Keating TA, Patel HM, Walsh CT (1999) Assembly of the Pseudomonas aeruginosa nonribosomal peptide siderophore pyochelin: in vitro reconstitution of aryl-4, 2-bisthiazoline synthetase activity from PchD, PchE, and PchF. Biochemistry 38(45): 14941–14954

81. Ronnebaum TA, Lamb AL (2018) Nonribosomal peptides for iron acquisition: pyochelin biosynthesis as a case study. Curr Opin Struct Biol 53:1–11. https://doi.org/10.1016/j.sbi.2018.01.015
82. Patel HM, Tao J, Walsh CT (2003) Epimerization of an L-cysteinyl to a D-cysteinyl residue during thiazoline ring formation in siderophore chain elongation by pyochelin synthetase from Pseudomonas aeruginosa. Biochemistry 42(35):10514–10527. https://doi.org/10.1021/bi034840c
83. Meneely KM, Lamb AL (2012) Two structures of a thiazolinyl imine reductase from Yersinia enterocolitica provide insight into catalysis and binding to the nonribosomal peptide synthetase module of HMWP1. Biochemistry 51(44):9002–9013. https://doi.org/10.1021/bi3011016
84. Ronnebaum TA, McFarlane JS, Prisinzano TE, Booker SJ, Lamb AL (2019) Stuffed methyltransferase catalyzes the penultimate step of Pyochelin biosynthesis. Biochemistry 58(6):665–678. https://doi.org/10.1021/acs.biochem.8b00716
85. Wang J, Li D, Chen L, Cao W, Kong L, Zhang W, Croll T, Deng Z, Liang J, Wang Z (2022) Catalytic trajectory of a dimeric nonribosomal peptide synthetase subunit with an inserted epimerase domain. Nat Commun 13(1):592. https://doi.org/10.1038/s41467-022-28284-x
86. Yuwen L, Zhang FL, Chen QH, Lin SJ, Zhao YL, Li ZY (2013) The role of aromatic L-amino acid decarboxylase in bacillamide C biosynthesis by bacillus atrophaeus C89. Sci Rep 3:1753. https://doi.org/10.1038/srep01753
87. Fortinez CM, Bloudoff K, Harrigan C, Sharon I, Strauss M, Schmeing TM (2022) Structures and function of a tailoring oxidase in complex with a nonribosomal peptide synthetase module. Nat Commun 13(1):548. https://doi.org/10.1038/s41467-022-28221-y
88. Jumper J, Evans R, Pritzel A, Green T, Figurnov M, Ronneberger O, Tunyasuvunakool K, Bates R, Zidek A, Potapenko A, Bridgland A, Meyer C, Kohl SAA, Ballard AJ, Cowie A, Romera-Paredes B, Nikolov S, Jain R, Adler J, Back T, Petersen S, Reiman D, Clancy E, Zielinski M, Steinegger M, Pacholska M, Berghammer T, Bodenstein S, Silver D, Vinyals O, Senior AW, Kavukcuoglu K, Kohli P, Hassabis D (2021) Highly accurate protein structure prediction with AlphaFold. Nature 596(7873):583–589. https://doi.org/10.1038/s41586-021-03819-2
89. Baek M, DiMaio F, Anishchenko I, Dauparas J, Ovchinnikov S, Lee GR, Wang J, Cong Q, Kinch LN, Schaeffer RD, Millan C, Park H, Adams C, Glassman CR, DeGiovanni A, Pereira JH, Rodrigues AV, van Dijk AA, Ebrecht AC, Opperman DJ, Sagmeister T, Buhlheller C, Pavkov-Keller T, Rathinaswamy MK, Dalwadi U, Yip CK, Burke JE, Garcia KC, Grishin NV, Adams PD, Read RJ, Baker D (2021) Accurate prediction of protein structures and interactions using a three-track neural network. Science 373(6557):871–876. https://doi.org/10.1126/science.abj8754

Part II

In Vivo and In Vitro Methods

Chapter 3

Using NMR Titration Experiments to Study *E. coli* FAS-II- and AcpP-Mediated Protein–Protein Interactions

Desirae A. Mellor, Javier O. Sanlley, and Michael Burkart

Abstract

Acyl carrier proteins (ACPs) are central to many primary and secondary metabolic pathways. In *E. coli* fatty acid biosynthesis (FAB), the central ACP, AcpP, transports intermediates to a suite of partner proteins (PP) for iterative modification and elongation. The regulatory protein–protein interactions that occur between AcpP and the PP in FAB are poorly understood due to the dynamic and transient nature of these interactions. Solution-state NMR spectroscopy can reveal information at the atomic level through experiments such as the 2D heteronuclear single quantum coherence (HSQC). The following protocol describes NMR HSQC titration experiments that can elucidate biomolecular recognition events.

Key words Fatty acid biosynthesis, FAS-II, Carrier proteins, AcpP, NMR spectroscopy, Titration experiments, HSQC

1 Introduction

Protein–protein interactions (PPIs) regulate essential biosynthetic pathways such as fatty acid biosynthesis, polyketide synthesis, and non-ribosomal peptide synthesis [1, 2]. These pathways are carrier protein dependent, as they rely on a central carrier protein that shuttles a covalently bound substrate to various partner proteins (PP) for modification and elongation. The PPIs between carrier protein's and the PPs in their respective pathways are crucial for proper function and product formation; however, the transient and dynamic nature of these interactions has left the molecular basis of these recognition events poorly understood. Here we will focus on the application of solution-phase protein NMR titration as a tool to provide restraints for *in silico* docking analysis, with a demonstration of these techniques upon fatty acid biosynthesis (FAB) in *Escherichia coli*.

FAB in *E. coli* has been extensively studied and these studies have provided much of the foundational knowledge for bacterial

Michael Burkart and Fumihiro Ishikawa (eds.), *Non-Ribosomal Peptide Biosynthesis and Engineering: Methods and Protocols*, Methods in Molecular Biology, vol. 2670, https://doi.org/10.1007/978-1-0716-3214-7_3,

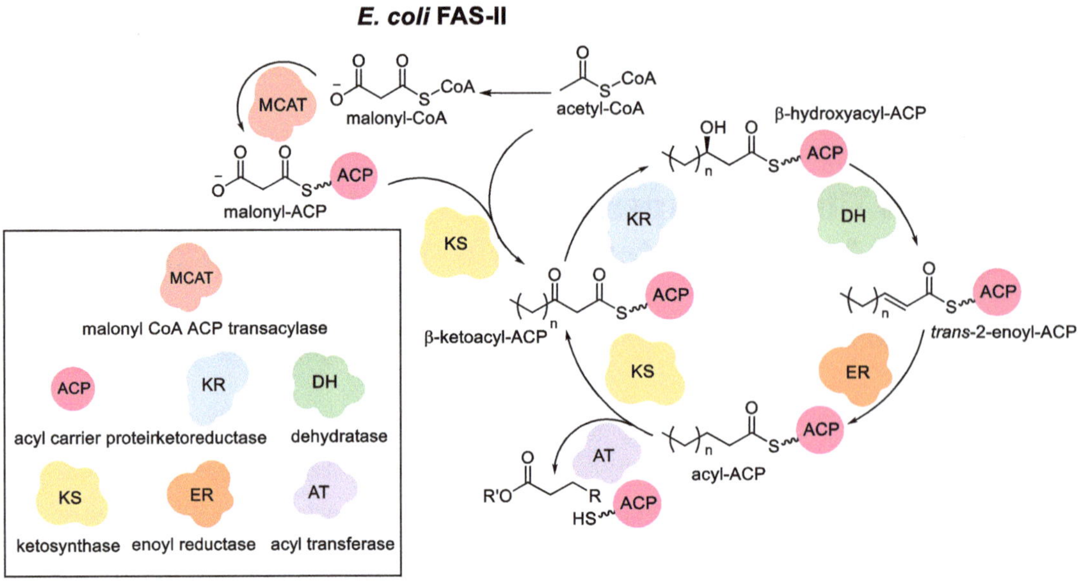

Fig. 1 Diagram of *E. coli* fatty acid biosynthesis (FAS-II) highlighting the varied PPI that are mediated by ACPs. In *E.coi* FAS-II, the ACP will iteratively interact with a suite of PP to produce the final product

FAB [3]. FAB occurs via two distinct organizations. Eukaryotes predominantly use the modular type I fatty acid synthase (FAS-I), which contains all proteins as domains on a large megasynthase. Typically, bacteria, such as *E. coli*, deploy the type II fatty acid synthase (FAS-II) which is comprised of discrete individual proteins that iteratively interact with a central carrier protein bearing a substrate or intermediate (Fig. 1).

In *E. coli* FAS-II, the central acyl carrier protein (ACP), AcpP, is the ACP responsible for transporting the growing substrate to each of the PP in the pathway for iterative modification and elongation [4]. A key feature of AcpP is its ability to sequester the substrate into the hydrophobic pocket to prevent premature hydrolysis. When AcpP comes into the appropriate proximity of a PP, the substrate undergoes "chain-flipping" into the pocket of the PP which has been found to be more hydrophobic than the pocket of AcpP. After the substrate has been modified, it is resequestered back into the AcpP pocket [5–7]. In addition to "chain-flipping," AcpP and the PP in FAS-II must orchestrate a series of complex events, and the full spectrum of the PPI in these events remains elusive. NMR Titration studies have provided valuable insight into the biomolecular recognition in *E. coli* PPI [8–13], but further studies are needed to fully understand the complete choreography.

NMR titration experiments have been validated for studying protein–small molecule and protein–ligand interactions. Additional methods can be used to study binding and kinetics, but NMR titration experiments elucidate PPI at the atomic level providing a more in-depth, comprehensive analysis [14, 15]. In conventional NMR titration experiments, the concentration of the receptor protein is held constant, and the ligand or small molecule concentration is steadily altered. This gradient alteration reveals the dynamic changes that occur at the surface residues of the receptor protein that are induced by the ligand or small molecule. These changes are made apparent through the mapping of complexation-induced changes in the chemical environment of the proteins, also known as chemical shift perturbations (CSPs). By mapping the CSP, the locations of binding sites, allosteric regulation, surface interactions, and residue-specific information can be revealed [16]. In ACP-mediated interactions, the ACP carrier shuttles the substrate that is covalently attached to a conserved serine residue to the various PP in the pathway. In these interactions, the PP functions as the receptor and the ACP functions as the ligand. The provided method is reversed from traditional methods and holds the AcpP concentration constant while varying the PP concentration. This change in convention allows a refined focus that elucidates AcpP-residue-specific information that has been previously inaccessible.

2 Materials and Methods for Proteins

2.1 M9 Minimal Media

Isotopically labeled proteins are grown in minimal media with ^{15}N and/or ^{13}C isotopes added to control the carbon and nitrogen sources. This protocol for M9 minimal media with ^{15}N is as follows:

1. Weigh out the following reagents:
   (a) 6 g Na_2HPO_4.
   (b) 3 g KH_2PO_4.
   (c) 0.5 g NaCl.
   (d) 0.25 g $MgSO_4*7H_2O$.
   (e) 0.015 g $CaCl_2*2H_2O$.
   (f) 0.1 g thiamine.
   (g) 0.003 g $FeSO_4*7H_2O$.
   (h) 8 g D-glucose.
   (i) 1 g ^{15}N-NH_4Cl.
2. Dissolve all reagents in 1 L of DI water.
3. Sterile filter the prepared media into a clean, dry, autoclaved reagent bottle with a lid.

2.2 Lysis Buffer

1. Weigh out the following reagents
   (a) 6.057 g Tris Base.
   (b) 8.766 g NaCl.
   (c) 100 mL glycerol.
2. Add all reagents to 500 mL of ultra-pure water in a 1 L reagent bottle.
3. Adjust the pH to 7.4.
4. Fill to 1 L with ultra-pure water.
5. Final composition is 50 mM Tris Base, 150 mM NaCl, 10% glycerol, pH 7.4.

2.3 Urea-PAGE (Polyacrylamide Gel Electrophoresis)

Urea-PAGE is a conformationally sensitive electrophoresis method that allows the resolution of proteins that exist in a multitude of states. This method is particularly useful for monitoring the substrate modification of carrier proteins.

1. Set up a gel casting stand such as the Mini-PROTEAN Tetra Cell Casting Stand.
2. Prepare 20% Urea Gels.
   (a) Prepare Resolving layer.
      (i) In a 15 mL falcon tube, combine the following:
         1. 3.1 mL 10 M Urea.
         2. 6.1 mL 40% acrylamide.
         3. 3.2 mL 1.5 M Tris Base, pH 8.8.
         4. 0.125 mL ice cold 10% ammonium persulfate (APS) w/v.
         5. 0.005 mL *N, N, N, N′*-Tetramethyl-ethylenediamine (TEMED).
      (ii) After all reagents are combined, invert the falcon tube 2–3 times to thoroughly mix. Quickly pour into the casting gel plates up to approximately 1.5 cm from the top.
      (iii) Fill the remaining 1.5 cm with isopropanol (IPA).
      (iv) Allow to polymerize for 45 minutes or until set. ***There will be a small remainder in the falcon tube. When the remainder has polymerized, the gels will be set***.
      (v) After polymerization, decant off the IPA layer.
   (b) Prepare the Stacking layer.

(i) In a 15 mL falcon tube, combine the following:
1. 2.9 mL ultra-pure water.
2. 0.5 mL 40% acrylamide.
3. 0.5 mL 1.0 M Tris, pH 6.8.
4. 0.04 mL 10% APS.
5. 0.005 mL TEMED.

(ii) After all reagents are combined, invert the falcon tube 2–3 times to thoroughly mix. Quickly pour on top of the set resolving layer and fill to the top.

(iii) Insert combs with desired number of lanes.

(iv) Allow to polymerize for 45 minutes or until set. ***There will be a small remainder in the falcon tube. When the remainder has polymerized, the gels will be set***.

(c) Gels can be stored in urea buffer, lying flat, at 4 °C.

3. Prepare 1 L of 10X Urea Running Buffer.
   (a) Fill 1 L bottle with 700 mL of ultra-pure water.
   (b) Add 30.3 g Tris Base.
   (c) Add 144.10 g glycine. ***add in batches while using a magnetic spin bar to allow faster solubilization***.
   (d) Fill to 1 L with ultra-pure water.
   (e) Final composition 250 mM Tris Base, 1.92 M glycine, pH 8.3 ***pH will naturally be between 8.3 and 8.9. Do not adjust pH***.
4. Prepare Samples.
   (a) Prepare 5X Urea Loading Dye ***Must be stored at – 20 °C due to DTT. If DTT is omitted, stock solution can be stored at room temperature and DTT added when preparing the individual samples***.
      (i) 250 mM Tris Base pH 6.8.
      (ii) 0.5% Bromophenol blue.
      (iii) 500 mM DTT.
      (iv) 50% glycerol.
   (b) Prepare samples.
      (i) For each sample, prepare a total volume of 15uL.
         1. 3uL 5X Urea Loading dye.
         2. 12uL of sample ***For more pure protein samples, ~2ug of protein is ideal per well. For less pure samples ~ 20ug of protein is ideal. If the protein sample is purer, reduce volume of protein sample as necessary and fill to 12uL with lysis buffer***.

5. Running the gel.
   (a) Set up a 1-D vertical gel electrophoresis system such as the Bio-Rad Mini-PROTEAN Tetra Cell.
   (b) Secure the 20% urea gel and plastic dummy plate in the electrode assembly plate.
   (c) Fill the electrode assembly with 1X urea running buffer and remove the comb from the gel.
   (d) Load 8-15uL of sample into a well. Repeat for all samples.
   (e) Fill the chamber with 1X urea buffer to the designated fill line and top off the chamber of the electrode assembly as needed if the volume was decreased during sample loading.
   (f) Run at 200 V for 120–150 minutes. ***For urea-gels, the dye front will run completely off the gel. Do not stop to prevent the dye front from running off or there will not be sufficient resolution***.
6. Visualizing the gel.
   (a) When the run is complete, carefully remove the gel from the gel cassette and transfer to a container. Rinse lightly with DI water.
   (b) "Fix" the gel by soaking in a solution of 10% glacial acetic acid, 50% methanol, and 40% water for 30 minutes to overnight.
   (c) Decant off the "fix" solution and rinse lightly with DI water.
   (d) Stain 30 minutes to overnight in Coomassie Brilliant Blue.
   (e) Decant off Coomassie Brilliant Blue and rinse with DI water.
   (f) Soak in DI water with trace amounts of ethanol to remove background staining. If gel is over stained, repeat step b and stain again for a shorter duration.

2.4 SDS-PAGE (Sodium Dodecyl Sulfate Polyacrylamide Gel Electrophoresis)

1. Set up a gel casting stand such as the Mini-PROTEAN Tetra Cell Casting Stand
2. Prepare 12% SDS Gels
   (a) Prepare Resolving layer
      (i) In a 15 mL falcon tube, combine the following:
         1. 3.7 mL 40% acrylamide.
         2. 3.1 mL 1.5 M Tris Base, pH 8.8.
         3. 5.4 mL ultra-pure water.
         4. 0.120 uL 10% SDS.

5. 0.120 mL ice cold 10% ammonium persulfate (APS) w/v.
6. 0.005 mL *N, N, N, N'*-Tetramethyl-ethylenediamine (TEMED).

(ii) After all reagents are combined, invert the falcon tube 2–3 times to thoroughly mix. Quickly pour into the casting gel plates up to approximately 1.5 cm from the top.

(iii) Fill the remaining 1.5 cm with isopropanol (IPA).

(iv) Allow to polymerize for 45 minutes or until set. ***There will be a small remainder in the falcon tube. When the remainder has polymerized, the gels will be set***.

(v) After polymerization, decant off the IPA layer.

(b) Prepare the Stacking layer

(i) In a 15 mL falcon tube, combine the following:

1. 2.9 mL ultra-pure water.
2. 0.5 mL 40% acrylamide.
3. 0.5 mL 1.0 M Tris Base, pH 6.8.
4. 0.04 mL 10% SDS.
5. 0.04 mL 10% APS.
6. 0.005 mL TEMED.

(ii) After all reagents are combined, invert the falcon tube 2–3 times to thoroughly mix. Quickly pour on top of the set resolving layer and fill to the top.

(iii) Insert combs with desired number of lanes

(iv) Allow to polymerize for 45 minutes or until set. ***There will be a small remainder in the falcon tube. When the remainder has polymerized, the gels will be set***

(c) Gels can be stored in SDS buffer, lying flat, at 4 °C

3. Prepare 1 L of 10X SDS Running Buffer

(a) Fill 1 L bottle with 700 mL of ultra-pure water

(b) Add 30.3 g Tris Base

(c) Add 144.10 g glycine. ***add in batches while using a magnetic spin bar to allow faster solubilization***

(d) Add 10 g SDS

(e) Fill to 1 L with ultra-pure water

(f) Final composition 250 mM Tris Base, 1.92 M glycine, pH 8.3 ***pH will naturally be between 8.3–8.9. Do not adjust pH***

4. Sample preparation
   (a) Prepare 5X SDS Loading Dye ***Must be stored at – 20 °C due to DTT. If DTT is omitted, stock solution can be stored at room temperature and DTT added when preparing the individual samples***
      (i) 10% SDS.
      (ii) 250 mM Tris Base pH 6.8.
      (iii) 0.5% Bromophenol blue.
      (iv) 500 mM DTT.
      (v) 50% glycerol.
   (b) Prepare samples
      (i) For each sample prepare a total volume of 15uL
         1. 3uL 5X SDS Loading dye
         2. 12uL of sample ***For more pure protein samples, ~2ug of protein is ideal per well. For less pure samples ~ 20ug of protein is ideal. If the protein sample is purer, reduce volume of protein sample as necessary and fill to 12uL with lysis buffer***
         3. Heat samples at 90–100 °C for 5 minutes
         4. Remove from heat and proceed to "Running the gel"
5. Running the gel
   (a) Set up a 1-D vertical gel electrophoresis system such as the Bio-Rad Mini-PROTEAN Tetra Cell
   (b) Secure the 12% SDS gel and plastic dummy plate in the electrode assembly plate
   (c) Fill the electrode assembly with 1X SDS running buffer and remove the comb from the gel
   (d) Load 8–15uL of appropriate protein ladder into the first lane
   (e) Load 8–15uL of sample into a subsequent well. Repeat for all samples
   (f) Fill the chamber with 1X SDS buffer to the designated fill line and top off the chamber of the electrode assembly as needed if the volume was decreased during sample loading
   (g) Run at 200 V for 30–45 minutes. ***Do not allow the dye front to run off the gel. Watch carefully and stop the run when the dye front is 0.5–1 cm from the base of the gel***
6. Visualizing the gel
   (a) When the run is complete, carefully remove the gel from the gel cassette and transfer to a container. Rinse lightly with DI water

(b) "Fix" the gel by soaking in a solution of 10% glacial acetic acid, 50% methanol, and 40% water for 30 minutes to overnight

(c) Decant off the "fix" solution and rinse lightly with DI water

(d) Stain 30 minutes to overnight in Coomassie Brilliant Blue

(e) Decant off Coomassie Brilliant Blue and rinse with DI water

(f) Soak in DI water with trace amounts of ethanol to remove background staining. If gel is over stained, repeat step b and stain again for a shorter duration

2.5 Isotopically Labeled ^{15}N-AcpP

2.5.1 Growth and Protein Expression

1. Using pet29a C-terminal His_6-tagged AcpP that has been transformed into BL21 cells, prepare a 5 mL starter culture in Luria–Bertani (LB) media with 5 uL of 100 mg/mL ampicillin. Allow to incubate with rotation overnight at 37 °C for a minimum of 12–18 hours.
2. Centrifuge the overnight culture at 12,000 rpm and 4 °C for 2 minutes to get a cell pellet. Discard the supernatant.
3. Resuspend the cell pellet it 1 mL of sterile ultra-pure water. Centrifuge under the same conditions in step 2. Discard supernatant. Repeat once more to ensure all LB has been removed. Resuspend a final time in 1 mL of ultra-pure water ** *Washing the cells in ultra-pure water will induce the cytolysis of some cells. This can result in longer growth times. This cytolysis can be reduced by replacing ultra-pure water with M9 minimal media in this step***.
4. Transfer 1 L of sterile filtered M9 minimal media with 1 g ^{15}N-NH_4Cl to a clean, dry, autoclaved 2 L baffled flask.
5. Add the resuspended starter culture and 1 mL of 100 mg/mL ampicillin to the baffled flask with M9 minimal media.
6. Grow at 37 °C with shaking of 150 rpm until OD_{600} reaches 0.8.
7. When OD_{600} is reached, induce protein expression with 500uL of 1 M IPTG for a final concentration of 0.5 mM IPTG.
8. Leave shaking at 150 rpm and 37 °C overnight, for 12–18 hours.
9. The following day, harvest the cell pellet with centrifugation for 45 minutes at 4 °C and 2000 rpm. Discard the supernatant. ***harvested pellets can be stored at − 20 °C or used immediately***.

2.5.2 Protein Purification

1. If cell pellet was previously frozen, thaw on ice.
2. Resuspend cell pellet in lysis buffer.
3. Sonicate the resuspended pellet for 6–8 minutes with intervals of 1 s on and 4 s off. Sonication should be performed over an ice bath and closely monitored to ensure the sample does not warm.
4. Collect the lysate through centrifugation for 45 min at 12,000 rpm and 4 °C. Transfer clarified lysate for protein purification. Discard pelleted membranes and insoluble materials.
5. Batch-bind the lysate with Ni-NTA resin that has been equilibrated with lysis buffer. Use approximately 0.5 mL of resin for 1 L of growth. Allow to batch-bind for 30 minutes with spinning at 4 °C.
6. Prepare the following buffers and keep on ice.
   (a) Wash 1: 50 mL of lysis buffer.
   (b) Wash 2: 50 mL of lysis buffer with 10 mM imidazole (49.5 mL of lysis buffer with 500uL of 1 M imidazole added).
   (c) Elution buffer: 50 mL of lysis buffer with 250 mM imidazole (37.5 mL lysis buffer with 12.5 mL 1 M imidazole added).
7. Transfer lysate to an empty gravity flow column fitted with a stop cock. Collect the flow-through. The Ni-NTA will form the bed volume as the flow through passes through the column. Do not allow the bed to go dry and stop the flow just above the Ni-NTA resin. Keep on ice after collection.
8. Wash with 40 mL of Wash 1 buffer. Collect, wash, and stop flow just above the Ni-NTA resin. Keep on ice after collection.
9. Wash with 40 mL of Wash 2 buffer. ***Bradford reagent can be used here to denote when to move to the elution buffer. The volume of Wash 2 can be increased or decreased as needed. Collect 20uL of sample and add 180uL of Bradford reagent. When no color change is detected, this indicates the majority of contaminant proteins have been removed and purification can proceed to the next step***.
10. Elute protein with Elution buffer in 5 mL fractions testing by Bradford reagent as explained in the previous step.
11. Verify the presence and purity of the protein by testing the flow-through, wash 1, wash 2, and elution fractions by 12% SDS-PAGE gel.
12. To remove imidazole, dialyze fractions containing ^{15}N-AcpP in 2 L of 50 mM Tris Base, 150 mM NaCl, 10% glycerol, 1 mM DTT, and pH 7.4 overnight at 4 °C.

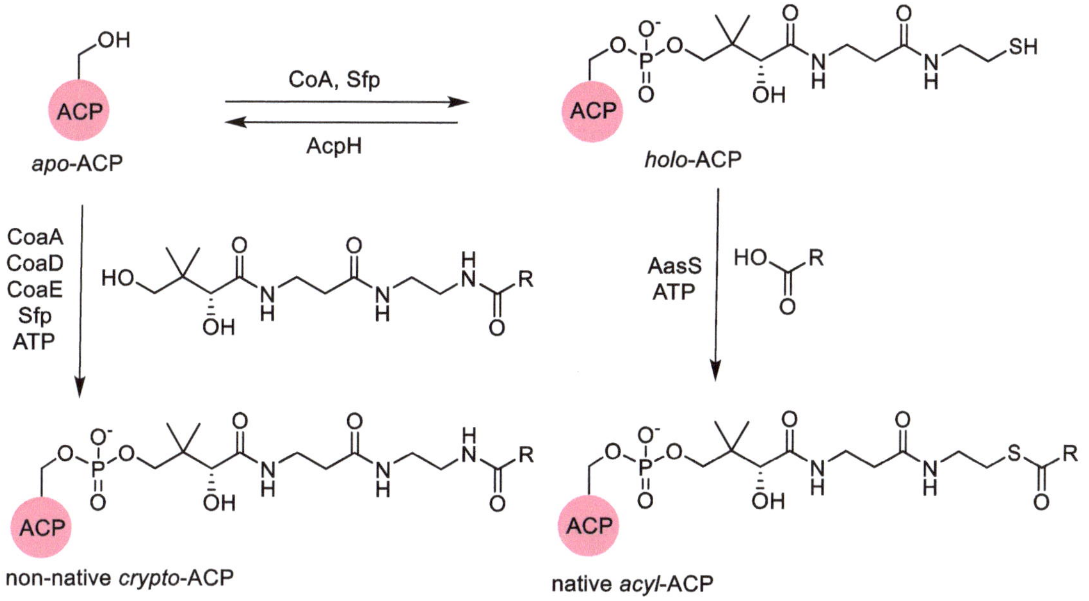

Fig. 2 AcpP is produced as a mixture of the inactive *apo*-ACP form and the active *holo*-ACP form. Homogenous *holo* samples are achieved using the phosphopantetheine transferase (Sfp) where homogenous *apo* samples are created with the phosphopantetheine hydrolase (AcpH). Synthesized phosphopantetheine mimetics can be loaded on to *apo* samples via a chemoenzymatic one-pot to produce non-native *crypto*-ACP where the native thioester bond has been replaced with an amide bond. The acyl-acyl carrier protein AasS can be used to attach free fatty acids to the *holo*-ACPs to produce *acyl*-ACPs that retain the native thioester bond [7, 8, 15, 16]

2.6 Chemoenzymatic Modification of ^{15}N-AcpP

In vivo expression of AcpP produces the ACP in two forms, *apo*- and *holo*-. Previous research has developed chemical biology tool kits to elucidate ACP-mediated interactions via the loading of substrate mimetics to form *crypto*-AcpP, or by loading fatty acids to form *acyl*-AcpP [2, 7–13, 17–20]. It is first necessary to produce a homogenous sample of the carrier protein in the desired state, and then one of two methods can by applied for either cryptofication or acylation (Fig. 2). Both methods are described below.

2.6.1 Holofication and Acylation

1. Prepare a uniformly *holo*-^{15}N-AcpP sample
   (a) Combine the following
      (i) *apo/holo*-^{15}N-AcpP mixture.
      (ii) 0.005 equivalents of *B. subtilis* 4′-phosphopantetheinyl transferase (Sfp) ***expression and purification as previously described*** [17].
      (iii) 12.5 mM $MgCl_2$.
      (iv) 1 mM Coenzyme A.
   (b) Rotate overnight at 37 °C
   (c) Confirm complete holofication by Urea-PAGE analysis

2. Prepare *acyl*- ^{15}N-AcpP
   (a) Combine the following
      (i) *holo*-^{15}N-AcpP prepared in the previous step.
      (ii) 0.005 equivalents of *V. harveyi* Acyl-acyl carrier protein synthetase (AasS) ***expression and purification as previously reported*** [7].
      (iii) 12.5 mM $MgCl_2$.
      (iv) 8 mM ATP.
      (v) 0.5 mM TCEP.
      (vi) 1.5 mM fatty acid.
   (b) Rotate overnight at 37 °C
   (c) Confirm complete acylation by Urea-PAGE analysis

2.6.2 Apofication and Cryptofication

1. Prepare a uniformly apo-^{15}N-AcpP sample
   (a) Combine the following
      (i) *apo/holo*-^{15}N-AcpP mixture.
      (ii) 0.1 equivalents of *P. aeruginosa* AcpH ***expressed and purified as previously reported*** [18].
      (iii) 12.5 mM $MgCl_2$.
      (iv) 5 mM $MnCl_2$.
      (v) 5 mM TCEP.
   (b) Rotate overnight at 37 °C
   (c) Confirm complete apofication by Urea-PAGE analysis
2. Prepare *crypto*- ^{15}N-AcpP
   (a) Combine the following
      (i) *apo*-^{15}N-AcpP prepared in the previous step.
      (ii) 0.1 equivalents each of *E. coli* CoaA, CoaD, and CoaE **expressed and purified as previously described** [19].
      (iii) 0.1 equivalents of *B. subtilis* 4′-phosphopantetheinyl transferase (Sfp) ***expression and purification as previously described*** [17].
      (iv) 12.5 mM $MgCl_2$.
      (v) 8 mM ATP.
      (vi) 0.2% Triton-X.
      (vii) 0.5 mM TCEP.
      (viii) 1.5 mM substrate mimetic probe.
   (b) Rotate overnight at 37 °C
   (c) Confirm complete loading by Urea-PAGE analysis

2.7 Partner Proteins

Partner proteins should be expressed and purified as previously reported with the incorporation of the following modifications as necessary:

For NMR Titrations experiments using 2.6a, the active site of the partner protein may need to be mutated to prevent unwanted substrate modification. Further discussion can be found as previously reported [1, 7, 16, 20].

2.7.1 FPLC Purification and NMR Buffers

After preparing the substrate-loaded ^{15}N-AcpP, it will need to be isolated and transferred into a NMR suitable buffer solution.

2.7.2 ^{15}N-AcpP

1. Separate by size exclusion chromatography. As AcpP is an 8.64 kDa protein, the Superdex S75 provides sufficient separation.
2. Buffer composition: 50 mM phosphate, 0.5 mM TCEP, 0.1% azide, pH 7.4.

2.7.3 Partner Protein

It may be necessary to test different buffer conditions for the chosen partner protein. The following should be taken into consideration when preparing the NMR buffer:

1. Stability: Titration experiments can require the sample to stay at experiment conditions for several hours to several days. The protein must be stable in the chosen buffer, under experimental conditions. Examples can be found as previously reported [7]
2. Salt: KCl and NaCl are often used to increase stability and solubility of proteins. However, the conductivity of these components interferes with the instrument's signal. Phosphate 10 mM–50 mM is typically used to avoid conductivity interference with the signal to noise ratio.
3. pH: Acidic pH is ideal (4–7) but the stability and longevity of the sample should be prioritized.
4. Reducing agents: 0.5–5 mM DTT or TCEP can be used to prevent dimerization and unwanted activity due to the presence of free thiols.
5. Microbial growth: Sodium azide <50uM can be used to prevent microbial growth in samples over the course of the experiments.

** *Note: Both the ^{15}N-AcpP and the partner protein must be in the SAME buffer. The given ^{15}N-AcpP buffer will need to be tested for compatibility with the partner protein. If adjustments are necessary to maintain protein stability, the determined buffer conditions should be uniformly applied to both ^{15}N-AcpP and the partner protein.* **.

3 Materials and Methods for NMR Analysis

3.1 NMR Sample Preparation

The ^{15}N-AcpP and partner protein samples must be prepared, then transferred to a high-quality NMR tube with a frequency rating of at least 500 MHz, though 600 MHz is preferable.

3.1.1 Sample Requirements

1. ^{15}N-AcpP concentration: No less than 100uM.
2. Partner protein concentrations: molar equivalents of 0.5, 1, 1.5, 2 with respect to ^{15}N-AcpP.
3. 10% D_2O in each sample.
4. The total sample volume should be 500uL for a 5 mm tube.

3.2 NMR Spectra Collection

Spectra should be collected on a 600 or 800 MHz instrument such as the Bruker Advance spectrometer. The following protocol contains commands and parameters for use with TopSpin. TopSpin commands are shown in bold. All NMR experiments should be prepared, parameterized, and collected by, or under the supervision of, a person who has the appropriate expertise in NMR spectroscopy and the facilities being used.

1. Set the temperature to 37 °C by using the command **edta.**
2. Load the sample of onto the instrument. ***The "zero-point" sample of homogenous ^{1}H-^{15}N-AcpP should be collected first. Experiments consisting of ^{1}H-^{15}N-AcpP with increasing concentrations of partner protein will be conducted subsequently.***
3. Use the **lock** command to lock the sample on the deuterium signal frequency.
4. Tuning and matching the signal frequency are important for NMR experiments. If the instrument and software allow automated tuning, use the command **atma** for tuning and matching of the H, N, and C channels as necessary for the experiment. If manual tuning is required, display the wobble curve with the **wobb** command.
5. Shimming is a process that optimizes field homogeneity. The TopSpin command for shimming is **topshim**, though it can vary with instruments so it is important to verify the appropriate command.
6. Create a dataset to store output files with spectra data and specify parameters
7. Run a proton 1D experiment with water suppression to check if protein is folded correctly

(a) Parameters that need to be reviewed. The command **ased** will display the summary of acquisition parameters. The command **eda** will show all parameters.

(i) Recycle delay: 1.3 s to 1.5 s is recommended.

(ii) 90-degree pulse sequence calibrated to properly acquire signal from sample: **getprosol** command will populate the values. To find the accurate 90-degree pulse, use the **pulsecal** command.

(iii) Set O1 to water frequency.

(iv) Number of scans: You will need to specify both dummy scans (DS) and number of scans (NS). 16 scans should be sufficient, though number of scans will depend on the concentration of the sample. More dilute samples will need more scans for better sensitivity.

8. After 1D experiment is complete, create a new dataset to prepare the 2D HSQC

(a) Parameters that need to be reviewed. The command **ased** will display the summary of acquisition parameters. The command **eda** will show all parameters.

(i) Recycle delay: 1.3 s to 1.5 s is recommended.

(ii) 90-degree pulse sequence calibrated to properly acquire signal from sample: **getprosol** command will populate the values. To find the accurate 90-degree pulse, use the **pulsecal** command.

(iii) Set O1 to water frequency.

(iv) Data points: 2048 data points for the ^{1}H dimension and 128 data points for ^{15}N and ^{13}C dimensions are recommended. More data points can be used to achieve better resolution.

(v) Number of scans: You will need to specify both dummy scans (DS) and number of scans (NS). 16 scans for the zero-point experiment is recommended and 24 scans is recommended for the samples with both the carrier protein and the partner protein. Number of scans will depend on the concentration of the sample. More dilute samples will need more scans for better sensitivity.

(vi) Set your spectral widths (**sw**) and adjust the offset frequency as necessary for both proton and either N or C

9. The command **rga** sets the receiver gain.

10. The command **zg** starts the data acquisition.

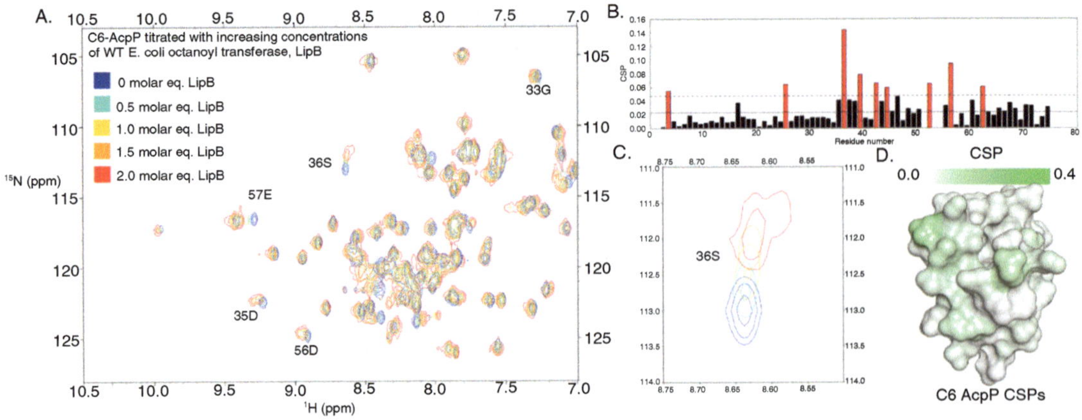

Fig. 3 Example figure of titration data of AcpP with *E. coli* LipB. (**a**) The ^{1}H-^{15}N-HSQC spectra were overlayed and individual peaks were highlighted. (**b**) The CSP of each residue in the titration showing key residue-specific interactions. (**c**) Focus of the conserved Ser of AcpP. (**d**) CSP of AcpP colored by magnitude to show key surface areas [10]

11. The command **xfb** will process the 2D spectrum after acquisition.
12. Further processing can be performed though NMRPipe [21].

3.3 NMR Data Analysis

3.3.1 Chemical Shift Perturbations

Analysis of multiple HSQC spectra can be achieved using additional software designed to visualize and process different types of NMR data. As shown in Fig. 3, by looking at chemical shift perturbations (CSP), information on individual residues, particularly residue-specific interactions (Fig. 3b and c) and key surface areas (Fig. 3d), can be obtained. For titration experiments, there is a generalized workflow that ultimately calculates the chemical shift perturbations (CSP) per residue using the following formula:

$$\Delta\delta_{\mathrm{NH}} = \sqrt{\Delta\delta H^2 + \left(\frac{\Delta\delta N}{5}\right)^2}$$

where $\Delta\delta_{\mathrm{NH}}$ is the chemical shift for the NH bonds in the protein, while $\Delta\delta H$ and$\Delta\delta N$ represent the change in chemical shift for H and N nuclei, respectively. Backbone assignments, or peak lists, of the labeled protein are required for the proper assignment of residues to chemical shifts in the spectra. Commonly obtained through triple resonance assignments beyond the scope of this chapter, the peak list can be downloaded from an online repository such as the Biological Magnetic Resonance Bank (BMRB) [22].

1. Obtain all necessary files pertaining to the HSQC spectra and convert them to an appropriate format for processing. Topspin saves the spectra in a particular format which can be converted to more accessible versions using a given NMR software.

Popular software such as SPARKY, POKY (a SPARKY extension), or CCPN can be used for these purposes [22–25].

2. Using either of the programs mentioned previously, load and visualize the zero-point spectra, modify contour as needed to obtain a clean view that only shows the NH signals and reduces noise.
3. Download the list of peaks from BMRB and map them to the spectra. Given slight differences in the buffers and other conditions, the peaks may appear slightly shifted from where they are supposed to be. Regardless, the overall fingerprint should appear the same. Manually adjust the peaks to center the signals observed that match the backbone.
4. NMR analysis software may be equipped with built-in functions that center the peaks using mathematical algorithms; this can help clear out bias once the peaks have been aligned with their corresponding chemical shifts.
5. Once all the peaks have been assigned to their corresponding residues, save the new peak list that contains the coordinates in the H and N dimensions for all the residues in the protein for the zero point.
6. Proceed to overlay the saturated spectra over the zero-point spectra and repeat the same steps for obtaining the coordinates that correspond to the saturated peaks. You may benefit by assigning different colors to each spectrum and go over each peak sequentially.
7. Save the saturated point spectra peak list as a separate file, and using excel or other numerical processing software, proceed to calculate the CSP for each residue in the protein.

Note: Due to their lack of backbone NH bonds, prolines do not show up in the spectra. Histidine tags can also potentially interfere with HSQC spectra. Additionally, signal loss can be observed for some residues during data acquisition, which may cause their peaks to be invisible.

3.3.2 Line Shape Analysis

NMR titration and binding experiments offer the ability to study structural and dynamical biomolecular interactions. Line shape analysis can be applied to two-dimensional experiments as unique chemical shifts are associated with each dimension of the experiment [26]. Line shape analysis allows the extrapolation of kinetic data such as exchange processes, k_{ex}, the dissociation constant, k_d, as well as the k_{on} and k_{off} rates [26]. This kinetic information can be determined for specific residues providing valuable insight into regulatory PPIs. Programs such as TITAN 2D have been designed for these purposes and can be used with the data from the acquired spectra [7, 26, 27].

3.3.3 Computational Analysis and Applications

Disrupting regulatory protein–protein interactions has become an enticing new target for rational drug discovery and design. High-throughput screening (HTS) is commonly employed in the pharmaceutical industry to screen large libraries of possible compounds, though its shortfalls include a high false-positive rate. This high failure rate combined with the complexity, time constraints, and high cost of drug discovery has contributed to an overall decline in the number of identified "possible" compounds successfully making it to market [28]. The development of methods, such as those described here, have unlocked valuable information that had previously remained elusive. NMR titration experiments on ^{15}N-labeled carrier proteins provide detailed structural information about the backbone NH bonds that are perturbed upon binding to partner proteins [12]. In addition to binding affinities, NMR titration experiments provide valuable insight into allosteric interactions, interfacial interactions, and confirmational changes that occur upon binding and complexation [29]. Each of these interactions offers new potential druggable sites. The quantitative amide proton CSP information provided though NMR titration experiments can be exploited to refine and improve docking simulations. By incorporating experimental restraints, an increased sampling of native-like confirmations can be generated, which in turn reduces the failure of scoring functions that rely on near-native conformation sampling. The augmentation of scoring/docking function with experimental restraints provides additional improvements in predictive power by overcoming deficiencies in current scoring schemes [30]. Unlike HTS, virtual screening (VS) and computer-aided drug discovery (CADD) allow many molecules to be screened in a short amount of time though deficiencies in scoring functions still present limitations. Structure-based virtual screening (SBVS) incorporates experimental and structural information allowing more refined screens with more optimized scoring functions [31]. Lead compounds or fragments identified through *in silico* methods that have been optimized with experimental restraints can then be tested in target-based and ligand-based NMR screens through additional NMR titration experiments [31].

References

1. Chen A, Re RN, Burkart MD (2018) Type II fatty acid and polyketide synthases: deciphering protein-protein and protein-substrate interactions. Nat Prod Rep 35(10):1029–1045. https://doi.org/10.1039/c8np00040a. PMID: 30046786
2. Jaremko MJ, Davis TD, Corpuz JC, Burkart MD (2020) Type II non-ribosomal peptide Synthetase proteins: structure, mechanism, and protein-protein interactions. Nat Prod Rep 37(3):355–379. https://doi.org/10.1039/c9np00047j. PMID: 31593192
3. Cronan JE, Thomas J (2009) Bacterial fatty acid synthesis and its relationships with polyketide synthetic pathways. Methods Enzymol 459:395–433. https://doi.org/10.1016/S0076-6879(09)04617-5. PMID: 19362649

4. Nguyen C, Haushalter RW, Lee DJ, Markwick PRL, Bruegger J, Caldara-Festin G, Finzel K, Jackson DR, Ishikawa F, O'Dowd B, McCammon JA, Opella SJ, Tsai S-C, Burkart MD (2014) Trapping the dynamic acyl carrier protein in fatty acid biosynthesis. Nature 505(7483):427–431. https://doi.org/10.1038/nature12810. PMID: 24362570
5. Cronan JE (2014) The chain-flipping mechanism of ACP (acyl carrier protein)-dependent enzymes appears universal. Biochem J 460(2): 157–163. https://doi.org/10.1042/BJ20140239. PMID: 24825445
6. Roujeinikova A, Simon WJ, Gilroy J, Rice DW, Rafferty JB, Slabas AR (2007) Structural studies of fatty acyl-(acyl carrier protein) thioesters reveal a hydrophobic binding cavity that can expand to fit longer substrates. J Mol Biol 365(1):135–145. https://doi.org/10.1016/j.jmb.2006.09.049. PMID: 17059829
7. Finzel K, Lee DJ, Burkart MD (2015) Using modern tools to probe the structure-function relationship of fatty acid synthases. Chembiochem 16(4):528–547. https://doi.org/10.1002/cbic.201402578. PMID: 25676190, Sztain, T.; Bartholow, T. G.; Lee, D. J.; Casalino, L.; Mitchell, A.; Young, M. A.; Wang, J.; McCammon, J. A.; Burkart, M. D. Decoding aAllosteric rRegulation by the aAcyl cCarrier pProtein. Proc Natl Acad Sci U S A 2021, 118 (16), e2025597118. 10.1073/pnas.2025597118. PMID: 33846262
8. Sztain T, Bartholow TG, McCammon JA, Burkart MD (2019) Shifting the hydrolysis equilibrium of substrate loaded acyl carrier proteins. Biochemistry 58(34):3557–3560. https://doi.org/10.1021/acs.biochem.9b00612. PMID: 31397556. Sztain, T.; Patel, A.; Lee, D. J.; Davis, T. D.; McCammon, J. A.; Burkart, M. D. One aAtom mMatters: mModifying the tThioester lLinkage aAffects the sStructure of the aAcyl cCarrier pProtein. Angew Chem Int Ed Engl 2019, 58 (32), 10888–10892. 10.1002/anie.201903815. PMID: 31140212
9. Bartholow TG, Sztain T, Young MA, Lee DJ, Davis TD, Abagyan R (2022) Burkart. M. D. Control of Unsaturation in De Novo Fatty Acid Biosynthesis by FabA, Biochemistry. https://doi.org/10.1021/acs.biochem.2c00094. PMID 35255690
10. Bartholow TG, Sztain T, Young MA, Davis TD, Abagyan R, Burkart MD Protein–Protein interaction based substrate control in the E. Coli octanoic acid transferase, LipB. RSC Chem Biol 2(5):1466–1473. https://doi.org/10.1039/d1cb00125f. PMID 34705040
11. Milligan JC, Lee DJ, Jackson DR, Schaub AJ, Beld J, Barajas JF, Hale JJ, Luo R, Burkart MD, Tsai S-C (2019) Molecular basis for interactions between an acyl carrier protein and a Ketosynthase. Nat Chem Biol 15(7): 669–671. https://doi.org/10.1038/s41589-019-0301-y. PMID: 31209348
12. Purslow JA, Khatiwada B, Bayro MJ, Venditti V (2020) NMR methods for structural characterization of protein-protein complexes. Front Mol Biosci 7:9. PMID: 32047754
13. Maity S, Gundampati RK, Suresh Kumar TK (2019) NMR methods to characterize protein-ligand interactions. Nat Prod Commun 14(5): 1934578X19849296. https://doi.org/10.1177/1934578X19849296. PMID: 33865181
14. Williamson MP (2013) Using chemical shift perturbation to characterise ligand binding. Prog Nucl Magn Reson Spectrosc 73:1–16. https://doi.org/10.1016/j.pnmrs.2013.02.001. PMID 23962882
15. Worthington AS, Burkart MD (2006) One-pot chemo-enzymatic synthesis of reporter-modified proteins. Org Biomol Chem 4(1): 44–46. https://doi.org/10.1039/b512735a. PMID: 16357994
16. Mindrebo JT, Patel A, Misson LE, Kim WE, Davis TD, Ni QZ, La Clair JJ, Burkart MD (2020) 1.04 – Structural basis of acyl-carrier protein interactions in fatty acid and polyketide biosynthesis. In: Liu HW (Ben), Begley TP (eds) Comprehensive natural products III. Oxford, Elsevier, pp 61–122. https://doi.org/10.1016/B978-0-12-409547-2.14662-1
17. Lambalot RH, Gehring AM, Flugel RS, Zuber P, LaCelle M, Marahiel MA, Reid R, Khosla C, Walsh CT (1996) A new enzyme superfamily — the Phosphopantetheinyl transferases. Chem Biol 3(11):923–936. https://doi.org/10.1016/S1074-5521(96)90181-7. PMID: 8939709
18. Kosa NM, Haushalter RW, Smith AR, Burkart MD (2012) Reversible Chemoenzymatic labeling of native and fusion carrier protein motifs. Nat Methods 9(10):981–984. https://doi.org/10.1038/nmeth.2175. PMID: 22983458
19. Charov K, Burkart MD (2019) A single tool to monitor multiple protein–protein interactions of the Escherichia Coli acyl carrier protein. ACS Infect Dis 5(9):1518–1523. https://doi.org/10.1021/acsinfecdis.9b00150. PMID: 31317739

20. Mindrebo JT, Patel A, Kim WE, Davis TD, Chen A, Bartholow TG, La Clair JJ, McCammon JA, Noel JP, Burkart MD (2020) Gating mechanism of elongating β-Ketoacyl-ACP synthases. Nat Commun 11(1):1727. https://doi.org/10.1038/s41467-020-15455-x. PMID: 32265440
21. Delaglio F et al (1995) NMRPipe: a multidimensional spectral processing system based on UNIX pipes. J Biomol NMR 6:277–293
22. BioMagResBank, Eldon L. Ulrich; Hideo Akutsu; Jurgen F. Doreleijers; Yoko Harano; Yannis E. Ioannidis; Jundong Lin; Miron Livny; Steve Mading; Dimitri Maziuk; Zachary Miller; Eiichi Nakatani; Christopher F. Schulte; David E. Tolmie; R. Kent Wenger; Hongyang Yao; John L. Markley (2008) Nucleic Acids Res 36:D402–D408. https://doi.org/10.1093/nar/gkm957
23. Lee W, Tonelli M, Markley JL (2015) NMRFAM-SPARKY: enhanced software for biomolecular NMR spectroscopy. Bioinformatics 31(8):1325–1327. https://doi.org/10.1093/bioinformatics/btu830. Epub 2014 Dec 12. PMID: 25505092; PMCID: PMC4393527
24. Lee W, Rahimi M, Lee Y, Chiu A (2021) POKY: a software suite for multidimensional NMR and 3D structure calculation of biomolecules. Bioinformatics 37(18): 3041–3042
25. Skinner SP, Fogh RH, Boucher W et al (2016) CcpNmr Analysis Assign: a flexible platform for integrated NMR analysis. J Biomol NMR 66: 111–124. https://doi.org/10.1007/s10858-016-0060-y
26. Waudby CA, Ramos A, Cabrita LD, Christodoulou J (2016) Two-dimensional NMR Lineshape analysis. Sci Rep 6(1):24826. https://doi.org/10.1038/srep24826. (PMID: 27109776)
27. Waudby CA, Ouvry M, Davis B, Christodoulou J (2020) Two-dimensional NMR Lineshape analysis of single, multiple, zero and double quantum correlation experiments. J Biomol NMR 74(1):95–109. https://doi.org/10.1007/s10858-019-00297-7. (PMID: 3191580)
28. Stark JL, Powers R (2012) Application of NMR and molecular docking in structure-based drug discovery. Top Curr Chem 326:1–34. https://doi.org/10.1007/128_2011_213
29. Maity S, Gundampati RK, Suresh Kumar TK (**2019**) NMR methods to characterize protein-ligand interactions. Nat Product Commun 14(5):1934578X19849296. https://doi.org/10.1177/1934578X19849296
30. González-Ruiz D, Gohlke H (2009) Steering protein–ligand docking with quantitative NMR chemical shift perturbations. J Chem Inf Model 49(10):2260–2271. https://doi.org/10.1021/ci900188r
31. Maia EHB, Assis LC, de Oliveira TA, da Silva AM, Taranto AG (2020) Structure-based virtual screening: from classical to artificial intelligence. Front Chem 8:343

Chapter 4

Chemoproteomic Profiling of Adenylation Domain Functions in Gramicidin S-Producing Non-ribosomal Peptide Synthetases

Fumihiro Ishikawa and Genzoh Tanabe

Abstract

Many amino acid-containing natural products are biosynthesized by large, multifunctional enzymes known as non-ribosomal peptide synthetases (NRPSs). Adenylation (A) domains in NRPSs are responsible for the incorporation of amino acid building blocks and can be considered as engineering domains; therefore, advanced techniques are required to not only rapidly verify expression and folding, but also accelerate the functional prediction of the A-domains in lysates from native and heterologous systems. We recently developed activity-based protein profiling (ABPP) of NRPSs that offers a simple and robust analytical platform for A-domains and provides insights into their enzyme–substrate specificity. In this chapter, we describe the design and synthesis of these ABPP probes and provide a summary of our work on the development of a series of protocols for labeling, visualizing, and analyzing endogenous NRPSs in complex biological systems.

Key words Non-ribosomal peptide synthetase (NRPS), Adenylation domain, Activity-based protein profiling, Photoaffinity labeling, *Aneurinibacillus migulanus* ATCC 9999, Gramicidin S-NRPSs

1 Introduction

Amino-acid-containing natural products, including the antibiotic vancomycin, immunosuppressant cyclosporine, and virulence factor pyoverdine, are promising resources for use in medicine because of their structural diversity and wide range of biological activities [1]. Biosynthesis of these amino-acid-containing natural products involves the large, modular, and multifunctional non-ribosomal peptide synthetases (NRPSs). Each NRPS module is responsible for the selection, loading, and condensation of a single amino acid building block and its incorporation into the peptide backbone. The typical NRPS module comprises three enzyme domains: adenylation (A), thiolation (T, also known as the peptidyl carrier protein, PCP), and condensation (C) (Fig. 1a) [2]. First, the

Michael Burkart and Fumihiro Ishikawa (eds.), *Non-Ribosomal Peptide Biosynthesis and Engineering: Methods and Protocols*, Methods in Molecular Biology, vol. 2670, https://doi.org/10.1007/978-1-0716-3214-7_4,

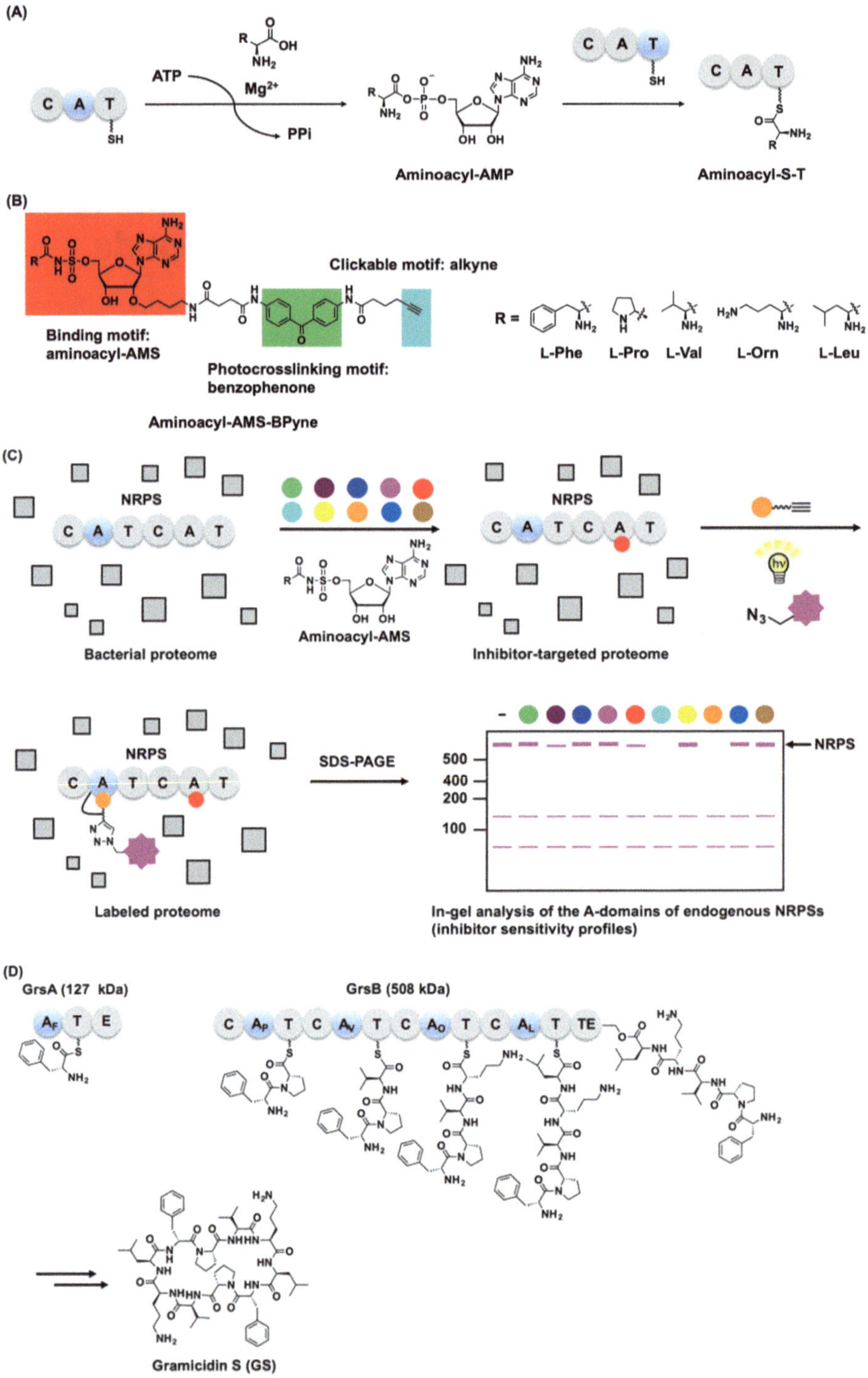

Fig. 1 (**a**) Adenylation reaction and amino acid transfer catalyzed by the adenylation (A) domain. (**b**) Structures of activity-based protein profiling (ABPP) probes. (**c**) Functional profiling of A-domains of the NRPSs using competitive ABPP. In ABPP, proteomes from producer organisms were treated with probes, exposed to UV light (365 nm) to perform cross-linking, and reacted with 5/6-TAMRA-peg$_3$-azide under copper(I)-catalyzed azide-alkyne cycloaddition (CuAAC). Labeled proteomes were visualized using SDS-PAGE and in-gel fluorescence imaging. The substrate candidates of the A-domains of NRPSs can be imaged by pre-incubating with a series

A-domain catalyzes the formation of aminoacyl adenylate in the presence of adenosine triphosphate and Mg^{2+} (Fig. 1a). Next, the adenylated amino acid is transferred into the 4′--phosphopantetheine group of a downstream T domain (Fig. 1a). Finally, the C domain catalyzes amide bond formation between the two aminoacyl-S-T domains in both the upstream and downstream T domains.

Metabolic engineering is an attractive technique for the biosynthesis of non-native products. However, natural product biosynthesis depends on many regulatory mechanisms and their orchestrated regulation of processes including the transcription of relevant genes into mRNAs, translation of mRNAs into proteins, and post-translational protein modifications [3]. Indeed, the complex cellular physiology frequently makes it difficult to produce the target compounds [4]. Our knowledge of NRPSs is dependent on studies in which the full NRPS protein or each enzyme domain is produced in a heterologous system. Therefore, strategies to directly analyze the endogenous NRPSs in native systems are needed. Furthermore, NRPSs are responsible for the biosynthesis of numerous metabolites that function as virulence factors in pathogenic bacteria. For example, pyoverdine, the major siderophore produced by *Pseudomonas aeruginosa*, is a non-ribosomal peptide and a key virulence determinant [5]. Compromising pyoverdine, whether by inhibiting its production with fluoropyrimidines [6, 7] or its function using gallium (Ga^{3+}) or pyoverdine inhibitors [8, 9], is sufficient to mitigate *P. aeruginosa* pathogenesis, demonstrating that pyoverdine-related pathways can be potential therapeutic targets. Therefore, understanding the activity, transcriptional control, and dynamics of pyoverdine-NRPS in the bacterial pathogenesis is essential.

Based on these considerations, we developed activity-based protein profiling (ABPP) probes that can be used for labeling, visualization, enrichment, identification, and tracking of endogenous NRPSs in proteomes (Fig. 1b) [10, 11]. We and others have described 5′-*O*-*N*-(aminoacyl)sulfamoyladenosines (aminoacyl-AMSs) that are capable of binding A-domains with substrate specificity that corresponds to a conjugated amino acid (Figs. 1b and 3) [10–12]. Aminoacyl-AMSs discriminate the differential active-site

Fig. 1 (continued) of aminoacyl-AMS compounds and subsequent probe labeling. Inhibitor-bound A-domains are evaluated by monitoring the reduction in probe labeling intensity (purple and red circles). The A-domains of endogenous NRPSs that are completely inhibited by aminoacyl-AMSs can be visualized as a complete loss of fluorescence signal following in-gel fluorescence analysis (light blue and orange circles). In the competitive ABPP for the A-domains of NRPSs, the inhibitor sensitivity profiles correspond to their substrate specificities. The target A-domain for the probe is colored blue. The rectangles display non-specific proteins. (**d**) Non-ribosomal peptide synthesis of gramicidin S. Modules are comprised of T, A (A_F, L-Phe; A_P, L-Pro; A_V, L-Val; A_O, L-Orn; A_L, L-Leu), C, and TE domains. The target A-domains for probes are colored blue

structures of the A-domains of NRPSs. The ABPP probes consist of a terminal alkyne (yne) for Cu(I)-catalyzed azide-alkyne [3 + 2] click chemistry (CuAAC), a photoreactive benzophenone (BP) for crosslinking of protein targets, and an aminoacyl-AMS moiety for selective binding toward the A-domains of NRPSs (Fig. 1b) [10, 11]. It is important to note that the clickable benzophenone linker was incorporated into the aminoacyl-AMS moiety at the 2′-OH group of ribose, which has no effect on the binding affinity to the NRPS A-domain (Fig. 1b) [10, 11]. To expand the utility of the ABPP probes, an advanced assay platform that enables the functional prediction and assessment of the A-domains of endogenous NRPSs in crude cellular lysates has been developed (Fig. 1c) [13]. Recently, we reported an in-cell ABPP system that enabled the labeling and imaging of an endogenous NRPS with remarkable selectivity and sensitivity in the gram-positive bacterium *Bacillus subtilis* [14]. We refer to excellent review articles that explain the basic technology of ABPP [15, 16].

In this chapter, we describe the competitive format of the ABPP strategy using a combined library of aminoacyl-AMS compounds, which offers a simple assay system for the A-domains of endogenous NRPSs and provides insights into enzyme substrate candidates and enzyme active-site microenvironments (Fig. 1c) [13]. The inhibition profiles correlate with the known substrate promiscuity of the A-domains [13]. We describe the method used for the synthesis of ABPP probes, L-Phe-AMS-BPyne, L-Pro-AMS-BPyne, L-Val-AMS-BPyne, L-Orn-AMS-BPyne, and L-Leu-AMS-BPyne and the synthesis of a series of aminoacyl-AMS compounds, L-Phe-AMS, L-Pro-AMS, L-Orn-AMS, Gly-AMS, L-AlaAMS, L-Val-AMS, L-Leu-AMS, L-Ile-AMS, L-Asn-AMS, L-Gln-AMS, L-Ser-AMS, L-Thr-AMS, L-Met-AMS, L-Tyr-AMS, L-Trp-AMS, L-Asp-AMS, L-Glu-AMS, L-Lys-AMS, L-Arg-AMS, and L-His-AMS, as well as a protocol for ABPP of NRPSs to label, visualize, and analyze endogenous NRPSs from the crude cellular lysate of the gramicidin S producer, *Aneurinibacillus migulanus* ATCC 9999, as a model bacterium of the overall utility of this technique (Fig. 1b, c, and d). This technique could serve as a powerful diagnostic platform to analyze the A-domains of endogenous NRPSs, facilitating metabolic engineering and combinatorial biosynthesis in native and heterologous systems.

2 Materials

All materials and reagents described below can be used without further purification.

1. The organic solvents used in the following experiments, including ethyl acetate (EtOAc), chloroform ($CHCl_3$), methanol

2.1 Synthetic Procedure of Aminoacyl-AMS-BPyne Labeling Reagents

(MeOH), and *N*,*N*-dimethylformamide, can be obtained from commercial suppliers.

2.2 Synthetic Procedure of Aminoacyl-AMS Compounds

All materials and reagents described below can be used without further purification.

1. The organic solvents used in the following experiments, including EtOAc, $CHCl_3$, MeOH, acetonitrile, and *N*,*N*-dimethylformamide, can be obtained from commercial suppliers.

2.3 Bacterial Culture

1. *Aneurinibacillus migulanus* ATCC 9999.
2. Petri dishes.
3. Nutrient agar: Add 3.0 g beef extract, 5.0 g peptone, and 15 g agar to 1 L of Milli-Q water and sterilize the resulting mixture at 121 °C for 20 min before pouring into Petri dishes and allowing to cool to room temperature. Store the plates at 4 °C.
4. Nutrient broth: Add 3.0 g beef extract and 5.0 g peptone to 1 L of Milli-Q water and sterilize the resulting mixture at 121 ° C for 20 min. Store at room temperature.
5. YPG medium: Add 50 g yeast extract, 50 g bacto-peptone, and 5 g glucose to 1 L of Milli-Q water and sterilize the resulting mixture at 121 °C for 20 min. Store it at room temperature.
6. Phosphate-buffered saline (PBS): Add 9.6 g of Dulbecco's PBS to 1 L of Milli-Q water and sterilize the resulting mixture at 121 °C for 20 min. Store it at room temperature.
7. 80% (v/v) glycerol solution: Add 80 mL glycerol to 20 mL of Milli-Q water and sterilize the resulting mixture at 121 °C for 20 min. Store at room temperature.
8. Milli-Q water.
9. Screw cap tubes, 2 mL.
10. Culture tubes, glass.
11. Plastic tubes, 50 mL.
12. Eppendorf tubes, 1.5 mL.
13. Baffled flasks, 1 L.
14. Constant temperature incubator shaker.
15. Cell density meter.
16. High speed refrigerated microcentrifuge.

2.4 Sample Preparation for Labeling Studies

1. 20 mM Tris–HCl (pH 8.0): Add 3.15 g tris(hydroxymethyl)-aminomethane to 900 mL Milli-Q water. Adjust pH to 8.0 with an appropriate volume of aqueous hydrochloric acid (HCl) solution and make up to 1 L with Milli-Q water. Store at room temperature.
2. Protease inhibitor cocktail, EDTA-free (100×): Dissolve the powdered product in 1 mL Milli-Q water in a glass bottle. Aliquot and store it at −80 °C.
3. 1 M Magnesium chloride ($MgCl_2$) solution: Add 10.2 g $MgCl_2{\cdot}6H_2O$ to 50 mL Milli-Q water. Store at room temperature.
4. 1 M Tris(2-carboxyethyl)phosphine (TCEP) solution: Add 14.3 g TCEP hydrochloride to 50 mL Milli-Q water. Use a syringe filter to remove any insoluble material, aliquot, and store at 4 °C.
5. 10% (v/v) Nonidet(R) P40 substitute: Add 50 mL Nonidet (R) P40 substitute to 450 mL Milli-Q water. Store it at 4 °C.
6. 20 mM Tris–HCl (pH 8.0), 1 mM $MgCl_2$, 1 mM TCEP, and 0.05%(v/v) Nonidet(R) P40 substitute: For example, add 50 μL of 1 M $MgCl_2$, 50 μL of 1 M TCEP, and 250 μL of 10%(v/v) Nonidet(R) P40 substitute to 50 mL of 20 mM Tris–HCl (pH 8.0). Store at 4 °C.
7. 20 mM Tris–HCl (pH 8.0), 1 mM $MgCl_2$, 1 mM TCEP, 0.05%(v/v) Nonidet(R) P40 substitute, and protease inhibitor cocktail: For example, add 10 μL of 1 M $MgCl_2$, 10 μL of 1 M TCEP, and 50 μL of 10%(v/v) Nonidet(R) P40 substitute to 10 mL of 20 mM Tris–HCl (pH 8.0) and 100 μL of protease inhibitor cocktail (100×), which, to prevent the decomposition of the protease inhibitors, was cooled on ice prior to use. Prepare fresh at the time of use.
8. 10 mg/mL lysozyme solution: Add 10 mg lysozyme from egg white to 1 mL of 20 mM Tris–HCl (pH 8.0), 1 mM $MgCl_2$, 1 mM TCEP, and 0.05% (v/v) Nonidet(R) P40 substitute. Prepare fresh at the time of use.
9. 2 mg/mL bovine serum albumin (BSA) solution for Bradford protein assay: Add 2 mg BSA to 1 mL Milli-Q water. Store at 4 °C.
10. Coomassie Brilliant Blue (CBB) solution for Bradford protein assay: Add 10 mL of protein assay CBB solution (5×) to 40 mL Milli-Q water.
11. Recombinant DNase I (RNase-free).
12. Milli-Q water.
13. Eppendorf tubes, 1.5 mL.

14. Tissue culture plate (96 well), flat bottom with low evaporation.
15. pH meter.
16. Constant temperature incubator.
17. High speed refrigerated microcentrifuge.
18. Microplate reader.

2.5 Activity-Based Protein Profiling

1. *Aneurinibacillus migulanus* ATCC 9999 proteome.
2. 20 mM Tris–HCl (pH 8.0), 1 mM $MgCl_2$, 1 mM TCEP, and 0.05% (v/v) Nonidet(R) P40 substitute: For example, add 50 μL of 1 M $MgCl_2$, 50 μL of 1 M TCEP, and 250 μL of 10%(v/v) Nonidet(R) P40 substitute to 50 mL of 20 mM Tris–HCl (pH 8.0). Store at 4 °C.
3. Dimethyl sulfoxide (DMSO) for molecular biology research.
4. 100 mM DMSO solution of L-Phe-AMS: Add 4.9 mg L-Phe-AMS to 100 μL DMSO. Store at −30 °C.
5. 100 mM DMSO solution of L-Pro-AMS: Add 4.4 mg L-Pro-AMS to 100 μL DMSO. Store at −30 °C.
6. 100 mM DMSO solution of Gly-AMS: Add 4.0 mg Gly-AMS to 100 μL DMSO. Store at −30 °C.
7. 100 mM DMSO solution of L-Ala-AMS: Add 4.2 mg L-Ala-AMS to 100 μL DMSO. Store at −30 °C.
8. 100 mM DMSO solution of L-Val-AMS: Add 4.5 mg L-Val-AMS to 100 μL DMSO. Store at −30 °C.
9. 100 mM DMSO solution of L-Leu-AMS: Add 4.6 mg L-Leu-AMS to 100 μL DMSO. Store at −30 °C.
10. 100 mM DMSO solution of L-Ile-AMS: Add 4.6 mg L-Ile-AMS to 100 μL DMSO. Store at −30 °C.
11. 100 mM DMSO solution of L-Asn-AMS: Add 4.6 mg L-Asn-AMS to 100 μL DMSO. Store at −30 °C.
12. 100 mM DMSO solution of L-Gln-AMS: Add 4.7 mg L-Gln-AMS to 100 μL DMSO. Store at −30 °C.
13. 100 mM DMSO solution of L-Ser-AMS: Add 4.3 mg L-Ser-AMS to 100 μL DMSO. Store at −30 °C.
14. 100 mM DMSO solution of L-Thr-AMS: Add 4.5 mg L-Thr-AMS to 100 μL DMSO. Store at −30 °C.
15. 100 mM DMSO solution of L-Met-AMS: Add 4.8 mg L-Met-AMS to 100 μL DMSO. Store at −30 °C.
16. 100 mM DMSO solution of L-Tyr-AMS: Add 5.1 mg L-Tyr-AMS to 100 μL DMSO. Store at −30 °C.
17. 100 mM DMSO solution of L-Trp-AMS: Add 5.3 mg L-Trp-AMS to 100 μL DMSO. Store at −30 °C.

18. 100 mM DMSO solution of L-Asp-AMS: Add 4.6 mg L-Asp-AMS to 100 μL DMSO. Store at −30 °C.
19. 100 mM DMSO solution of L-Glu-AMS: Add 4.8 mg L-Glu-AMS to 100 μL DMSO. Store at −30 °C.
20. 100 mM DMSO solution of L-Lys-AMS: Add 4.7 mg L-Lys-AMS to 100 μL DMSO. Store at −30 °C.
21. 100 mM DMSO solution of L-Arg-AMS: Add 5.0 mg L-Arg-AMS to 100 μL DMSO. Store at −30 °C.
22. 100 mM DMSO solution of L-His-AMS: Add 5.0 mg L-His-AMS to 100 μL DMSO. Store at −30 °C.
23. 100 mM DMSO solution of L-Orn-AMS: Add 4.6 mg L-Orn-AMS to 100 μL DMSO. Store at −30 °C.
24. 10 mM DMSO solution of L-Phe-AMS-BPyne: Add 1 mg L-Phe-AMS-BPyne 100 μL DMSO. Store at −30 °C.
25. 10 mM DMSO solution of L-Pro-AMS-BPyne: Add 1 mg L-Pro-AMS-BPyne to 100 μL DMSO. Store at −30 °C.
26. 10 mM DMSO solution of L-Val-AMS-BPyne: Add 1 mg L-Val-AMS-BPyne to 100 μL DMSO. Store at −30 °C.
27. 10 mM DMSO solution of L-Orn-AMS-BPyne: Add 1 mg L-Orn-AMS-BPyne to 100 μL DMSO. Store at −30 °C.
28. 10 mM DMSO solution of L-Leu-AMS-BPyne: Add 1 mg L-Leu-AMS-BPyne to 100 μL DMSO. Store at −30 °C.
29. 5 mM DMSO solution of 5/6-TAMRA-peg$_3$-azide: Add 1 mg 5/6-TAMRA-peg$_3$-azide to 317 μL of DMSO. Aliquot and store at −80 °C.
30. 5 mM DMSO solution of Tris[(1-benzyl-1H-1,2,3-triazol-4-yl)methyl]amine (TBTA): Add 2.7 mg TBTA to 1 mL DMSO. Aliquot and store at −80 °C.
31. 50 mM Tris(2-carboxyethyl)phosphine HCl (TCEP) solution: Add 14.3 mg TCEP-HCl to 1 mL Milli-Q water. Aliquot and store at −80 °C.
32. 50 mM Copper(II) sulfate ($CuSO_4$) solution: Add 12.5 mg $CuSO_4{\cdot}5H_2O$ to 1 mL Milli-Q water. Aliquot and store at −80 °C.
33. 0.25 M Tris–HCl (pH 6.8): Add 15.0 g Tris(hydroxymethyl)-aminomethane to 900 mL Milli-Q water. Adjust pH to 6.8 with an appropriate volume of aqueous HCl solution and make up to 1 L with Milli-Q water. Store at room temperature.
34. 10% (w/v) sodium dodecyl sulfate (SDS) solution: Add 5 g SDS to 50 mL of Milli-Q water. Store at room temperature.
35. 0.5 M dithiothreitol (DTT) solution: Add 771 mg DTT to 10 mL Milli-Q water. Aliquot and store at −30 °C.

36. 50% (v/v) glycerol solution: Add 50 mL glycerol to 50 mL Milli-Q water. Store at room temperature.
37. 0.1% (w/v) bromophenol blue (BPB) solution: Add 10 mg BPB to 10 mL Milli-Q water. Store at room temperature.
38. SDS-PAGE gel loading buffer (5×): Mix 0.25 M Tris–HCl (pH 6.8) with 10% (w/v) SDS, 0.5 M DTT, 50% (v/v) glycerol, and 0.1% (w/v) BPB. Store one aliquot of the resulting mixture at 4 °C for immediate use. Store the remaining aliquots at −30 °C until further use.
39. SDS-PAGE running buffer (10×): Add 30.2 g Tris (hydroxymethyl)aminomethane, 144 g glycine, and 10 g SDS to 1 L Milli-Q water and dilute the resulting solution to 1× for running the gels. Store at room temperature.
40. CBB fixing solution: Add 400 mL methanol and 150 mL acetic acid to 1450 mL Milli-Q water. Store at room temperature.
41. CBB staining solution: Add 100 g ammonium sulfate, 1 g Coomassie Brilliant Blue G-250, 30 mL phosphoric acid, and 200 mL of ethanol (99.5%) to 1 L Milli-Q water. Store at room temperature.
42. 40% (w/v)-acrylamide/bis mixed solution (29:1).
43. Wide range gel preparation buffer (4×) for PAGE.
44. 10% (w/v) ammonium persulfate solution: Add 1.0 g ammonium persulfate to 10 mL Milli-Q water. Aliquot and store at −80 °C.
45. *N,N,N′,N′*-Tetramethylethylenediamine.
46. SDS-PAGE gel (6%).
47. HiMark unstained protein standard.
48. ECL plex fluorescent rainbow markers.
49. Plastic container for SDS-PAGE.
50. Eppendorf tubes, 1.5 mL.
51. Tissue culture plate (96 well), flat bottom with low evaporation.
52. UVP crosslinker.
53. High speed refrigerated microcentrifuge.
54. Typhoon laser-scanner platform.

3 Methods

3.1 Synthetic Procedures of Aminoacyl-AMS-BPyne Labeling Reagents

This section describes the synthesis of a series of **aminoacyl-AMS-BPyne** probes (Fig. 2). The probes were synthesized from commercially available adenosine, 4-dibromobutane, 4-methoxy-4-oxobutanoic acid, 4,4′-diaminobenzophenone, and 5-hexynoic chloride (Fig. 2). Compounds **1** and **5** were synthesized as previously described in Ref. [10, 17], respectively (Fig. 2).

3.1.1 Synthesis of 2a

1. Dissolve compound **1** in *N,N*-dimethylformamide under an atmosphere of nitrogen at room temperature.
2. Add 1.5 eq of Boc-Pro-OSu and 3 eq of cesium carbonate and stir the solution at room temperature for 12 h.

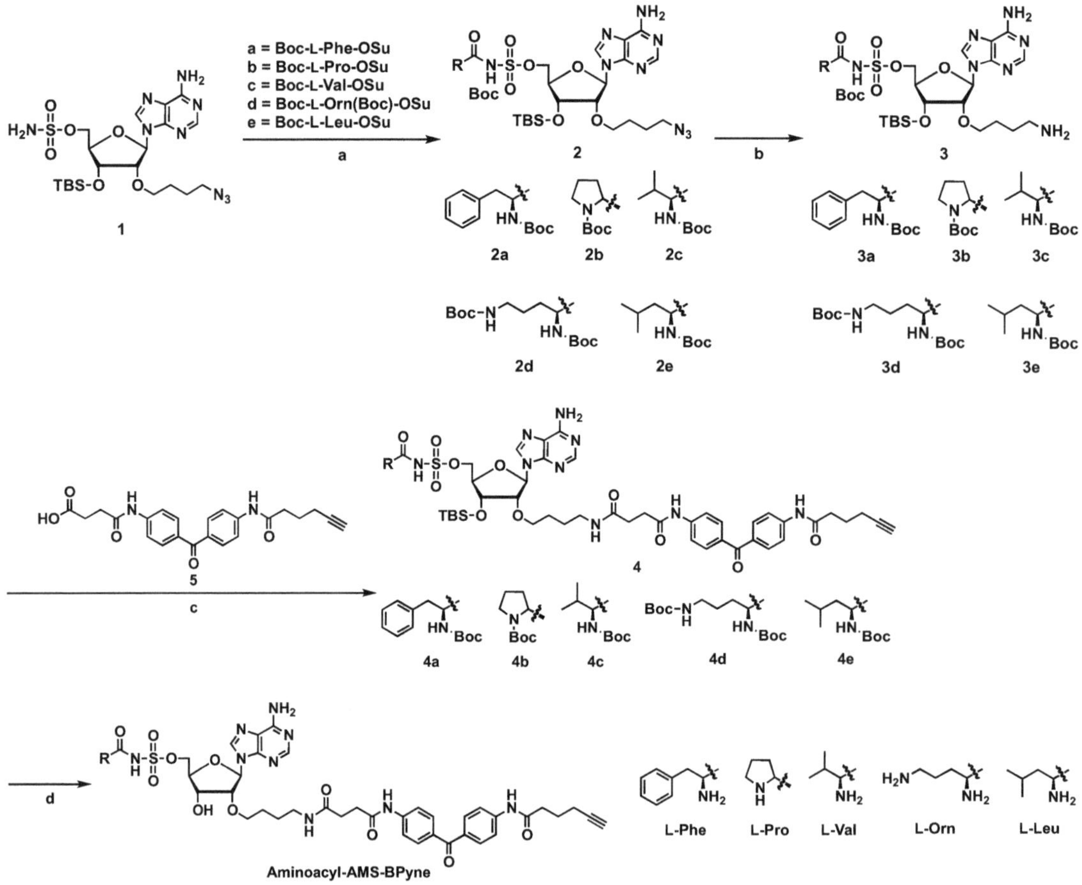

Fig. 2 Synthetic routes to **aminoacyl-AMS-BPyne** [10, 11, 17]. *Reagents and conditions*: [a] Cs_2CO_3, DMF, rt.: 67% (**2a**); 73% (**2b**); 78% (**2c**); 83% (**2d**); 78% (**2e**). [b] Pd/C, H_2, MeOH, rt.: 25% (**3a**); 64% (**3b**); 66% (**3c**); 67% (**3d**); 52% (**3e**). [c] **5**, EDC, HOBt, DMF, rt.: 86% (**4a**); 50% (**4b**); 58% (**4c**); 60% (**4d**); 42% (**4e**). [d] 80% aqueous TFA, rt.: 56% (**L-Phe-AMS-BPyne**); 91% (**L-Pro-AMS-BPyne**); 92% (**L-Val-AMS-BPyne**); 88% (**L-Orn-AMS-BPyne**); 74% (**L-Leu-AMS-BPyne**)

3. Filtrate the solution through a pad of Celite.
4. Evaporate the filtrate *in vacuo*.
5. Purify the residue by flash chromatography (90:10 $CHCl_3$/MeOH).

3.1.2 Synthesis of 3a

1. Dissolve compound **2a** in MeOH under an atmosphere of nitrogen at room temperature.
2. Add 10% Pd/C, and hydrogenate the solution under an atmosphere of hydrogen at room temperature for 12 h.
3. Filtrate the solution through a pad of Celite.
4. Evaporate the filtrate *in vacuo*.
5. Purify the residue by flash chromatography (80:20 $CHCl_3$/MeOH).

3.1.3 Synthesis of 4a

1. Dissolve 1.2 eq of compound **5** in *N,N*-dimethylformamide under an atmosphere of nitrogen at room temperature.
2. Add 1.2 eq of 1-Ethyl-3-(3-dimethylaminopropyl) carbodiimide hydrochloride and 1.2 eq of 1-hydroxybenzotriazole and stir the solution at room temperature for 10 min.
3. Add compound **3a** and stir the solution at room temperature for 12 h.
4. Dilute the solution with EtOAc and wash it with 5% aqueous citric acid, saturated aqueous sodium bicarbonate, and brine.
5. Dry the organic layer over anhydrous sodium sulfate.
6. Evaporate the solution to dryness.
7. Purify the residue by flash chromatography (86:14 $CHCl_3$/MeOH).

3.1.4 Synthesis of L-Phe-AMS-BPyne

1. Dissolve compound **4a** in a mixture of trifluoroacetic acid and water [80:20 (v/v)].
2. Stir the solution at room temperature for 8 h.
3. Evaporate the solution *in vacuo*.
4. Purify the residue by flash chromatography (80:20 $CHCl_3$/MeOH).

3.1.5 Synthesis of 2b

1. Dissolve compound **1** in *N,N*-dimethylformamide under an atmosphere of nitrogen at room temperature.
2. Add 1.5 eq of Boc-Pro-OSu and 3 eq of cesium carbonate and stir the solution at room temperature for 1 h.
3. Filtrate the solution through a pad of Celite.
4. Evaporate the filtrate *in vacuo*.
5. Purify the residue by flash chromatography (91:9 to 83:17 $CHCl_3$/MeOH).

3.1.6 Synthesis of **3b**

1. Dissolve compound **2b** in MeOH under an atmosphere of nitrogen at room temperature.
2. Add 10% Pd/C, and hydrogenate the solution under an atmosphere of hydrogen at room temperature for 24 h.
3. Filtrate the solution through a pad of Celite.
4. Evaporate the filtrate *in vacuo.*
5. Purify the residue by flash chromatography (80:20 to 75:25 $CHCl_3$/MeOH).

3.1.7 Synthesis of **4b**

1. Dissolve 1.2 eq of compound **5** in *N,N*-dimethylformamide under an atmosphere of nitrogen at room temperature.
2. Add 1.2 eq of 1-Ethyl-3-(3-dimethylaminopropyl) carbodiimide hydrochloride and 1.2 eq of 1-hydroxybenzotriazole and stir the solution at room temperature for 10 min.
3. Add compound **3b** and stir the solution at room temperature for 18 h.
4. Dilute the solution with ethyl acetate and wash it with 5% aqueous citric acid, saturated aqueous sodium bicarbonate, and brine.
5. Dry the organic layer over anhydrous sodium sulfate.
6. Evaporate the solution *in vacuo.*
6. Purify the residue by flash chromatography (91:9 to 83:17 $CHCl_3$/MeOH).

3.1.8 Synthesis of **L-Pro-AMS-BPyne**

1. Dissolve compound **4b** in a mixture of trifluoroacetic acid and water [80:20 (v/v)].
2. Stir the solution at room temperature for 6 h.
3. Evaporate the solution *in vacuo.*
4. Purify the residue by flash chromatography (75:25 $CHCl_3$/MeOH).

3.1.9 Synthesis of **2c**

1. Dissolve compound **1** in *N,N*-dimethylformamide under an atmosphere of nitrogen at room temperature.
2. Add 1.5 eq of Boc-Val-OSu and 3 eq of cesium carbonate and stir the solution at room temperature for 1 h.
3. Filtrate the solution through a pad of Celite.
4. Evaporate the filtrate *in vacuo.*
5. Purify the residue by flash chromatography (91:9 to 83:17 $CHCl_3$/MeOH).

3.1.10 Synthesis of 3c

1. Dissolve compound **2c** in MeOH under an atmosphere of nitrogen at room temperature.
2. Add 10% Pd/C and hydrogenate the solution under an atmosphere of hydrogen at room temperature for 24 h.
3. Filtrate the solution through a pad of Celite.
4. Evaporate the filtrate *in vacuo.*
5. Purify the residue by flash chromatography (80:20 $CHCl_3$/MeOH).

3.1.11 Synthesis of 4c

1. Dissolve 1.2 eq of compound **5** in *N,N*-dimethylformamide under an atmosphere of nitrogen at room temperature.
2. 2. Add 1.2 eq of 1-Ethyl-3-(3-dimethylaminopropyl) carbodiimide hydrochloride and 1.2 eq of 1-hydroxybenzotriazole and stir the solution at room temperature for 10 min.
3. Add compound **3c** and stir the solution at room temperature for 12 h.
4. Dilute the reaction mixture with EtOAc and wash it with 5% aqueous citric acid, saturated aqueous sodium bicarbonate, and brine.
5. Dry the organic layer over anhydrous sodium sulfate.
6. Evaporate the solution *in vacuo.*
7. Purify the residue by flash chromatography (83:17 $CHCl_3$/MeOH).

3.1.12 Synthesis of L-Val-AMS-BPyne

1. Dissolve compound **4c** in a mixture of trifluoroacetic acid and water [80:20 (v/v)].
2. Stir the solution at room temperature for 6 h.
3. Evaporate the solution *in vacuo.*
4. Purify the residue by flash chromatography (75:25 $CHCl_3$/MeOH).

3.1.13 Synthesis of 2d

1. Dissolve compound **1** in *N,N*-dimethylformamide under an atmosphere of nitrogen at room temperature.
2. Add 1.5 eq of Boc-Orn(Boc)-OSu and 3 eq of cesium carbonate and stir the solution at room temperature for 2 h.
3. Filtrate the solution through a pad of Celite.
4. Evaporate the filtrate *in vacuo.*
5. Purify the residue by flash chromatography (91:9 to 83:17 $CHCl_3$/MeOH).

3.1.14 Synthesis of 3d

1. Dissolve compound **2d** in MeOH under an atmosphere of nitrogen at room temperature.

2. Add 10% Pd/C and hydrogenate the solution under an atmosphere of hydrogen at room temperature for 24 h.
3. Filtrate the solution through a pad of Celite.
4. Evaporate the filtrate *in vacuo*.
5. Purify the residue by flash chromatography (80:20 $CHCl_3$/MeOH).

*3.1.15 Synthesis of **4d***

1. Dissolve 1.2 eq of compound **5** in *N,N*-dimethylformamide under an atmosphere of nitrogen at room temperature.
2. Add 1.2 eq of 1-Ethyl-3-(3-dimethylaminopropyl) carbodiimide hydrochloride and 1.2 eq of 1-hydroxybenzotriazole and stir the solution at room temperature for 10 min.
3. Add compound **3d** and stir the solution at room temperature for 3 h.
4. Dilute the solution with ethyl acetate and wash it with 5% aqueous citric acid, saturated aqueous sodium bicarbonate, and brine.
5. Dry the organic layer over anhydrous sodium sulfate.
6. Evaporate the solution *in vacuo*.
7. Purify the residue by flash chromatography (91:9 to 83:17 $CHCl_3$/MeOH).

3.1.16 Synthesis of L-Orn-AMS-BPyne

1. Dissolve compound **4d** in a mixture of trifluoroacetic acid and water [80:20 (v/v)].
2. Stir the solution at room temperature for 3 h.
3. Evaporate the solution *in vacuo*.
4. Purify the residue by flash chromatography (80:20 $CHCl_3$/MeOH).

*3.1.17 Synthesis of **2e***

1. Dissolve compound **1** in *N,N*-dimethylformamide under an atmosphere of nitrogen at room temperature.
2. Add 1.5 eq of Boc-Lue-OSu and 3 eq of cesium carbonate and stir the solution at room temperature for 1 h.
3. Filtrate the solution through a pad of Celite.
4. Evaporate the filtrate *in vacuo*.
5. Purify the residue by flash chromatography (91:9 $CHCl_3$/MeOH).

*3.1.18 Synthesis of **3e***

1. Dissolve compound **2e** in MeOH under an atmosphere of nitrogen at room temperature.
2. Add 10% Pd/C and hydrogenate the solution under an atmosphere of hydrogen at room temperature for 24 h.

3. Filtrate the solution through a pad of Celite.
4. Evaporate the filtrate *in vacuo*.
5. Purify the residue by flash chromatography (80:20 to 75:25 $CHCl_3$/MeOH).

3.1.19 Synthesis of **4e**

1. Dissolve 1.2 eq of compound **5** in *N,N*-dimethylformamide under an atmosphere of nitrogen at room temperature.
2. Add 1.2 eq of 1-Ethyl-3-(3-dimethylaminopropyl) carbodiimide hydrochloride and 1.2 eq of 1-hydroxybenzotriazole and stir the solution at room temperature for 10 min.
3. Add compound **3e** and stir the solution at room temperature for 12 h.
4. Dilute the solution with EtOAc and wash it with 5% aqueous citric acid, saturated aqueous sodium bicarbonate, and brine.
5. Dry the organic layer over anhydrous sodium sulfate.
6. Evaporate the solution *in vacuo*.
7. Purify the residue by flash chromatography (91:9 to 83:17 $CHCl_3$/MeOH).

3.1.20 Synthesis of L-Leu-AMS-BPyne

1. Dissolve compound **4e** in a mixture of trifluoroacetic acid and water [80:20 (v/v)].
2. Stir the solution at room temperature for 6 h.
3. Evaporate the solution *in vacuo*.
3. Purify the residue by flash chromatography (83:17 $CHCl_3$/MeOH).

3.2 Synthetic Procedures of Aminoayl-AMS Compounds

This section describes the synthesis of a series of aminoacyl-AMS compounds **8a**–**8t** (Fig. 3). Compound **6** was synthesized as previously described in Ref. [17] (Fig. 3).

3.2.1 Synthesis of **7a**

1. Dissolve compound **6** in *N,N*-dimethylformamide under an atmosphere of nitrogen at room temperature.
2. Add 1.5 eq of Boc-Phe-OSu and 3.0 eq of cesium carbonate and stir the solution at room temperature for 2 h.
3. Filtrate the solution through a pad of Celite.
4. Evaporate the filtrate *in vacuo*.
5. Purify the residue by flash chromatography (100:1 EtOAc/triethylamine (Et_3N)) (*see* **Note 1**).

3.2.2 Synthesis of L-Phe-AMS **8a**

1. Dissolve compound **7a** in a mixture of trifluoroacetic acid and water [80:20 (v/v)].
2. Stir the solution at room temperature for 8 h.

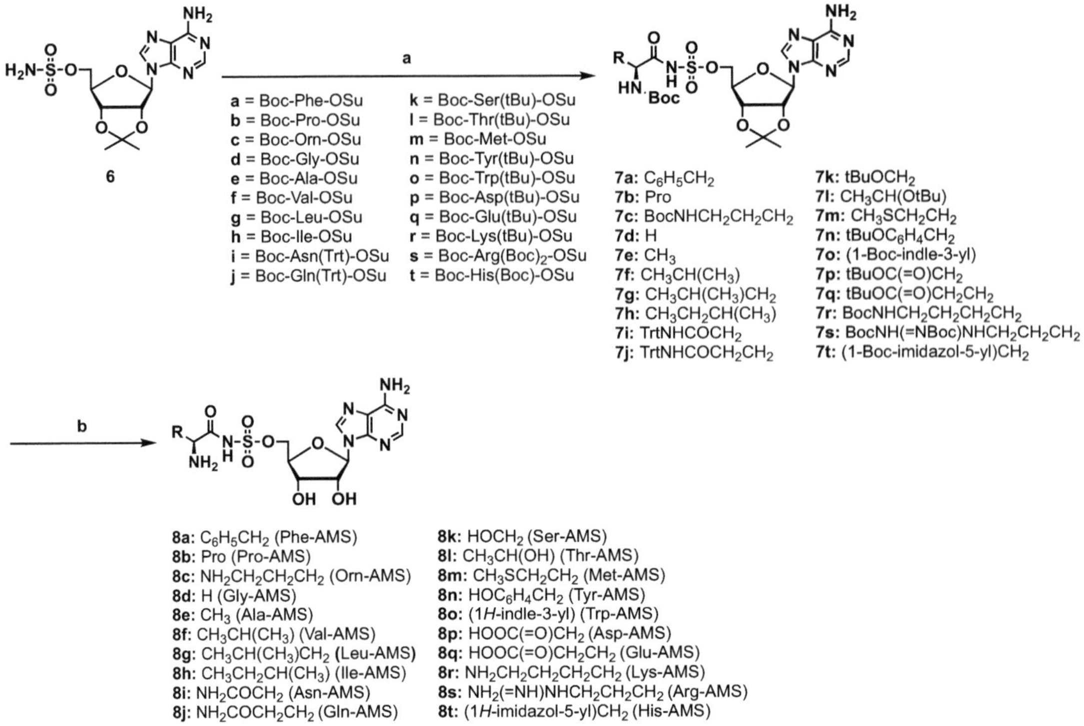

Fig. 3 Synthetic routes to a series of sulfamoyloxy-linked aminoacyl-AMP analogs **8a–8t** [10, 13, 17]. *Reagents and conditions*: [a] *N*-hydroxysuccinimide esters **a-t**, Cs_2CO_3, DMF, rt.: 80% (**7a**); 87% (**7b**); 22% (**7c**); 92% (**7d**); 53% (**7e**); 75% (**7f**); 61% (**7g**); 90% (**7h**); 88% (**7i**); 77% (**7j**); 65% (**7k**); 63% (**7l**); 75% (**7m**); 98% (**7n**); 57% (**7o**); 66% (**7p**); 76% (**7q**); 98% (**7r**); 78% (**7s**); 59% (**7t**). [b] 80% aqueous TFA or a mixture of 90:5:5 (v/v) of TFA, H_2O, and triisopropylsilane, rt.: 91% (**8a**); 74% (**8b**); 98% (**8c**); 52% (**8d**); 77% (**8e**); 82% (**8f**); 69% (**8g**); 70% (**8h**); 66% (**8i**); 40% (**8j**); 23% (**8k**); 49% (**8l**); 57% (**8m**); 74% (**8n**); 90% (**8o**); 83% (**8p**); 98% (**8q**); 61% (**8r**); 70% (**8s**); 74% (**8t**). (Reproduced from Kasai et al. [13])

3. Evaporate the solution *in vacuo*.
4. Purify the residue by flash chromatography (75:25:1 $CHCl_3$/MeOH/Et_3N) (*see* **Note 1**).

3.2.3 Synthesis of ***7b***

1. Dissolve compound **6** in *N,N*-dimethylformamide under an atmosphere of nitrogen at room temperature.
2. Add 1.5 eq of Boc-Pro-OSu and 3.0 eq of cesium carbonate and stir the solution at room temperature for 1 h.
3. Filtrate the solution through a pad of Celite.
4. Evaporate the filtrate *in vacuo*.
5. Purify the residue by flash chromatography (91:9:1 to 86:14:1 EtOAc/MeOH/Et_3N) (*see* **Note 1**).

3.2.4 Synthesis of L-Pro-AMS ***8b***

1. Dissolve compound **7b** in a mixture of trifluoroacetic acid and water [80:20 (v/v)].
2. Stir the solution at room temperature for 2 h.

3. Evaporate the solution *in vacuo*.
4. Purify the residue by flash chromatography (67:33:1 to 50:50:1 $CHCl_3$/MeOH/Et_3N) (*see* **Note 1**).

3.2.5 Synthesis of **7c**

1. Dissolve compound **6** in *N,N*-dimethylformamide under an atmosphere of nitrogen at room temperature.
2. Add 1.1 eq of Boc-Orn(Boc)-OSu and 3.0 eq of cesium carbonate and stir the solution at room temperature for 3 h.
3. Filtrate the solution through a pad of Celite.
4. Evaporate the filtrate *in vacuo*.
5. Purify the residue by flash chromatography (95:5:1 to 83:17:1 $CHCl_3$/MeOH/Et_3N) (*see* **Note 1**).

3.2.6 Synthesis of L-Orn-AMS **8c**

1. Dissolve compound **7c** in a mixture of trifluoroacetic acid and water [80:20 (v/v)].
2. Stir the solution at room temperature for 3 h.
3. Evaporate the solution *in vacuo*.
4. Purify the residue by high-performance liquid chromatography (HPLC) (*see* **Note 2**).

3.2.7 Synthesis of **7d**

1. Dissolve compound **6** in *N,N*-dimethylformamide under an atmosphere of nitrogen at room temperature.
2. Add 1.5 eq of Boc-Gly-OSu and 3.0 eq of cesium carbonate and stir the solution at room temperature for 12 h.
3. Filtrate the solution through a pad of Celite.
4. Evaporate the filtrate *in vacuo*.
5. Purify the residue by flash chromatography (95:10:1 $CHCl_3$/MeOH/Et_3N) (*see* **Note 1**).

3.2.8 Synthesis of Gly-AMS **8d**

1. Dissolve compound **7d** in a mixture of trifluoroacetic acid and water [80:20 (v/v)].
2. Stir the solution at room temperature for 10 h.
3. Evaporate the solution *in vacuo*.
4. Purify the residue by flash chromatography (50:50:1 $CHCl_3$/MeOH/Et_3N) (*see* **Note 1**).

3.2.9 Synthesis of **7e**

1. Dissolve compound **6** in *N,N*-dimethylformamide under an atmosphere of nitrogen at room temperature.
2. Add 1.5 eq of Boc-Ala-OSu and 3.0 eq of cesium carbonate and stir the solution at room temperature for 12 h.
3. Filtrate the solution through a pad of Celite.

4. Evaporate the filtrate *in vacuo*.
5. Purify the residue by flash chromatography (95:5:1 to 95:5:1 $CHCl_3$/MeOH/Et_3N) (*see* **Note 1**).

3.2.10 Synthesis of L-Ala-AMS 8e

1. Dissolve compound **7e** in a mixture of trifluoroacetic acid and water [80:20 (v/v)].
2. Stir the solution at room temperature for 3 h.
3. Evaporate the solution *in vacuo*.
4. Purify the residue by flash chromatography (67:33:1 $CHCl_3$/MeOH/Et_3N to 100:1 MeOH/Et_3N) (*see* **Note 1**).

3.2.11 Synthesis of 7f

1. Dissolve compound **6** in *N,N*-dimethylformamide under an atmosphere of nitrogen at room temperature.
2. Add 1.5 eq of Boc-Val-OSu and 3.0 eq of cesium carbonate and stir the solution at room temperature for 12 h.
3. Filtrate the solution through a pad of Celite.
4. Evaporate the filtrate *in vacuo*.
5. Purify the residue by flash chromatography (95:5:1 to 90:10:1 EtOAc/MeOH/Et_3N) (*see* **Note 1**).

3.2.12 Synthesis of L-Val-AMS 8f

1. Dissolve compound **7f** in a mixture of trifluoroacetic acid and water [80:20 (v/v)].
2. Stir the solution at room temperature for 6 h.
3. Evaporate the solution *in vacuo*.
4. Purify the residue by flash chromatography (83:17:1 to 67:33:1 $CHCl_3$/MeOH/Et_3N) (*see* **Note 1**).

3.2.13 Synthesis of 7g

1. Dissolve compound **6** in *N,N*-dimethylformamide under an atmosphere of nitrogen at room temperature.
2. Add 1.5 eq of Boc-Leu-OSu and 3.0 eq of cesium carbonate and stir the solution at room temperature for 4 h.
3. Filtrate the solution through a pad of Celite.
4. Evaporate the filtrate *in vacuo*.
5. Purify the residue by flash chromatography (97:3:1 $CHCl_3$/MeOH/Et_3N) (*see* **Note 1**).

3.2.14 Synthesis of L-Leu-AMS 8g

1. Dissolve compound **7g** in a mixture of trifluoroacetic acid and water [80:20 (v/v)].
2. Stir the solution at room temperature for 5 h.
3. Evaporate the solution *in vacuo*.
4. Purify the residue by flash chromatography (83:17:1 to 67:33:1 $CHCl_3$/MeOH/Et_3N) (*see* **Note 1**).

*3.2.15 Synthesis of **7h***

1. Dissolve compound **6** in *N,N*-dimethylformamide under an atmosphere of nitrogen at room temperature.
2. Add 1.5 eq of Boc-Ile-OSu and 3.0 eq of cesium carbonate and stir the solution at room temperature for 12 h.
3. Filtrate the solution through a pad of Celite.
4. Evaporate the filtrate *in vacuo*.
5. Purify the residue by flash chromatography (90:10:1 to $CHCl_3$/MeOH/Et_3N) (*see* **Note 1**).

*3.2.16 Synthesis of L-Ile-AMS **8h***

1. Dissolve compound **7h** in a mixture of trifluoroacetic acid and water [80:20 (v/v)].
2. Stir the solution at room temperature for 4 h.
3. Evaporate the solution *in vacuo*.
4. Purify the residue by flash chromatography (50:50:1 $CHCl_3$/MeOH/Et_3N) (*see* **Note 1**).

*3.2.17 Synthesis of **7i***

1. Dissolve Boc-Asn(Trt)-OH in *N,N*-dimethylformamide under an atmosphere of nitrogen at room temperature.
2. Add 1.5 eq of 1-Ethyl-3-(3-dimethylaminopropyl)carbodiimide hydrochloride and 2.5 eq of *N*-hydroxysuccinimide and stir the solution at room temperature for 3 h.
3. Dilute the solution with EtOAc and wash it with 5% aqueous citric acid, saturated aqueous sodium bicarbonate, and brine.
4. Dry the organic layer over anhydrous sodium sulfate.
5. Evaporate the solution *in vacuo*.
6. Dissolve compound **6** in *N,N*-dimethylformamide under an atmosphere of nitrogen at room temperature.
7. Add 1.5 eq of Boc-Asn(Trt)-OSu and 3.0 eq of cesium carbonate and stir the solution at room temperature for 1 h.
8. Filtrate the solution through a pad of Celite.
9. Evaporate the filtrate *in vacuo*.
10. Purify the residue by flash chromatography (95:5:1 to 90:10:1 $CHCl_3$/MeOH/Et_3N) (*see* **Note 1**).

*3.2.18 Synthesis of L-Asn-AMS **8i***

1. Dissolve compound **7i** in a mixture of trifluoroacetic acid and water [80:20 (v/v)].
2. Stir the solution at room temperature for 2 h.
3. Evaporate the solution *in vacuo*.
4. Purify the residue by flash chromatography (90:10:1 to 85:15:1 $CHCl_3$/MeOH/Et_3N) (*see* **Note 1**).

3.2.19 Synthesis of **7j**

1. Dissolve Boc-Gln(Trt)-OH in dichloromethane under an atmosphere of nitrogen at room temperature.
2. Add 1.5 eq of 1-Ethyl-3-(3-dimethylaminopropyl)carbodiimide hydrochloride and 1.5 eq of *N*-hydroxysuccinimide and stir the solution at room temperature for 5 h.
3. Evaporate the solution *in vacuo*.
4. Purify the residue by flash chromatography (20:80 to 33:67 EtOAc/hexane).
5. Dissolve compound **6** in *N,N*-dimethylformamide under an atmosphere of nitrogen at room temperature.
6. Add 1.5 eq of Boc-Gln(Trt)-OSu and 3.0 eq of cesium carbonate and stir the solution at room temperature for 1 h.
7. Filtrate the solution through a pad of Celite.
8. Evaporate the filtrate *in vacuo*.
9. Purify the residue by flash chromatography (95:5:1 to 90:10:1 $CHCl_3$/MeOH/Et_3N) (*see* **Note 1**).

3.2.20 Synthesis of L-Gln-AMS **8j**

1. Dissolve compound **7j** in a mixture of trifluoroacetic acid and water [80:20 (v/v)].
2. Stir the solution at room temperature for 30 min.
3. Evaporate the solution *in vacuo*.
4. Purify the residue by flash chromatography (67:33:1 $CHCl_3$/MeOH/Et_3N to 100:1 MeOH/Et_3N) (*see* **Note 1**).

3.2.21 Synthesis of **7k**

1. Dissolve Boc-Ser(tBu)-OH·DCHA in dichloromethane under an atmosphere of nitrogen at room temperature.
2. Add 1.2 eq of 1-Ethyl-3-(3-dimethylaminopropyl)carbodiimide hydrochloride, 1.2 eq of *N*-hydroxysuccinimide, 3.0 eq of *N,N*-diisopropylethylamine, and 0.1 eq of 4-dimethylaminopyridine and stir the solution at room temperature for 3 h.
3. Wash the solution with 5% aqueous citric acid, saturated sodium bicarbonate, and brine.
4. Dry the organic layer over anhydrous sodium sulfate.
5. Evaporate the solution *in vacuo*.
6. Purify the residue by flash chromatography (10:90 to 33:67 EtOAc/hexane).
7. Dissolve compound **6** in *N,N*-dimethylformamide under an atmosphere of nitrogen at room temperature.
8. Add 1.5 eq of Boc-Ser(tBu)-OSu and 3.0 eq of cesium carbonate and stir the solution at room temperature for 1 h.
9. Filtrate the solution through a pad of Celite.

10. Evaporate the filtrate *in vacuo*.
11. Purify the residue by flash chromatography (95:5:1 to 80:20:1 $CHCl_3$/MeOH/Et_3N) (*see* **Note 1**).

3.2.22 Synthesis of L-Ser-AMS ***8k***

1. Dissolve compound **7k** in a mixture of trifluoroacetic acid, water, and triisopropylsilane [90:5:5 (v/v)].
2. Stir the solution at room temperature for 30 min.
3. Evaporate the solution *in vacuo*.
4. Purify the residue by flash chromatography (80:20:1 to 50:50:1 $CHCl_3$/MeOH/Et_3N) (*see* **Note 1**).

3.2.23 Synthesis of ***7l***

1. Dissolve Boc-Thr(tBu)-OH in dichloromethane under an atmosphere of nitrogen at room temperature.
2. Add 1.2 eq of 1-Ethyl-3-(3-dimethylaminopropyl)carbodiimide hydrochloride, 1.2 eq of *N*-hydroxysuccinimide, 3.0 eq of *N,N*-diisopropylethylamine, and 0.1 eq of 4-dimethylaminopyridine and stir the solution at room temperature for 2 h.
3. Wash the solution with 5% aqueous citric acid, saturated aqueous sodium bicarbonate, and brine.
4. Dry the organic layer over anhydrous sodium sulfate.
5. Evaporate the solution *in vacuo*.
6. Purify the residue by flash chromatography (17:83 to 33:67 EtOAc/hexane).
7. Dissolve compound **6** in *N,N*-dimethylformamide under an inert atmosphere at room temperature.
8. Add 1.5 eq of Boc-Thr(tBu)-OSu and 3.0 eq of cesium carbonate and stir the solution at room temperature for 1 h.
9. Filtrate the solution through a pad of Celite.
10. Evaporate the filtrate *in vacuo*.
11. Purify the residue by flash chromatography (95:5:1 to 80:20:1 $CHCl_3$/MeOH/Et_3N) (*see* **Note 1**).

3.2.24 Synthesis of L-Thr-AMS ***8l***

1. Dissolve compound **7l** in a mixture of trifluoroacetic acid, water, and triisopropylsilane [90:5:5 (v/v)].
2. Stir the solution at room temperature for 30 min.
3. Evaporate the solution *in vacuo*.
4. Purify the residue by flash chromatography (90:10:1 to 60:40:1 $CHCl_3$/MeOH/Et_3N) (*see* **Note 1**).

3.2.25 Synthesis of ***7m***

1. Dissolve compound **6** in *N,N*-dimethylformamide under an atmosphere of nitrogen at room temperature.
2. Add 1.5 eq of Boc-Met-OSu and 3.0 eq of cesium carbonate and stir the solution at room temperature for 12 h.

3. Filtrate the solution through a pad of Celite.
4. Evaporate the filtrate *in vacuo*.
5. Purify the residue by flash chromatography (94:6:1 to EtOAc/MeOH/Et_3N) (*see* **Note 1**).

3.2.26 Synthesis of L-Met-AMS 8m

1. Dissolve compound **7m** in a mixture of trifluoroacetic acid and water [80:20 (v/v)].
2. Stir the solution at room temperature for 3 h.
3. Evaporate the solution *in vacuo*.
4. Purify the residue by flash chromatography (75:25:1 $CHCl_3$/MeOH/Et_3N) (*see* **Note 1**).

3.2.27 Synthesis of 7n

1. Dissolve Boc-Tyr(tBu)-OH in dichloromethane under an atmosphere of nitrogen at room temperature.
2. Add 2.2 eq of 1-Ethyl-3-(3-dimethylaminopropyl)carbodiimide hydrochloride and 2.2 eq of *N*-hydroxysuccinimide and stir the solution at room temperature for 12 h.
3. Wash the solution with 5% aqueous citric acid, saturated aqueous sodium bicarbonate, and brine.
4. Dry the organic layer over anhydrous sodium sulfate.
5. Evaporate the solution *in vacuo*.
6. Dissolve compound **6** in *N*,*N*-dimethylformamide under an atmosphere of nitrogen at room temperature.
7. Add 1.9 eq of Boc-Thr(tBu)-OSu and 6.0 eq of cesium carbonate and stir the solution at room temperature for 12 h.
8. Filtrate the solution through a pad of Celite.
9. Evaporate the filtrate *in vacuo*.
10. Purify the residue by flash chromatography (90:10:1 $CHCl_3$/MeOH/Et_3N) (*see* **Note 1**).

3.2.28 Synthesis of L-Tyr-AMS 8n

1. Dissolve compound **7n** in a mixture of trifluoroacetic acid, water, and triisopropylsilane [90:5:5 (v/v)].
2. Stir the solution at room temperature for 8 h.
3. Evaporate the solution *in vacuo*.
4. Purify the residue by HPLC (*see* **Note 3**).

3.2.29 Synthesis of 7o

1. Dissolve Boc-Trp(Boc)-OH in *N*,*N*-dimethylformamide under an atmosphere of nitrogen at room temperature.
2. Add 1.2 eq of 1-Ethyl-3-(3-dimethylaminopropyl)carbodiimide hydrochloride and 1.2 eq of *N*-hydroxysuccinimide and stir the solution at room temperature for 6 h.

3. Dilute the solution with EtOAc and wash it with 5% aqueous citric acid, saturated aqueous sodium bicarbonate, and brine.
4. Dry the organic layer over anhydrous sodium sulfate.
5. Evaporate the solution *in vacuo*.
6. Dissolve compound **6** in *N,N*-dimethylformamide under an atmosphere of nitrogen at room temperature.
7. Add 1.5 eq of Boc-Trp(Boc)-OSu and 3.0 eq of cesium carbonate and stir the solution at room temperature for 1 h.
8. Filtrate the solution through a pad of Celite.
9. Evaporate the filtrate *in vacuo*.
10. Purify the residue by flash chromatography (95:5:1 to 91:9:1 EtOAc/MeOH/Et_3N) (*see* **Note 1**).

3.2.30 Synthesis of L-Trp-AMS 8o

1. Dissolve compound **7o** in a mixture of trifluoroacetic acid and water [80:20 (v/v)].
2. Stir the solution at room temperature for 3 h.
3. Evaporate the solution *in vacuo*.
4. Purify the residue by flash chromatography (67:33:1 $CHCl_3$/MeOH/Et_3N) (*see* **Note 1**).

3.2.31 Synthesis of 7p

1. Dissolve Boc-Asp(OtBu)-OH in *N,N*-dimethylformamide under an atmosphere of nitrogen at room temperature.
2. Add 1.5 eq of 1-Ethyl-3-(3-dimethylaminopropyl)carbodiimide hydrochloride and 1.5 eq of *N*-hydroxysuccinimide and stir the solution at room temperature for 22 h.
3. Dilute the solution with EtOAc and wash it with 5% aqueous citric acid, saturated aqueous sodium bicarbonate, and brine.
4. Dry the organic layer over anhydrous sodium sulfate.
5. Evaporate the solution *in vacuo*.
6. Dissolve compound **6** in *N,N*-dimethylformamide under an atmosphere of nitrogen at room temperature.
7. Add 1.5 eq of Boc-Asp(OtBu)-OSu and 3.0 eq of cesium carbonate and stir the solution at room temperature for 3 h.
8. Filtrate the solution through a pad of Celite.
9. Evaporate the filtrate *in vacuo*.
10. Purify the residue by flash chromatography (94:6:1 EtOAc/MeOH/Et_3N) (*see* **Note 1**).

3.2.32 Synthesis of L-Asp-AMS 8p

1. Dissolve compound **7p** in a mixture of trifluoroacetic acid and water [80:20 (v/v)].
2. Stir the solution at room temperature for 6 h.

3. Evaporate the solution *in vacuo*.
4. Purify the residue by flash chromatography (67:33:1 $CHCl_3$/MeOH/Et_3N) (*see* **Note 1**).

*3.2.33 Synthesis of **7q***

1. Dissolve Boc-Glu(OtBu)-OH in *N,N*-dimethylformamide under an atmosphere of nitrogen at room temperature.
2. Add 1.2 eq of 1-Ethyl-3-(3-dimethylaminopropyl)carbodiimide hydrochloride and 1.2 eq of *N*-hydroxysuccinimide and stir the solution at room temperature for 24 h.
3. Dilute the solution with EtOAc and wash it with 5% aqueous citric acid, saturated aqueous sodium bicarbonate, and brine.
4. Dry the organic layer over anhydrous sodium sulfate.
5. Evaporate the solution *in vacuo*.
6. Dissolve compound **6** in *N,N*-dimethylformamide under an atmosphere of nitrogen at room temperature.
7. Add 1.5 eq of Boc-Glu(OtBu)-OSu and 3.0 eq of cesium carbonate and stir the solution at room temperature for 3 h.
8. Filtrate the solution through a pad of Celite.
9. Evaporate the filtrate *in vacuo*.
10. Purify the residue by flash chromatography (94:6:1 to 91:9:1 EtOAc/MeOH/Et_3N) (*see* **Note 1**).

*3.2.34 Synthesis of L-Glu-AMS **8q***

1. Dissolve compound **7q** in a mixture of trifluoroacetic acid and water [80:20 (v/v)].
2. Stir the solution at room temperature for 3 h.
3. Evaporate the solution *in vacuo*.
4. Purify the residue by flash chromatography (67:33:1 $CHCl_3$/MeOH/Et_3N to 100:1 MeOH/Et_3N) (*see* **Note 1**).

*3.2.35 Synthesis of **7r***

1. Dissolve compound **6** in *N,N*-dimethylformamide under an atmosphere of nitrogen at room temperature.
2. Add 1.5 eq of Boc-Lys(Boc)-OSu and 3.0 eq of cesium carbonate and stir the solution at room temperature for 12 h.
3. Filtrate the solution through a pad of Celite.
4. Evaporate the filtrate *in vacuo*.
5. Purify the residue by flash chromatography (90:10:1 to $CHCl_3$/MeOH/Et_3N) (*see* **Note 1**).

*3.2.36 Synthesis of L-Lys-AMS **8r***

1. Dissolve compound **7r** in a mixture of trifluoroacetic acid and water [80:20 (v/v)].
2. Stir the solution at room temperature for 10 h.

3. Evaporate the solution *in vacuo.*
4. Purify the residue by HPLC (*see* **Note 4**).

*3.2.37 Synthesis of **7s***

1. Dissolve Boc-Arg$(Boc)_2$-OH in tetrahydrofuran under an atmosphere of nitrogen at room temperature.
2. Add 1.2 eq of 1-Ethyl-3-(3-dimethylaminopropyl)carbodiimide hydrochloride and 1.2 eq of *N*-hydroxysuccinimide and stir the solution at room temperature for 15 h.
3. Evaporate the solution *in vacuo.*
4. Purify the residue by flash chromatography (90:10 to 67:33 EtOAc/hexane).
5. Dissolve compound **6** in *N,N*-dimethylformamide under an atmosphere of nitrogen at room temperature.
6. Add 1.5 eq of Boc-Arg$(Boc)_2$-OSu and 3.0 eq of cesium carbonate and stir the solution at room temperature for 1 h.
7. Filtrate the solution through a pad of Celite.
8. Evaporate the filtrate *in vacuo.*
9. Purify the residue by flash chromatography (95:5:1 to 90:10:1 EtOAc/MeOH/Et_3N) (*see* **Note 1**).

*3.2.38 Synthesis of L-Arg-AMS **8s***

1. Dissolve compound **8r** in a mixture of trifluoroacetic acid and water [80:20 (v/v)].
2. Stir the solution at room temperature for 2 h.
3. Evaporate the solution *in vacuo.*
4. Purify the residue by flash chromatography (50:50:1 $CHCl_3$/MeOH/Et_3N to 100:1 MeOH/Et_3N) (*see* **Note 1**).

*3.2.39 Synthesis of **7t***

1. Dissolve compound **6** in *N,N*-dimethylformamide under an atmosphere of nitrogen at room temperature.
2. Add 1.5 eq of Boc-His(Boc)-OSu and 3.0 eq of cesium carbonate and stir the solution at room temperature for 12 h.
3. Filtrate the solution through a pad of Celite.
4. Evaporate the filtrate *in vacuo.*
5. Purify the residue by flash chromatography (95:5:1 to 83:17:1 EtOAc/MeOH/Et_3N) (*see* **Note 1**).

*3.2.40 Synthesis of L-His-AMS **8t***

1. Dissolve compound **7t** in a mixture of trifluoroacetic acid and water [80:20 (v/v)].
2. Stir the solution at room temperature for 3 h.
3. Evaporate the solution *in vacuo.*
4. Purify the residue by flash chromatography (67:33:1 $CHCl_3$/MeOH/Et_3N to 100:1 MeOH/Et_3N) (*see* **Note 1**).

3.3 Bacterial Culture

This section describes the propagation and cultivation of the gramicidin S producer *A. migulanus* ATCC 9999.

3.3.1 Bacterial Propagation Procedure

1. Open vial containing culture lyophilizate of *A. migulanus* ATCC 9999.
2. Start with at least three culture tubes each containing 5 mL of nutrient broth.
3. Remove 1 mL nutrient broth using a 1 mL pipette and rehydrate the pellet.
4. Aseptically transfer this aliquot back into a culture tube and mix thoroughly.
5. Use several drops of this mixture to inoculate the remaining tubes of nutrient broth and nutrient agar plates.
6. Grow at 37 °C for 24 h.
7. Add 700 μL of liquid bacterial culture to each of the 2 mL screw cap vials.
8. Add 100 μL of 80% glycerol to each of the vials, mix well, and store at −80 °C.

3.3.2 Bacterial Culture Procedure

1. Prepare nutrient agar plates
2. Prepare a rich, complex medium (YPG) according to the literature procedure [17].
3. Streak *A. migulanus* ATCC 9999 for isolation of single colonies on nutrient agar plates and incubate the resulting plates at 37 °C for 24 h.
4. Pick a single colony.
5. Prepare a seed culture of 2–5 mL YPG medium in a culture tube and incubate the culture at 37 °C for 24 h on a reciprocating shaker at 200 rpm.
6. Inoculate 2 mL of the seed culture into 250 mL YPG in a 1 L baffled flask and grow at 37 °C for 24 h ($OD_{600} = 29.5$) on a reciprocating shaker at 200 rpm.
7. Determine growth at specific time points by measuring the absorbance at 600 nm on a spectrophotometer.
8. Harvest the cells into 50 mL plastic tubes and centrifuge them for 15 min at 15,000 × *g* at 4 °C.
9. Divide the cells into 1.5 mL Eppendorf tubes and centrifuge them for 10 min at 15,000 × *g* at 4 °C.
10. Discard the supernatant and store the cell pellets at −80 °C until use.

3.4 Sample Preparation for Labeling Studies

This section describes the preparation of proteomes from *A. migulanus* ATCC 9999.

1. Resuspend the frozen cell pellets in 0.5–1 mL of 20 mM Tris–HCl (pH 8.0) containing 1 mM $MgCl_2$, 1 mM TCEP, 0.05% Nonidet(R) P40 substitute, and the protease inhibitor cocktail on ice.
2. Add 10–20 μL of 10 mg/mL lysozyme solution to the cell suspension and incubate the resulting mixture on ice for 30 min with gentle shaking (*see* **Note 5**).
3. Transfer the mixture to a constant temperature incubator at 37 °C and incubate for 30 min with gentle shaking (*see* **Note 5**).
4. Centrifuge the mixture for 10 min at 15,000 × *g* at 4 °C and collect the supernatant (*see* **Note 6**).
5. Determine protein concentration with Bradford assay in a 96-well plate at an absorbance of 595 nm using bovine serum albumin as standard [18].

3.5 Activity-Based Protein Profiling

This section describes the labeling of endogenous GrsA- and GrsB-NRPSs *in vitro* (Figs. 4 and 5).

1. To label endogenous GrsA-NRPS, treat the resultant proteome (45 μL, 2.0–3.0 mg/mL) with DMSO (0.5 μL) and **L-Phe-AMS-BPyne** (0.5 μL of 100 μM stock in DMSO, final concentration: 1 μM) for 10 min at room temperature (*see* **Note 7**) (*see* **Note 8**).
2. To label endogenous GrsB-NRPS, treat the resultant proteomes (45 μL, 2.0–3.0 mg/mL) with DMSO (0.5 μL) and **L-Pro-AMS-BPyne**, **L-Val-AMS-BPyne**, **L-Orn-AMS-BPyne**, or **L-Leu-AMS-BPyne** (0.5 μL of 100 μM stock in DMSO, final concentration: 1 μM) for 10 min at room temperature (*see* **Note 7**) (*see* **Note 8**).
3. Perform UV irradiation for a 96-well plate using a UVP Crosslinker (Analytik Jena) (8 W, 365 nm, on ice) (*see* **Note 9**) (*see* **Note 10**).
4. Perform copper(I)-catalyzed alkyne-azide cycloaddition (CuAAC) click chemistry reaction on the 96-well plate by adding 4 μL CuAAC click chemistry mix (1 μL of 50 mM $CuSO_4$, 1 μL of 5 mM TBTA, 1 μL of 50 mM TCEP, and 1 μL of 5 mM 5/6-TAMRA-peg$_3$-azide) into each well (*see* **Note 11**).
5. Incubate each sample at room temperature in the dark for 1 h.
6. Take 40 μL of each sample for SDS-PAGE and in-gel fluorescence analysis.

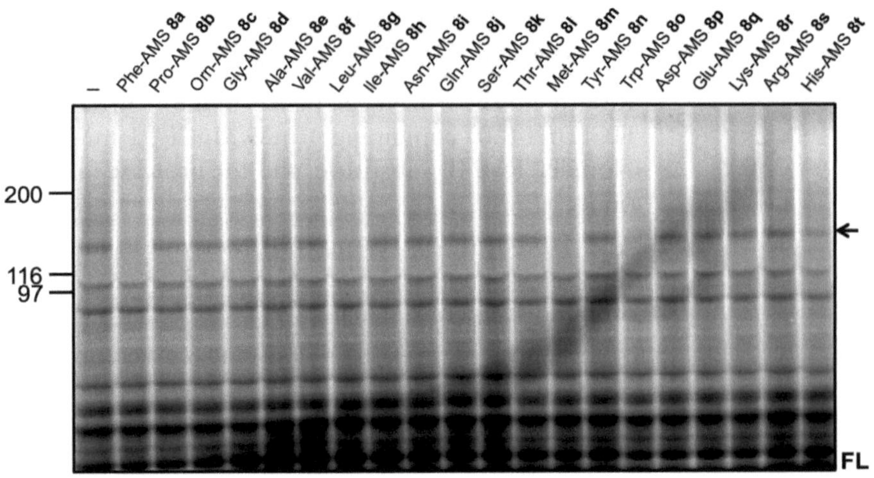

Fig. 4 Competitive ABPP experiment of **8a–8t** toward the A-domain of endogenous GrsA-NRPS [13]. Assessment of the inhibition potency (100 μM compound) in the *A. migulanus* ATCC 9999 proteome with **L-Phe-AMS-BPyne** (1 μM). The labeling of the $A_{L\text{-}Phe}$-domain of endogenous GrsA-NRPS by **L-Phe-AMS-BPyne** completely disappeared or reduced in the presence of L-Phe-AMS **8a**, L-Leu-AMS **8g**, L-Met-AMS **8m**, L-Trp-AMS **8o**, and L-His-AMS **8t**. This inhibitor sensitivity profile correlates with the substrate specificity of the $A_{L\text{-}Phe}$-domain of GrsA-NRPS. Indeed, the $A_{L\text{-}Phe}$-domain of recombinant GrsA catalyzed the adenylation of L-Phe, L-Leu, L-Met, L-Trp, and L-His [13]. The arrow indicates the endogenous GrsA-NRPS. The gel was visualized using in-gel fluorescence (FL). (Reproduced from Kasai et al. [13])

7. Add 10 μL of reducing 5× SDS-loading buffer to each sample, mix thoroughly, and heat at 95 °C for 5 min.
8. Load 20 μL of each sample on an SDS-PAGE gel.
9. After electrophoresis, wash thoroughly with Milli-Q water before scanning gels with a Typhoon 9410 Gel and Blot Imager (GE Healthcare) with 532 nm laser excitation and 580 nm emission using standard procedures.
10. Stain gels with CBB staining solution following standard procedures.

3.6 Competitive Activity-Based Protein Profiling

This section describes the functional profiling of endogenous GrsA- and GrsB-NRPSs *in vitro*.

1. For the functional profiling of an endogenous GrsA-NRPS, treat the resultant proteomes (45 μL, 2.0–3.0 mg/mL) with sulfamoyloxy-linked aminoacyl-AMP analogs **8a–8t** (0.5 μL of 10 mM stock in DMSO, final concentration: 100 μM) for 10 min at room temperature and incubate the mixture with **L-Phe-AMS-BPyne** (0.5 μL of 100 μM stock in DMSO, final concentration: 1 μM) for 10 min at room temperature (*see* **Note** 7) (*see* **Note 8**).
2. For the functional profiling of an endogenous GrsB-NRPS, treat the resultant proteomes (45 μL, 2.0–3.0 mg/mL) with

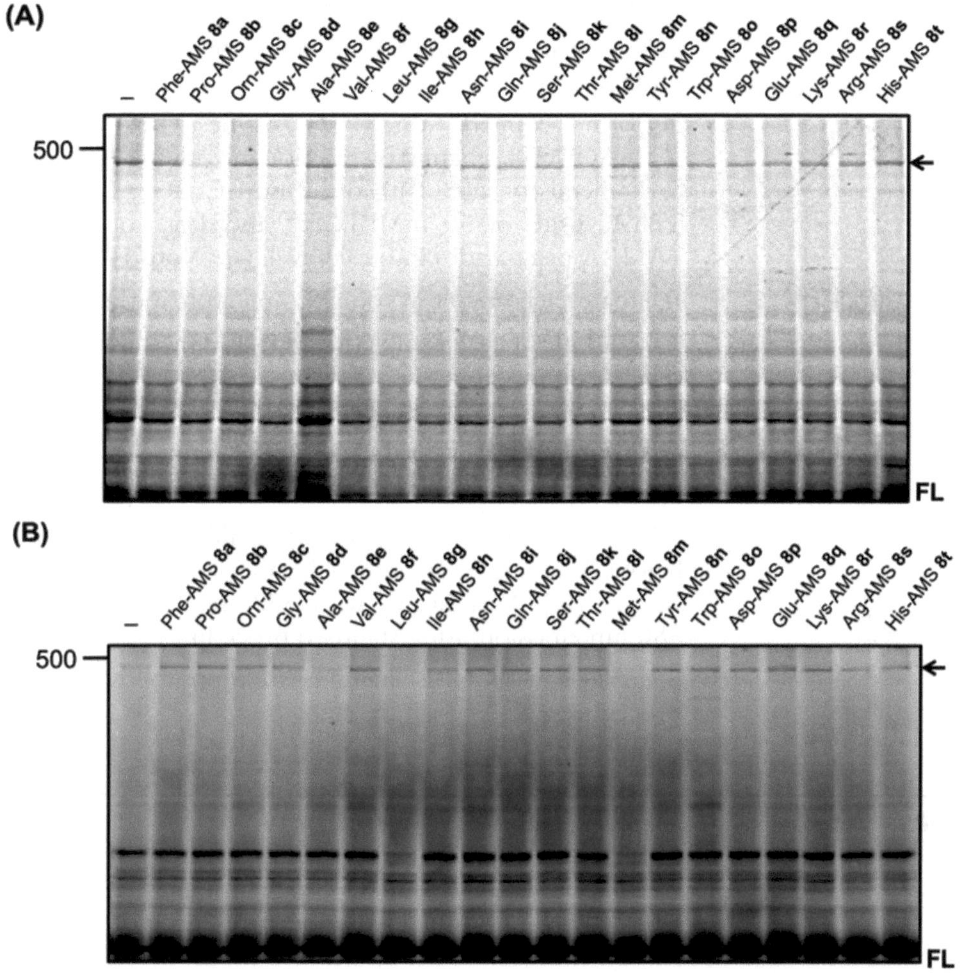

Fig. 5 Representative competitive ABPP experiments of **8a**–**8t** toward the A-domains of endogenous GrsB-NRPS [13]. (**a**) Assessment of the inhibition potency (100 μM compound) toward the $A_{L\text{-}Pro}$-domain of GrsB-NRPS with **L-Pro-AMS-BPyne** (1 μM). The labeling of the $A_{L\text{-}Pro}$-domain of GrsB-NRPS was lost completely only on adding L-Pro-AMS **8b**, corresponding identically to its substrate preferences [13]. (**b**) Assessment of the inhibition potency (100 μM compound) toward the $A_{L\text{-}Leu}$-domain of GrsB-NRPS with **L-Pro-AMS-BPyne** (1 μM). The labeling of the $A_{L\text{-}Leu}$-domain of GrsB-NRPS was inhibited on adding L-Ala-AMS **8e**, L-Leu-AMS **8g**, and L-Met-AMS **8m**. This inhibitor sensitivity profile suggests that the $A_{L\text{-}Leu}$-domain of GrsB-NRPS may accept L-Ala and L-Met as alternative substrates. The arrow indicates the endogenous GrsB-NRPS. The gel was visualized using in-gel fluorescence (FL). (Reproduced from Kasai et al. [13])

sulfamoyloxy-linked aminoacyl-AMP analogs **8a**–**8t** (0.5 μL of 10 mM stock in DMSO, final concentration: 100 μM) for 10 min at room temperature and incubate the mixture with **L-Pro-AMS-BPyne**, **L-Val-AMS-BPyne**, **L-Orn-AMS-BPyne**, or **L-Leu-AMS-BPyne** (0.5 μL of 100 μM stock in DMSO, final concentration: 1 μM) for 10 min at room temperature (*see* **Note** 7) (*see* **Note 8**).

3. Perform UV irradiation for a 96-well plate using a UVP Crosslinker (Analytik Jena) (8 W, 365 nm, on ice) (*see* **Note 9**) (*see* **Note 10**).
4. Perform copper(I)-catalyzed alkyne-azide cycloaddition (CuAAC) click chemistry reaction by adding 4 μL CuAAC click chemistry mix (1 μL of 50 mM $CuSO_4$, 1 μL of 5 mM TBTA, 1 μL of 50 mM TCEP, and 1 μL of 5 mM 5/6-TAMRA-peg_3-azide) into each well (*see* **Note 11**).
5. Incubate each sample at room temperature in the dark for 1 h.
6. Take 40 μL of each sample for SDS-PAGE and in-gel fluorescence analysis.
7. Add 10 μL reducing 5× SDS-loading buffer to each sample, mix thoroughly, and heat at 95 °C for 5 min.
8. Load 20 μL of each sample on an SDS-PAGE gel.
9. After electrophoresis, wash with Milli-Q water thoroughly before scanning gels with a Typhoon 9410 Gel and Blot Imager (GE Healthcare) with 532 nm laser excitation and 580 nm emission using standard procedures.
10. Stain gels with CBB staining solution following standard procedures.

4 Notes

1. As compounds **7a**–**7t** and **8a**–**8t** are unstable, flash chromatography was performed in the presence of 0.1% (v/v) triethylamine.
2. HPLC purification was performed under the following condition: COSMOSIL 5C_{18}-PAQ, φ 10 mm × 250 mm, aqueous trifluoroacetic acid (0.01%), 4.0 mL/min, 210 nm, t_R: 9.8 min. The trifluoroacetic acid was removed *in vacuo* and the water was removed via lyophilization to give L-Orn-AMS **8c**.
3. HPLC purification was performed under the following condition: COSMOSIL 5C_{18}-PAQ, φ 10 mm × 250 mm, acetonitrile/aqueous trifluoroacetic acid (0.1%, 10:90), 3.0 mL/min, 220 nm, t_R: 12.5 min. The trifluoroacetic acid was removed *in vacuo* and the water was removed via lyophilization to give L-Tyr-AMS **8n**.
4. HPLC purification was performed under the following condition: COSMOSIL 5C_{18}-PAQ, φ 10 mm × 250 mm, aqueous trifluoroacetic acid (0.01%), 3.0 mL/min, 220 nm, t_R: 9.5 min. The trifluoroacetic acid was removed *in vacuo* and the water was removed via lyophilization to give L-Lys-AMS **8r**.

5. Large NRPS proteins were found to be particularly sensitive to mechanical cell disruption processes, and it is, therefore, necessary to treat the bacterial cells gently to obtain the intracellular proteins [19].
6. If viscosity is observed in bacterial cell lysates because of their DNA content, add DNase I (RNase-free) (Takara), incubate on ice or 5 min, centrifuge the mixture for 10 min at 15,000 × *g* and 4 °C, and collect the supernatant.
7. A 96-well plate is used to obtain reproducible photo-labeling.
8. The final DMSO concentration is kept at 2.2% (v/v).
9. Removing the lid of the 96-well plate lid will provide more efficient irradiation and therefore higher labeling efficiency.
10. To obtain reproducible, efficient, and specific photoaffinity labeling, 1 cm distance from the top of a 96-well plate and 5 min labeling time were used.
11. $CuSO_4$, TBTA, TCEP, and 5/6-TAMRA-peg_3-azide were added to provide final concentrations of 1 mM, 100 μM, 1 mM, and 100 μM, respectively.

Acknowledgments

This work was partly supported by Grants-in-Aid for Research (C) (19 K05722 to F.I.) from JSPS and by grants from the Noda Institute for Scientific Research (F.I.), Research Foundation for Pharmaceutical Sciences (F.I.), Japan Foundation for Applied Enzymology (F.I.), and Institute Foundation for Science, Osaka (F.I.). We are also thankful for the financial support provided by the Antiaging Project for Private Universities.

References

1. Süssmuth RD, Mainz A (2017) Nonribosomal peptide synthesis-principles and prospects. Angew Chem Int Ed 56:3770–3821
2. Hur GH, Vickery CR, Burkart MD (2012) Explorations of catalytic domains in non- ribosomal peptide synthetase enzymology. Nat Prod Rep 10:1074–1098
3. Zhang B, Tian W, Wang S, Yan X, Jia X, Pierens GK, Chen W, Ma H, Deng Z, Qu X (2017) Activation of natural products biosynthetic pathways *via* a protein modification level regulation. ACS Chem Biol 12:1732–1736
4. Kwok R (2010) Five hard truths for synthetic biology. Nature 463:286–290
5. Minandri F, Imperi F, Frangipani E, Bonchi C, Visaggio D, Facchini M, Pasquali P, Bragonzi A, Visca P (2016) Role of iron systems in Pseudomonas aeruginosa virulenceand airway infection. Infect Immun 84: 2324–2335
6. Imperi F, Massai F, Facchini M, Frangipani E, Visaggio D, Leoni L, Bragonzi A, Visca P (2013) Repurposing the antimycotic drug flucytosine for suppression of *Pseudomonas aeruginosa* pathogenicity. Proc Natl Acad Sci U S A 110:7458–7463
7. Kirienko DR, Revtovich AV, Kirienko NV (2016) A high-content, phenotypic screen identifies fluorouridine as an inhibitor of pyoverdine biosynthesis and *Pseudomonas aeruginosa* virulence. mSphere 1:e00217–e00216
8. Kaneko Y, Thoendel M, Olakanmi O, Britigan BE, Singh PK (2007) The transition metal gallium disrupts *Pseudomonas aeruginosa* iron metabolism and has antimicrobial and

antibiofilm activity. J Clin Invest 117:877888–877888

9. Kirienko DR, Kang D, Kirienko NV (2019) Novel pyoverdine inhibitors mitigate Pseudomonas aeruginosa pathogenesis. Front Microbiol 9:3317
10. Konno S, Ishikawa F, Suzuki T, Dohmae N, Burkart MD, Kakeya H (2015) Active site-directed proteomic probes for adenylation domains in nonribosomal peptide synthetases. Chem Commun 51:2262–2265
11. Ishikawa F, Konno S, Suzuki T, Dohmae N, Kakeya H (2015) Profiling nonribosomal peptide synthetase activities using chemical proteomic probes for adenylation domains. ACS Chem Biol 10:1989–1997
12. Finking R, Neumüller A, Solsbacher J, Konz D, Kretzschmar G, Schweitzer M, Krumm T, Marahiel MA (2003) Aminoacyl adenylate substrate analogues for the inhibition of adenylation domains of nonribosomal peptide synthetases. Chembiochem 4:903–906
13. Kasai S, Konno S, Ishikawa F, Kakeya H (2015) Functional profiling of adenylation domains in nonribosomal peptide synthetases by competitive activity-based protein profiling. Chem Commun 51:15764–15767
14. Ishikawa F, Konno S, Uchida C, Suzuki T, Takashima K, Dohmae N, Kakeya H, Tanabe G (2022) Chemoproteomics profiling of surfactin-producing nonribosomal peptide synthetases in living bacterial cells. Cell Chem Biol 29:145–156.e8
15. Cravatt BF, Wright AT, Kozarich JW (2008) Activity-based protein profiling: from enzyme chemistry to proteomic chemistry. Annu Rev Biochem 77:383–414
16. Sanman LE, Bogyo M (2014) Activity-based profiling of proteases. Annu Rev Biochem 83: 249–273
17. Ishikawa F, Kakeya H (2014) Specific enrichment of nonribosomal peptide synthetase module by an affinity probe for adenylation domains. Bioorg Med Chem Lett 24:865–869
18. Bradford MM (1976) A rapid and sensitive method for the quantitation of microgram quantities of protein utilizing the principle of protein-dye binding. Anal Biochem 72:248–254
19. Augenstein DC, Thrasher KD, Sinskey AJ, Wang DI (1974) Optimization in the recovery of a labile intracellular enzyme. Biotechnol Bioeng 16:1433–1447

Chapter 5

In Vitro Biochemical Characterization of Excised Macrocyclizing Thioesterase Domains from Non-ribosomal Peptide Synthetases

Jordan T. Brazeau-Henrie, André R. Paquette, and Christopher N. Boddy

Abstract

Characterization of thioesterases (TEs) is an important step in understanding natural product biosynthesis. Studying non-ribosomal peptide synthetase (NRPS) TEs presents a unique set of challenges with specific cloning and expression issues as well as the challenging synthesis of the thioester peptides substrate required for characterization of the TE. In this method, we describe the cloning and expression of NRPS TEs, the synthesis of thioester peptides, and the *in vitro* biochemical characterization of the enzyme.

Key words Non-ribosomal peptides, Thioesterase, Solid-phase peptide synthesis, Thioester peptides, Biochemical characterization

1 Introduction

Thioesterases (TEs) are α/ß-hydrolases [1, 2] found at the C-terminus of many non-ribosomal peptide synthetases (NRPSs) protein complexes. They are responsible for the release of the completed non-ribosomal peptide (NRP) from the NRPS enzyme complex [3–5]. During NRP biosynthesis, the growing peptidyl intermediate is covalently linked via a thioester bond to carrier protein (CP) domains of the NRPS complex. The TE catalyzes the transfer of the completed NRP from the final CP onto its catalytic Ser, generating a peptidyl-TE intermediate. This intermediate can undergo release through macrocyclization, where an internal nucleophile of the NRP cleaves the peptidyl-TE bond, or through hydrolysis with water generating a free acid [2, 6, 7]. In some cases, an additional NRP on the final CP can act as the nucleophile, generating a dimer through transesterification. This transesterification process can be repeated to generate higher order oligomers that can be released as a macrocycle or free acid [8–11].

Michael Burkart and Fumihiro Ishikawa (eds.), *Non-Ribosomal Peptide Biosynthesis and Engineering: Methods and Protocols*, Methods in Molecular Biology, vol. 2670, https://doi.org/10.1007/978-1-0716-3214-7_5,

NRPS TEs are typically biochemically characterized as excised TE domains or CP-TE didomains, rather than as a part of their enormous native NRPS protein complexes [8, 11–16]. Excising the TE from the enormous NRPS significantly aids in the expression, purification, and assaying of the TE domain. PCR amplification of the DNA encoding the excised TE from the producing organism's genome provides the required sequence for generation of an expression vector. Alternatively, synthetic gene constructs can be designed, also enabling codon optimization. Either can be readily cloned into expression vectors compatible with a variety of heterologous hosts, most commonly the pET vector system for expression in *Escherichia coli*. N-terminal fusions with His-tag facilitate purification of the recombinant protein and do not typically disrupt biochemical function [8, 11–28]. In some examples C-terminal His-tags have also been successfully used [9, 10, 29–35]. Metal affinity chromatography is typically sufficient to obtain high purity TE domain for characterization, though in some cases, additional ion exchange or size exclusion chromatography is required [11, 12, 28, 36, 37].

A linear thioester activated peptide substrate is required to characterize NRPS TEs. This substrate mimics the CP-bound NRPS (Fig. 1). The most common thioester used in TE characterization is the N-acetylcysteamine thioester [8, 10, 11], which is homologous to the cysteamine terminus of the phosphopantenthienyl arm of the CP. In some cases, the more activated thiophenol thioester of the peptide is required [38].

Whether N-acetylcysteamaine or thiophenol, C-terminal thioester peptide synthesis is challenging. Because of their role in native chemical ligation [39], several solid-phase peptide synthesis (SPPS) methods have been developed to access thioesters. Boc SPPS [40] provides one of the most effective methods for accessing thioesters [41–43], however the demanding peptide cleavage methods have limited its use. Fmoc SPPS is the preferred method for peptide synthesis however the typical Fmoc deprotection conditions are incompatible with thioester bonds. While a handful of functionalized resins [44–47] have been developed to generate thioesters, these syntheses require additional peptide modifications or generate a non-standard thioester substrate. The most effective method for accessing N-acetylcysteamine peptide thioesters for the characterization of TEs is using hyper-sensitive SPPS resins such as 2-chlorotrityl chloride resins, which can be used to generate a free C-terminus while retaining acid-labile amino acid side chain protecting groups [48]. This allows for the off-resin thioester coupling and subsequent full deprotection to afford thioester peptides. The protocol to generate these thioester peptides is described in the methods below.

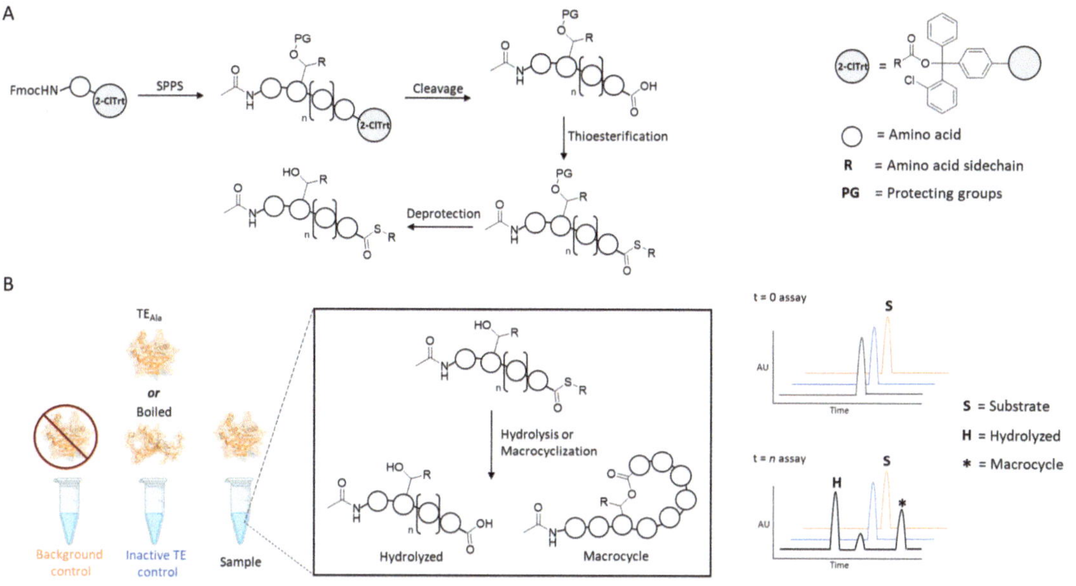

Fig. 1 Peptide synthesis and biochemical assay workflows. (**a**) General workflow for peptide synthesis, cleavage, thioesterification, and deprotection. (**b**) General workflow for TE product distribution characterization. The substrate is incubated without TE to probe background, with an inactivated TE for a negative control, and with functional TE for the test reaction. TE_{Ala} is a TE with the catalytic Ser mutated into Ala while boiled TE is heat denatured (both are inactive). Substrate is converted into free acid or macrocycle and detected by LCMS

Biochemical characterization of NRPS TEs typically involves evaluation of the product distribution. Assays containing recombinant purified TE and a synthetic thioester substrate are characterized by High-Performance Liquid Chromatography–Mass Spectrometry (LCMS), which can unambiguously differentiate the potential products from the starting thioester, providing key information on product distribution [8, 49–51]. HPLC analysis of TE assay is also suitable for analyzing product distribution although the use of authentic standards is required for product identification [49]. Relative product ratios can be calculated based on peak area. If authentic standards are available, absolute quantification of the products can be readily obtained [52]. TE-dependent formation of the macrocycle from the linear thioester confirms the biochemical function of the TE.

In addition to product distribution, kinetic characterization of any newly characterized TE is required. The most common strategy is to determine the Michaelis–Menten kinetic parameters (K_M and k_{cat}) [53, 54] for the formation of the acyl-enzyme intermediate. These parameters are obtained through determination of the initial velocities (v_0) for hydrolysis of the thioester at a fixed TE concentration with varying substrate concentrations [55]. Detection of the released free thiol from thioester hydrolysis is usually performed spectrophotometrically by supplementing assay mixes with DTNB

(Ellman's reagent), allowing for measurements of product formation in real-time [10]. Alternatively, consumption of thioester or production of products can be determined by LCMS or HPLC analysis [12, 50, 51]. Direct fitting of the initial velocities versus substrate concentrations to the Michaelis–Menten equation provides the kinetic parameters.

2 Materials

2.1 Thioesterase (TE) Expression

All glassware should be cleaned and autoclaved. Plasticware, such as pipette tips and Eppendorf tubes, are also autoclaved. For cloning of TE gene(s), all stock solutions and reagents were prepared exactly as previously described [71]. All solutions were prepared using deionized H_2O.

1. LB media: 5 g yeast extract, 10 g bactotryptone, 5 g NaCl dissolved in 1 L of H_2O. Sterilize by autoclave in a glass bottle.
2. Ampicillin stock: 1000× stock solution of ampicillin sodium (100 mg/mL). Weigh 1 g of ampicillin sodium salt in a 15 mL falcon tube. Dissolve with 10 mL of H_2O, mix using a vortex mixer. Sterilize the solution through a 0.22-micron filter, aliquot into Eppendorf tubes, and store at −20 °C.
3. Kanamycin stock: 1000× stock solution of kanamycin sulfate (50 mg/mL). Weigh 500 mg of kanamycin sulfate salt in a 15 mL falcon tube. Dissolve with 10 mL of H_2O and mix using a vortex mixer. Sterilize the solution through a 0.22-micron filter, aliquot into Eppendorf tubes, and store at −20 °C.
4. Bacterial strains: Any *E. coli* B strain that carries the gene for T7 RNA polymerase (DE3) can be used. The BL21(DE3) strain is used since it allows for over-expression of recombinant proteins and minimizes protein degradation through deletions of the Lon and OmpT proteases. BAP1 [56] can also be used in place of BL21(DE3).
5. Plasmids: pET-28a(+) and pET-21c(+) both contain the *lacI* gene and confer kanamycin and ampicillin resistance, respectively. Both plasmids contain a MCS flanked by the T7 cassette and encode a C-terminal hexa His tag. pET-28a(+) encodes an N-terminal hexa His tag in addition to the C-terminal hexa His tag.
6. IPTG: 1 M isopropyl-β-D-thiogalactopyranoside (IPTG). In a 15 mL falcon tube, dissolve 2.38 g of IPTG in 10 mL of H_2O. Filter-sterilize aliquots into sterile Eppendorf tubes and store at −20 °C.

7. Stock NaCl: 5 M NaCl. In a 500 mL glass bottle, weigh 87.60 g of NaCl. Dissolve in 300 mL of H_2O and mix using a magnetic stirrer. Store at room temperature.
8. Stock Tris: 1 M of Tris. In a 500 mL glass bottle, weigh 36.33 g of Tris base. Dissolve in 300 mL of H_2O and mix using a magnetic stirrer. Store at room temperature.
9. Stock glycerol: 60% *w/v* of glycerol. In a 500 mL glass bottle, weigh 180 g of glycerol solution. Add 140 mL of H_2O and mix vigorously using a magnetic stirrer. Store at 4 °C.
10. Stock imidazole: 1 M imidazole. In a 500 mL glass bottle, weigh 20.42 g of imidazole. Dissolve in 300 mL of H_2O and mix using a magnetic stirrer. Wrap the bottle in tin foil and store at 4 °C.
11. Pepstatin A: 1000× pepstatin A (1 mg/mL with 10% *v/v* acetic acid). In a 15 mL falcon tube, weigh 10 mg of pepstatin A. Dissolve in 9 mL of MeOH and mix using a vortex mixer. Add 1 mL of glacial acetic acid and mix once more using a vortex mixer. Aliquot into Eppendorf tubes and store at −20 °C.
12. Leupeptin: 1000× leupeptin (1 mg/mL). In a 15 mL falcon tube, weigh 10 mg of leupeptin. Dissolve in 10 mL of H_2O and mix using a vortex mixer. Aliquot into Eppendorf tubes and store at −20 °C.
13. Lysozyme: 100 mg/mL lysozyme. In a 15 mL falcon tube, weigh 1 g of lysozyme. Dissolve in 10 mL of H_2O and mix using a vortex mixer. Aliquot into Eppendorf tubes and store at −20 °C.
14. Lysis buffer: 300 mM NaCl, 100 mM Tris, 10% *w/v* glycerol, 1 mg/mL lysozyme, 1× pepstatin A, 1× leupeptin, pH 7.4. Fill a 15 mL falcon tube with H_2O up to the 5 mL marker. Add 600 μL of stock NaCl, 500 μL of stock Tris, and 1.67 mL of stock glycerol. Mix by inversion and adjust pH with 3 M HCl. Add 100 μL of 100 mg/mL lysozyme, 10 μL of 1000× pepstatin A, and 10 μL of 1000× leupeptin. Make up to 10 mL with H_2O, mix using a vortex mixer, and store at 4 °C.
15. Imidazole solutions: 300 mM NaCl, 100 mM Tris, varying concentrations of imidazole (0, 20, 100, and 250 mM), pH 7.4. In four separate 50 mL falcon tubes, fill tubes with H_2O to the 15 mL marker. Add to each tube 3 mL of stock NaCl and 5 mL of stock Tris. For the desired concentrations of imidazole, add 0 mL, 1 mL, 5 mL, or 12.5 mL of stock imidazole (for 0, 20, 100, and 250 mM imidazole respectively) to one of the four tubes. Mix tubes using a vortex mixer. Adjust the pH using 3 M HCl. Add H_2O up to the 50 mL marker, wrap tubes in tin foil, and store at 4 °C.

16. Dialysis buffer: 300 mM NaCl, 100 mM Tris, pH 7.4. In a 100 mL graduated cylinder, add 60 mL of H_2O. Add 6 mL of stock NaCl and 10 mL of stock Tris and mix using a magnetic stirrer. Adjust pH using 3 M HCl then make up volume to 100 mL using H_2O. Store at 4 °C in a sealed glass bottle.
17. Centrifuge buckets, Ni-NTA resin, low-pressure glass protein purification columns, and centrifugal concentrator tube (10 kDa MWCO).

2.2 Peptide Synthesis

Solvents used are ACS grade, while N,N-dimethylformamide (DMF) is HPLC grade. All pipettes used are long glass Pasteur pipettes with a pressure bulb. Syringes are 1–12 mL non-pyrogenic sterile equipped with metal 18-gauge needles. Fritted peptide vessels are 12 mL syringes with an inserted polyethylene frit and a Luer Plug (*see* **Note 1**). Appropriate 50 mL falcon tubes are used for reagents on the peptide synthesizer.

1. Capping mixture: 17:2:1 parts DCM:MeOH:DIEA. Measure 17 mL of dichloromethane (DCM) in a 25 mL graduated cylinder and add 2 mL of methanol (MeOH) and 1 mL of N, N-diisopropylethylamine (DIEA) using a pipette. Gently stir with a long glass rod.
2. DBU solution: 2% *v/v* DBU. Add 9.8 mL N,N-Dimethylformamide (DMF) in a 7 dram glass vial using a 12 mL syringe followed by 200 μL 1,8-Diazabicyclo[5.4.0]undec-7-ene (DBU) using a 1 mL syringe. The solution was capped and gently agitated.
3. Amino acid solutions: 0.2 M amino acid. Weigh out the desired amino acid (2 mmoL) and transfer to a 50 mL falcon tube. Add 10 mL of DMF and mix the solution (*see* **Note 2**).
4. Activator solution: 0.5 M HTBU. Weigh out 2-(1H-Benzotriazole-1-yl)-1,1,3,3-tetramethyluronium hexafluorophosphate (HBTU, 25 mmoL) and transfer to a 250 mL dark glass screw-on flask provided with the instrument. Add 50 mL of DMF and mix the solution (*see* **Note 3**).
5. Activator base solution: 2 M DIEA. Aspirate 17.42 mL DIEA with a 5 mL syringe and transfer to a 250 mL glass screw-on flask provided with the instrument. Add 50 mL of N-methyl-2-pyrrolidone (NMP) and gently stir the solution.
6. Fmoc deprotection solution: 20% *v/v* piperidine. Add 10 mL of piperidine to a 50 mL glass graduated cylinder. Fill to 50 mL with DMF and transfer to the glass screw-on flask provided with the instrument. Gently stir the flask.
7. Acetic anhydride solution: 10% *v/v* acetic anhydride. Add 2 mL of acetic anhydride to a 50 mL falcon tube. Fill to 20 mL with DMF and mix by inversion.

8. Resin cleavage solution: 20% *v/v* HFIP. Add 2 mL 1,1,1,3,3,3-Hexafluoro-2-propanol (HFIP) to a 10 mL plastic graduated cylinder. Fill to 10 mL with DCM and use a pipette to gently mix the solution.
9. Peptide deprotection cocktail: 95:2.5:2.5% TFA:H_2O:TIPS. Add 950 μL of Trifluoroacetic acid (TFA) into a glass vial with a 1 mL syringe followed by the addition of 25 μL of deionized H_2O and 25 μL of triisopropylsilane (TIPS). Gently agitate the mixture and use immediately after mixing.
10. LCMS solvents: LCMS Optima grade acetonitrile (MeCN) and H_2O, both supplemented with 0.05% *v/v* formic acid (FA). Add 200 mL of LCMS Optima grade MeCN or H_2O in 250 mL plastic graduated cylinders. Add 125 μL of LCMS grade FA to the solution with a 1 mL micropipette and fill to 250 mL with either Optima grade MeCN or H_2O. Gently agitate the graduated cylinder and pour the solution into the appropriate bottle.
11. Semi-preparative HPLC solvents: Optima grade MeCN or water, both supplemented with 0.1% *v/v* TFA. Add 200 mL of Optima grade MeCN or H_2O in 250 mL plastic graduated cylinders. Add 250 μL of TFA into the solution with a 1 mL micropipette and fill to 250 mL with the appropriate Optima grade solvent. Gently agitate the graduated cylinder and pour the solution into the appropriate bottle.

2.3 Biochemical Assays

All glassware should be cleaned and dried prior to use. All solutions were prepared using deionized water.

1. Assay buffer: 200 mM potassium phosphate buffer, pH 7.4. Weigh 5.7 g of potassium phosphate monobasic (KH_2PO_4) and 3.2 g of potassium phosphate dibasic (K_2HPO_4), then transfer solids to a 500 mL graduated cylinder containing a magnetic stir bar. Dissolve solids in 200 mL of H_2O using a magnetic stir plate. Adjust pH using 3 M HCl or 3 M of NaOH. Complete volume to 300 mL with H_2O then transfer into a 500 mL glass bottle. Store at room temperature.
2. Working TE stock: 50 μM functional TE. Dilute concentrated TE stock to a final volume of 250 μL using 50 mM of assay buffer. Store at 4 °C until needed (*see* **Note 4**).
3. Inactive TE stock: 50 μM inactivated TE. Dilute concentrated TE stock to a final volume of 20 μL using 50 mM of assay buffer. Heat-inactivate TE at 95 °C for 10 min. Store at −80 °C.
4. Substrate stock: *see* **step 9** of Subheading 3.2.5.
5. Quenching buffer: LCMS Optima grade MeCN with 0.05% *v/v* FA. Add 2 mL of LCMS Optima grade MeCN to a 2 mL

amber glass vial. Add 1 μL of LCMS grade FA directly into solution, cap vial, then mix by inversion. Store at room temperature.

6. LCMS solvents: same as Subheading 2.2.
7. DTNB stock: saturated DTNB solution. In an Eppendorf tube, add 1 mL of H_2O. Add 0.5 g portions of DTNB until the solution is saturated. Mix using a vortex mixer after each addition. Store at 4 °C.
8. DTNB working solution: 25% *v/v* DTNB. Mix the DTNB stock using a vortex mixer and let it rest on the benchtop for 1 min. In a separate Eppendorf tube, add 750 μL of H_2O then take 250 μL of the DTNB stock's top layer and add directly into the tube. Mix using a vortex mixer and store at 4 °C until needed (*see* **Note 5**).
9. N-Acetylcysteamine (SNAC) stock: 500 μM SNAC (*see* Subheading 3.2.5). Weigh 298 mg on an analytical balance and transfer into a 15 mL falcon tube. Dissolve in 5 mL of H_2O: DMSO mixture (75:25% *v/v*), mix using a vortex mixer, and store at 4 °C until needed (*see* **Note 6**).
10. HPLC autosampler vials (0.75 mL) with cap-septum, 2 mL amber glass vials, half-area clear flat bottom 96-well plate (190 μL well capacity) and multichannel pipettor.

3 Methods

3.1 TE Cloning and Expression

3.1.1 TE Cloning

The coding sequence for the excised TE of interest can be inserted into the multiple cloning region of pET-28a(+) or pET-21c(+) in the chosen reading frame by conventional methods [71]. To identify boundaries for the excision of the TE from the final NRPS module, the native 3′ end of the gene is used for the C-terminus. The N-terminal boundary is typically chosen to be close to the end of the last α-helix of the upstream CP domain. This ensures the full N-terminus of the TE is captured while minimizing the number of unstructured residues at the N-terminus. These unstructured N-terminal residues typically do not affect TE folding or activity [8, 11, 27]. To find this boundary, locate the CP and the TE domain using hidden Markov models (HMM) with tools such as MIBiG [57], antiSMASH [58], Pfam [59], or HMMER [60]. Use this as a guide to position your forward primer to encompass the last few residues (1–2 turns) of the last α-helix of the CP domain. The reverse primer should be positioned at the 3′ end of the NRPS gene encoding the TE (*see* **Note 8**). Alternatively, a synthetic gene construct can be made and codon-optimized for *E. coli*. For most TEs, N-terminal His tags produce functional proteins, and it is recommended to generate the N-terminal construct first (*see*

Note 9). Transform 1–50 ng of the generated plasmid in *E. coli* BL21 (DE3) following standard protocols [71]. Plate various amounts of the transformed bacteria on LB agar plates containing the appropriate antibiotic for the plasmid used.

3.1.2 Recombinant TE Expression

Fresh transformants are inoculated and tested for their ability to express soluble His-tagged TE. Often a test expression is performed using two single colonies of the desired TE encoding clones to confirm that soluble TE is produced under the desired expression conditions. As a positive control for test expression of a soluble TE, a single colony transformed with the plasmid pRSG33 can be used. This construct will express the soluble 6-deoxyerythronolide B (DEBS) TE, which has been extensively characterized in the literature [36, 61–63]. To generate a seed culture for TE expression, each clone is grown in 5 mL of LB containing the appropriate antibiotic at 37 °C with agitation (200 rpm) to stationary phase. A 2 mL aliquot of the overnight culture is added to a flask containing 400 mL of LB with the appropriate antibiotic.

For test expression, the protocol below can be scaled down to 25 mL scale to optimize expression parameters (i.e., media, temperature, incubation time, IPTG concentration). Soluble and insoluble protein fractions are obtained using commercially available bacterial lysis reagents. Fractions are analyzed via Western Blotting [72] using a monoclonal anti-His antibody conjugated to HRP. If expression conditions were optimized during test expression and are different from those described in this section, use the optimized conditions instead.

1. Periodically measure the OD_{600} of the bacterial cultures.
2. When an OD_{600} of 0.6–0.8 is reached (*see* **Note 10**), let culture sit on ice for 5 min.
3. Induce protein expression by adding 400 μL of IPTG (1 M) directly into the culture (final concentration 1 mM). For test expression, induce one of the two cultures with IPTG. The other culture will be used as a no-induction control.
4. Allow protein to express for 16 h in a shaker (200 rpm) set at 16 °C.
5. Transfer bacterial culture into pre-chilled centrifuge buckets. Harvest cells by centrifugation at 4 °C and 3500 g for 20 min.
6. Carefully decant supernatant (*see* **Note 11**), then leave tube upside down on paper towel to dry pellet for 5 min.

3.1.3 Recombinant TE Purification

This protocol is used for purification of recombinant TE from 400 mL scale expression experiments. Depending on the yield of recombinant TE, larger scale expressions may be required. Scale reagents as appropriate for larger scale expression experiments. All solutions and tubes used are pre-chilled and kept on ice.

1. Resuspend bacterial pellet in 10 mL of lysis buffer (*see* **Note 12**).
2. Lyse cells via sonication (*see* **Note 13**). Make sure to always perform sonication on ice and avoid the generation of foam.
3. Centrifuge resulting lysate at 4 °C and 12,000 g for 1 h.
4. Carefully transfer the supernatant to a new falcon tube and add fresh Ni-NTA resin (1 mL captures 5 mg of protein).
5. Incubate the protein extract tube at 4 °C with constant agitation for 1.5 h (*see* **Note 14**).

   All subsequent steps should be performed on ice or in a cold room (<10 °C) to maximize protein yields. All elution steps are performed by gravity but can be performed by carefully increasing the pressure. Ensure that the pressure is kept low to prevent shearing of Ni-NTA resin. During elution, leave a small aliquot of solution at the bottom of the protein purification column to prevent drying of Ni-NTA resin (*see* **Note 15**).
6. Transfer contents of the tube into a protein purification column. Fill only about 75% of the column's capacity and place an empty falcon tube underneath the column to collect flow through.
7. Cap the column and let the resin settle to the bottom by gravity (*see* **Note 16**). This process takes approximately 5–10 min.
8. Elute and collect eluent into a new falcon tube. If some protein extract from **step 4** is left in the tube, repeat **steps 6–8**.
9. Add 10 column bed volumes (*see* **Note 17**) of 0 mM imidazole solution. Let the resin settle to the bottom by gravity. Elute and collect eluent into a new falcon tube.
10. Add 10 column bed volumes of 20 mM imidazole solution. Let the resin settle to the bottom by gravity. Elute and collect eluent into a new falcon tube.
11. Add 5 column bed volumes of 100 mM imidazole solution. Let the resin settle to the bottom by gravity. Elute and collect eluent into a new falcon tube.
12. Repeat **step 11** one more time.
13. Add 5 column bed volumes of 250 mM imidazole solution. Let the resin settle to the bottom by gravity. Elute and collect eluent into a new falcon tube.
14. Repeat **step 13** one more time.
15. Prepare and run 15 μL aliquots of eluents via SDS-PAGE [64].
16. Pool eluents containing the desired protein (*see* **Note 18**). Properly discard the remaining eluents.

3.1.4 Recombinant TE Concentration and Ultrafiltration

Purified TE will be in buffer with high imidazole concentration. This can affect protein stability and activity. It is highly recommended to remove imidazole via ultrafiltration. Ultrafiltration will not only enable the buffer to be exchanged but will also concentrate the purified TE. Alternatively, the TE can be further polished via FPLC (either ion exchange or size exclusion), which will further purify the protein and remove imidazole. Consider this option if proteolytic fragments of TE or other impurities are still present after the metal affinity chromatography step. All steps are performed on ice or in a cold room (<10 °C) to maximize protein yields. All solutions and tubes used are pre-chilled and kept on ice.

1. Transfer pooled eluents into a centrifugal concentrator tube. Centrifuge at 4 °C and 3500 g for 45 min or until the solvent line reaches the 0.5 mL marker on the filter (*see* **Note 19**).
2. Carefully remove filter from tube and decant flowthrough into a 50 mL falcon tube (*see* **Note 20**). Keep the falcon tube on ice throughout the entire process.
3. Add 30 volumes of dialysis buffer (or up to the maximal volume line of the filter) and mix gently by inversion. Centrifuge again as previously described.
4. Carefully remove filter from tube and decant flowthrough into the same 50 mL falcon tube.
5. Repeat **steps 3** and **4** two more times.
6. Add 30 volumes of dialysis buffer (or up to the maximal volume line of the filter) and mix gently by inversion. Centrifuge at 4 °C and 3500 g for 1 h or until the solvent line reaches the 0.25 mL marker on the filter.
7. Transfer solution from filter (*see* **Note 21**) into an Eppendorf tube and measure volume by reverse pipetting.
8. Add 1 volume of 60% *w/v* of glycerol to the tube. Mix gently with a pipette then aliquot into Eppendorf tubes (final composition: 30% *w/v* glycerol, 150 mM NaCl, 50 mM Tris, pH 7.4).
9. Quantify protein concentration by Bradford assay or any other suitable protein quantitation method.
10. Flash freeze tubes in liquid nitrogen or an acetone-dry ice bath. Store at −80 °C until needed.

3.2 Peptide Synthesis

Solid-phase peptide synthesis (SPPS) with the Fmoc protecting group strategy is the standard for syntheses of peptides and can be performed with either hand-shaking or with an automated microwave assisted peptide synthesizer. The following method is adaptable to the synthesis of most peptide sequences, where the N-terminus is either Fmoc protected or acylated, the side chains are protected with acid labile groups, and the C-terminus is a free carboxylic acid (*see* **Note 22**). For substrates where the N-terminal alpha amino group is the nucleophile, Boc protection is required at the N-terminus. For substrates with side chain nucleophiles, the N-terminus is typically acylated. The following method outlines the synthesis of a peptide with an internal sidechain hydroxyl group nucleophile [65, 66]. The loading and substitution level estimation of these acid sensitive resins, followed by peptide synthesis, cleavage, and deprotection, are described in detail below.

3.2.1 Resin Loading

Hyper-sensitive 2-chlorotrityl chloride resins can be purchased pre-loaded with your desired C-terminal residue; however, its preparation is straightforward. This step is performed off the automated peptide synthesizer in fritted peptide vessels.

1. Swell 2-chlorotrityl chloride resin (100–200 mesh, 0.75 mmoL/g) in dry DCM (10 mL/g of resin) in a fritted peptide vessel for 30 min then remove the solvent by flushing.
2. Dissolve C-terminus amino acid (2 eq.) and DIEA (4 eq.) in fresh DCM (10 mL/g of resin) and add to swelled resin. Agitate at room temperature for 2 h.
3. Flush the solution and wash the resin with DCM twice. Add the capping mixture and agitate for 45 min.
4. Flush the solution and wash twice with DCM and DMF. Dry the resin under vacuum and store at −20 °C.

3.2.2 Substitution Level Estimation

1. Weigh out dry loaded resin (5 mg) and add 2 mL of DBU solution and agitate for 30 min. Dilute to 10 mL with MeCN, aspirate 2 mL of the solution, and dilute to 25 mL with MeCN [67].
2. A reference is prepared the same way without adding resin.
3. Quartz cuvettes are filled with either 3 mL of resin solution or reference solution, letting the resin settle to the bottom.
4. Record the absorbance in a spectrophotometer at 304 nM. The estimated loading of the first Fmoc residue is determined from Eq. (1).

$$\text{Fmoc loading}\left(\frac{\text{mmol}}{\text{g}}\right) = \left(A_{\text{Sample}} - A_{\text{Ref}}\right) \text{ x } \frac{16.4}{\text{mg of resin}} \quad (1)$$

5. Repeat the procedure at least twice to determine an average loading in the same range.
6. Resin loading should be from 0.3 to 0.8 mmoL/g for these hypersensitive resins.

3.2.3 Peptide Synthesis

The following protocol describes the method for automated microwave assisted peptide synthesis using a CEM Liberty Blue. Peptide synthesis can be performed on any instrument, and the appropriate conditions for hypersensitive resins should be adjusted as per the manufacturer's guidelines (*see* **Note 23**).

1. Amino acids (0.2 M), HBTU (0.5 M), DIEA (2 M), piperidine (20% *v/v*), and acetic anhydride (10% *v/v*) solutions were prepared and placed onto the respective CEM Liberty Blue peptide synthesizer ports (*see* **Note 23**).
2. Weigh out loaded resin for a 0.1 mmol scale on a weigh paper and add to the reaction vessel.
3. Perform coupling for 10 min at 50 °C, followed by Fmoc deprotections for each amino acid. The N-terminus is Fmoc deprotected and acetylated with the acetic anhydride capping protocol (*see* **Note 24**).
4. Transfer the resin linked peptide to a fritted peptide vessel with DCM and flush dry. Dry the resin by vacuum and store at −20 °C.

3.2.4 Peptide Cleavage from Resin

1. Place approximately 100 mg of dried resin with bound peptide in a fritted peptide reaction flask. Add 5 mL of resin cleavage solution with a pipette. Cap the reaction vessel (*see* **Note 25**) and agitate for 45 min.
2. Flush the peptide vessel directly into a clean-dried glass round bottom flask. Rinse the resin twice with DCM, collecting and polling the washes into the round bottom flask.
3. Add 5 mL of resin cleavage solution to the resin again and agitate for another 45 min. Flush the peptide vessel into the round bottom flask and rinse the resin twice with DCM.
4. Place the round bottom flask on a rotary evaporator and remove the solvent under reduced pressure to afford a clear oil.
5. Purify the residue by flash column chromatography (*see* **Note 26**), with an appropriate MeOH:DCM mixture as the mobile phase.
6. Combine the fractions containing the peptide and remove the solvent under reduced pressure to afford pure free C-terminus protected peptide.
7. Characterize the peptide by LCMS (*see* **Note 27**) and NMR (*see* **Note 28**).

3.2.5 Thiol Synthesis, Thioester Synthesis, and Peptide Deprotection

Most standard carboxylic acid-activating reagents can be used for thioester formation; however, HOBt/EDC is recommended to ensure solubility and reduced epimerization of the peptide [68]. The synthesis of N-acetylcysteamine will also be described as this is a standard thiol for thioester reagent preparation.

1. N-acetylcysteamine synthesis [69]: Dissolved cysteamine hydrochloride (1.14 g, 10.0 mmoL, 1.0 eq.), KOH (0.56 g, 10.0 mmoL, 1.0 eq.), and $NaHCO_3$ (2.52 g, 30.0 mmoL, 3.0 eq.) in a round bottom flask with water (50 mL, 0.2 M) (*see* **Note 29**). Add acetic anhydride (0.65 mL, 10.0 mmoL, 1.0 eq.) dropwise to the solution and stir at room temperature for 10 min. Adjust the pH to 7.3 using concentrated HCl. Extract the solution with EtOAc (20 mL × 3) and dry the organic fractions over $MgSO_4$. Filter and remove the organic solvent under reduced pressure to afford clear oil.
2. Dissolve the thiol in deuterated chloroform and characterize by NMR (*see* **Note 30**).
3. Thioester synthesis [70]: Dissolve the free C-terminal cleaved peptide (25 mg, 0.021 mmoL, 1 eq.) and HOBt (3 mg, 0.022 mmoL, 1.05 eq.) in DCM (0.1 M) and stir in a 0 °C ice-bath for 10 min. Add EDC (3.4 mg, 0.022 mmoL, 1.05 eq.) to the solution and stir for a further 30 min at 0 ° C. Add N-acetylcysteamine (2.6 mg, 0.024 mmoL, 1.15 eq) and stir the solution at room temperature for 16 h. Wash the solution with $NaHCO_3$ saturated solution and extract with DCM (5 mL × 3). Wash the organic fractions with water and brine then dry over $MgSO_4$. Remove the organic solvents under reduced pressure in a rotary evaporator to afford clear oil. Purify the residue by flash column chromatography, with an appropriate MeOH:DCM mixture as the mobile phase.
4. Characterize the peptide by LCMS (*see* **Note 27**) and NMR (*see* **Note 28**).
5. Treat the protected thioester peptide with an appropriate TFA cocktail (*see* **Note 31**) for amino acid side chain protecting group removal.
6. Stir the solution at room temperature for 2 h in a flask capped with a septum with an exhaust needle.
7. Add cold ether to the crude residue to precipitate the peptide or if the peptide is highly hydrophobic, directly remove the volatiles by rotary evaporator.
8. Perform peptide purification on a reversed phase HPLC, using 20-min methods, with MeCN and H_2O supplemented with 0.1% *v/v* TFA (*see* **Note 32**).
9. Store the final purified compound as a 50 mM stock in DMSO at −20 °C.

3.3 Biochemical Assays

3.3.1 Assay Preparation and Incubation

Assay preparation should be performed in Eppendorf tubes and on ice. Thaw a fresh tube of protein concentrate on ice and thaw a substrate tube on the benchtop. It is also recommended to prepare two sets of controls: one to probe background reactions with peptide substrate and a negative control. For the background reaction, an assay without the presence of TE is prepared. For the negative control, the TE is replaced with an inactivated TE denatured at 95 °C. Alternatively, you can mutate the catalytic Ser to Ala via a site-directed mutagenesis method. Assay tubes contain the following components at the indicated final concentrations: 50 mM phosphate buffer, pH 7.4, 1 mM substrate, 5% *v/v* DMSO, and 5 μM TE or inactivated TE. Incubation time, substrate and TE concentrations may need to be optimized for detectable signals. Scale up reaction volumes appropriately and take periodic aliquots to determine proper incubation time. Concentrations ranging from 0.1 to 5 mM for substrate and 5–25 μM for TE can be used to optimize these conditions.

1. Prepare in triplicate a 20 μL reaction assay by adding 12 μL of H_2O and 5 μL of assay buffer (200 μM, pH 7.4). Mix using a pipette.
2. Add 2 μL of working TE stock (50 μM) to assay tubes (*see* **Note 33**). Mix gently using a pipette.
3. Prepare in triplicate the inactive enzyme control assay by adding 2 μL of inactive TE stock (50 μM) to 12 μL of H_2O and 5 μL of assay buffer (200 μM, pH 7.4). Mix gently using a pipette.
4. Prepare in triplicate the substrate control assay by adding 14 μL of H_2O to 5 μL of assay buffer (200 μM, pH 7.4) and mix gently using a pipette.
5. Add 1 μL of substrate stock (50 mM) to every tube. Mix gently with a pipette then incubate at room temperature for 16 h (*see* **Note 34**).

3.3.2 LCMS Analysis of TE Product Distribution

1. Once incubation is complete, quench reactions with 3 volumes of quenching buffer. Mix the tube by inversion and collect all droplets to the bottom of the tube using a microcentrifuge.
2. Incubate tube on benchtop for 15 min at room temperature.
3. Centrifuge at room temperature and max speed in a microcentrifuge for 10 min.
4. Transfer supernatant into an HPLC autosampler vial and insert into the autosampler rack adapter.
5. Analyze by reversed phase LCMS (or HPLC) using 15-min methods with MeCN and H_2O supplemented with 0.05% *v/v* FA in ESI positive mode (*see* **Note 35**).

3.3.3 Kinetic Characterization of TE

Kinetic assays are prepared at room temperature and arrayed in triplicate into a 96-well plate. A range of substrate final concentrations from 0.1 to 1.5 mM is usually sufficient for determination of k_{cat} and K_M from NRPS TEs. In some cases where the TE has high affinity for the substrate, more dilute substrate final concentrations may be needed to determine the K_M. Some thioester substrates may not be soluble in assay mixtures at the higher concentrations. In such cases, determine the maximum solubility of your substrate and use final substrate concentrations in the assay starting at the maximum soluble concentration and decrease by twofold for 6–8 serial dilutions. Depending on the magnitude of the K_M and substrate solubility, it may only be possible to calculate the specificity constant (k_{cat}/K_M). Thaw a fresh tube of protein concentrate on ice and thaw a substrate tube on the benchtop. Each well will contain 50 μL as a final reaction volume (50 mM phosphate buffer, pH 7.4, 0–1.5 mM substrate, 5% *v/v* DMSO, 4% *v/v* DTNB and 5 μM TE). The protocol described below is a continuous assay and can be adapted to be discontinuous for TEs that are inactivated by DTNB. For discontinuous assays, reaction mixtures are scaled up to 200 μL in triplicate and 25 μL aliquots are taken every two minutes for 10 min and quenched with 25 μL of 8% *v/v* DTNB mix for analysis via a microplate reader (single read at 412 nM).

1. In 8 Eppendorf tubes, add 25.6 μL of DTNB working solution (25% *v/v*) then prepare 80 μL of substrate mixes as shown in Table 1 in the same Eppendorf tubes.

2. Array 25 μL of substrate mixes in a 96-well plate as shown in Fig. 2.
3. Take absorbance measurements at 412 nM every 15 s for 2.5 min at room temperature.
4. In an Eppendorf tube, prepare a 750 μL TE master mix by adding 375 μL of assay buffer (200 mM, pH 7.4), 225 μL of H_2O and 150 μL of working TE stock (50 μM). Mix gently using a pipette.

Table 1
Substrate mixes for Michaelis–Menten kinetics of TE

C_f (mM)	H_2O (μL)	DMSO (μL)	Substrate stock [50 mM] (μL)	Substrate dilution [10 mM] (μL)
0	46.4	8.0	0	0
0.1	45.6	7.8	0	0.8
0.2	45.2	7.6	0	1.6
0.5	43.4	7.0	0	4.0
1.0	46.4	6.4	1.6	0
1.5	46.4	5.6	2.4	0
2.0	46.4	4.8	3.2	0
3.0	46.4	3.2	4.8	0

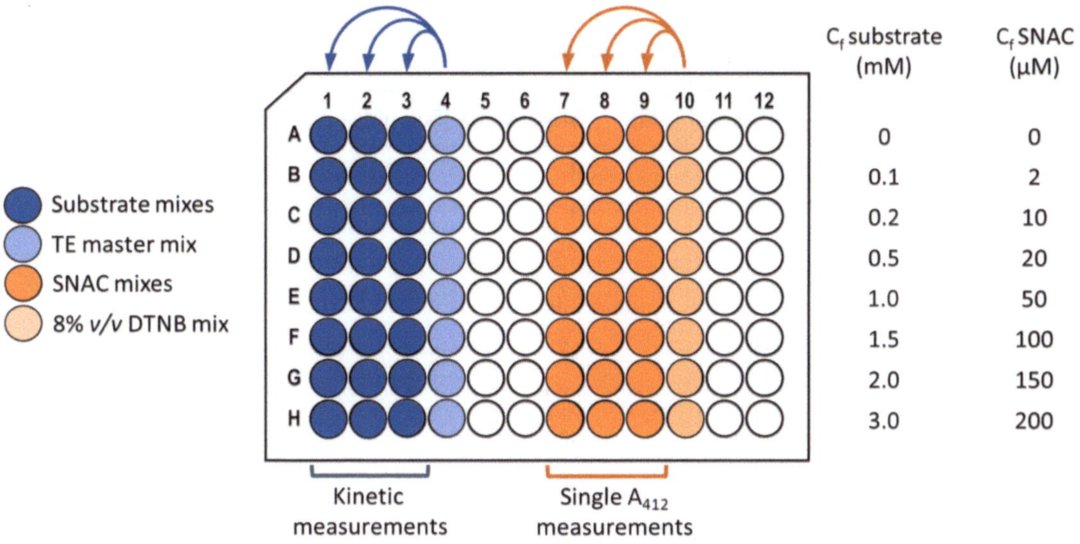

Fig. 2 TE Kinetics and SNAC standard curve 96-well plate layout. Typical 96-well plate layout for TE kinetics and SNAC standard curve measurements. Master mixes containing DTNB, substrate, and DMSO are arrayed in triplicate (shown in dark blue) for background measurements at 412 nM. TE master mix containing assay buffer and TE (shown in light blue) is arrayed beside substrate mixes and added into each substrate mix well for kinetic measurements at 412 nM. SNAC mixes containing SNAC and DMSO are also arrayed in triplicate (shown in dark orange). An 8% v/v DTNB mix containing DTNB and assay buffer (shown in light orange) is arrayed beside SNAC mixes and added into each SNAC mix well. A short incubation of 5 min followed by single measurements at 412 nM affords the SNAC standard curve needed for Michaelis–Menten kinetic modeling of data

5. Array 90 μL of TE master mix in the same 96-well plate as shown in Fig. 2.
6. With the use of a multichannel pipettor, start assay reactions by pipetting 25 μL of the arrayed TE master mix into the wells containing substrate master mixes (final substrate concentrations: 0, 0.05, 0.10, 0.25, 0.50, 0.75, 1.00, and 1.50 mM).
7. Quickly shake the plate to mix and take measurements at 412 nM every 15 s for 15 min at room temperature with periodic mixing (2 s) after every read.
8. After kinetic reads are done, prepare 175 μL of SNAC mixes in Eppendorf tubes as shown in Table 2.
9. In an Eppendorf tube, prepare 750 μL of an 8% *v/v* DTNB mix by adding 240 μL of DTNB working solution (25% *v/v*) with 375 μL of assay buffer (200 mM, pH 7.4) and 135 μL of H_2O. Mix using a vortex mixer.
10. Array 25 μL of SNAC mixes and 90 μL of 8% *v/v* DTNB mix into a new 96-well plate as shown in Fig. 2.

Table 2
SNAC mixes for standard curve

C_f SNAC (μM)	SNAC stock [500 μM] (μL)	DMSO (μL)	H_2O (μL)
0	0	17.5	157.5
2	0.7	17.3	157.0
10	3.5	16.6	154.9
20	7.0	15.8	152.2
50	17.5	13.1	144.4
100	35.0	8.8	131.2
150	52.5	4.4	118.1
200	70.0	0	105.0

11. With a multichannel pipettor, add 25 μL of the arrayed DTNB mix to each well containing SNAC mixes (final SNAC concentration: 0, 1, 5, 10, 25, 50, 75, and 100 μM). Mix by shaking the plate.
12. Incubate on the benchtop for 5 min.
13. Using a microplate reader, take a single absorbance reading at 412 nM and room temperature.
14. Generate SNAC standard curve by plotting average absorbance readings as a function of the SNAC concentration. Take note of the linear regression equation. If R^2 value is below 0.95, redo **steps 8–13**.
15. Convert each absorbance value collected in **step 3** using the SNAC standard curve's linear regression equation from **step 14**.
16. Measure the slope (mM/s) of the data set generated in **step 15**. This will correspond to the raw initial velocities (rv_0) for the background.
17. Convert each absorbance value collected in **step 7** using the same linear regression equation from **step 14**.
18. Measure the slope (mM/s) of the data set generated in **step 17**. This will correspond to the raw initial velocities (rv_0) for the TE reactions.
19. Subtract the rv_0 TE (**step 18**) with each background rv_0 (**step 16**) at each respective substrate concentration. This will give the corrected initial velocities (v_0).
20. Convert v_0 units (mM/s) into the appropriate velocity units (mM/min).

21. Using GraphPad Prism, or other similar software, plot v_0 as a function of substrate concentration.
22. Fit a non-linear curve (Michaelis-Menten, Eq. 2) onto the graph and take note of the kinetic parameters K_M and V_{max} (*see* **Note 36**).

$$v_0 = \frac{V_{max} \cdot [S]}{K_M + [S]} \tag{2}$$

23. Calculate k_{cat} using values of V_{max} (mM/min) and $[E_T]$ (mM) with the rate-law (Eq. 3).

$$k_{cat} = \frac{V_{max}}{[E_T]} \tag{3}$$

4 Notes

1. The syringe plunger is used for solvent flushing with the cap removed. Polyethylene frit for 12 mL syringe (Supelco, 57182-U) and a Visidry Female Luer Plug (Supelco, 57098) are effect options.
2. Cap the falcon tube and shake the solution. Non-polar amino acids such as leucine and isoleucine might require 10 s of sonication to completely dissolve.
3. Cap the flask and stir for 2 min, until the solution becomes clear. An amber flask is preferred if the solution is planned to be left at room temperature for a couple of days.
4. Prepare this working stock immediately prior to adding it into assay mixes. Do not use working stocks that are >2 h old. This limits variation with TE catalytic activity.
5. Prepare this mixture when needed. It can be stored for up to 2 months; however, it is recommended to prepare this mixture fresh every time.
6. Keep tube sealed to limit oxidation. Use only freshly prepared SNAC stock.
7. It is likely that the codons used by the producing organism alter the folding dynamics of the TE in *E. coli*, yielding misfolded TE. To fix this, build a synthetic gene construct and optimize codon usage for *E. coli*.
8. For an N-terminal hexa His tag, use NdeI and EcoRI restriction sites for unidirectional cloning into pET-28a(+) and include the stop codon before the EcoRI site. For a

C-terminal hexa His tag, use the same restriction sites for unidirectional cloning into pET-21c(+). Do not include the stop codon before the EcoRI site. EcoRI should be in frame 3 to ensure the His tag is in frame.

9. If the N-terminal construct does not produce soluble protein, re-clone the TE into pET-21c(+) with a C-terminal His tag and test for soluble protein production.
10. If culture OD_{600} exceeds 1.0, dilute culture to an OD_{600} of 0.15 using fresh LB media supplemented with the correct antibiotic. Let the culture incubate for 2 doubling times (0.5–1 h) in a shaker (200 rpm) set at 37 °C to allow cultures to adapt. Monitor OD_{600} carefully.
11. If residual media is found near the pellet, carefully use a pipette to aspirate the remaining liquid. At this step, pellets can be stored at −20 °C if needed.
12. Add pepstatin A and leupeptin to the lysis buffer before pellet resuspension. These protease inhibitors will not inhibit TE activity.
13. Other cell lysis protocols, such as a French press or high-pressure homogenization, can be used. It is not recommended to use detergents (Triton X-100, CHAPS, etc.) as these will affect downstream activity assays.
14. Incubation times up to 5 h will not negatively impact protein yields.
15. If the resin dries during elution, quickly add 0 mM imidazole solution to cover the resin.
16. If the column is clogged, leaving the resin floating in suspension, apply light pressure to the column. The resin should then settle within 5–10 min.
17. Ni-NTA resin is typically a 50% *v/v* slurry. Therefore, 50% of the volume added is equivalent to 1 column bed volume.
18. Pooled eluents can be stored in the fridge (<10 °C) for 24 h. Avoid storing for more than 3 days. Do not store eluents in the freezer.
19. Occasionally proteins can precipitate as the volume decreases. Shortening the spin time or supplementing the dialysis buffer with imidazole (20 mM final concentration) may solve this problem. If not, stop concentrating the protein just prior to precipitation.
20. Keep this falcon tube to verify that it does not contain any protein (SDS-PAGE gel or absorbance at 280 nm). If TE is present, a new centrifugal concentrator is required and the flowthrough must be buffer exchanged again.

21. Rinse the filter using the pipetted solution. This is to recover any protein stuck on the filter membrane.
22. As there is a wide variety of amino acid sidechains, consult a general peptide synthesis guide (https://www.peptide.com/resources/solid-phase-peptide-synthesis/) to understand the possible side-reactions for each residue in your sequence.
23. More acidic additives such as HOBt, HOAt, and Oxyma are not recommended for hypersensitive resins as premature resin cleavage might occur. DIEA activated reagents such as Benzotriazole coupling reagents HBTU or HATU are preferred over DIC/Oxyma (http://cem.com/media/contenttype/media/literature/600329-cemv6.pdf).
24. Acetylation is a standard in peptide synthesis; however, any N-terminal modification can be performed at this step.
25. It is important to allow the solution to release any gas prior to agitation.
26. Use thin layer chromatography (TLC) to determine appropriate solvent mixtures to purify your peptide. Most protected sidechain peptides with a free C-terminus will elute on normal phase silica. However, larger, and more polar peptides may require reverse phase liquid chromatography (HPLC).
27. Protected peptides tend to be hydrophobic. Expect the peptide to elute at high concentrations of ACN on a standard C18 column. Due to the hydrophobicity, they may also ionize poorly in ESI positive mode.
28. CD_3OD or $CDCl_3$ are often suitable solvents for these protected peptides.
29. It is important to degas the H_2O once components are dissolved. To degas, the solution is capped with a septum and equipped with a short exit needle. A second long needle is added onto the septum with the tip submerged in solution. A stream of nitrogen gas is applied to the long needle and a steady stream nitrogen is bubbled through the stirring solution for 10 min. The outlet needle is removed, the gas flow is stopped, and a nitrogen gas filled balloon is added to the septum.
30. The N,S-diacetylcysteamine and the disulfide products can sometimes be detected in small quantities; however, this reaction is high yielding (>95%).
31. TFA cocktails differ depending on the amino acid sequence and their protecting groups. The standard mixture is TFA:H_2O:TIPS (95:2.5:2.5%).
32. Determine an appropriate HPLC gradient to have sufficient separation for peak isolation. A 5–95% MeCN gradient over 10 min is usually sufficient.

33. Proteins are partly solvated by glycerol and therefore will sink to the bottom of the tube. It is highly recommended to mix tube before pipetting. When adding the protein, pipette directly into the solution to ensure it is added to the mix.
34. For most TEs, 16-h incubation is sufficient. However, incubation times may vary. For TEs that have been previously assayed, verify incubation time in the literature. If TE is uncharacterized, it is highly recommended to use a scaled-up assay and quenching aliquots at varying time points to probe this.
35. A 10–95% MeCN gradient over 15 min with 1 mL/min flow-rate is usually sufficient with C18 columns. However, solvent gradient and flow-rate may need to be adjusted for proper separation and peak resolution of TE products. The MS allows for unambiguous detection of products via their m/z ratio. Using the ion extraction feature can help find products easily. If using an HPLC, compare peaks to synthetically made or purchased standards for unambiguous detection of products.
36. Kinetic parameters can be calculated using Excel. To do this, plot v_0 as a function of $v_0/[S]$ (Eadie–Hofstee plot). V_{max} will correspond to the y-intercept and K_M the negative of the slope. You can then use this to generate a non-linear curve fit in Excel using the Michaelis–Menten equation (Eq. 2).

Acknowledgments

This work was funded by the Natural Sciences and Engineering Research Council of Canada (RGPIN 2019-06859).

References

1. Holmquist M (2000) Alpha/Beta-hydrolase fold enzymes: structures, functions and mechanisms. Curr Protein Pept Sci 1(2): 209–235
2. Horsman ME, Hari TPA, Boddy CN (2016) Polyketide synthase and non-ribosomal peptide synthetase thioesterase selectivity: logic gate or a victim of fate? Nat Prod Rep 33(2): 183–202
3. Martinez-Nunez MA, Lopez y Lopez VE (2016) Nonribosomal peptide synthetases and their applications in industry. Sustain Chem Proc 4(13)
4. Challis GL, Naismith JH (2004) Structural aspects of non-ribosomal peptide biosynthesis. Curr Opin Struct Biol 14(6):748–756
5. Finking R, Marahiel MA (2004) Biosynthesis of nonribosomal peptides1. Annu Rev Microbiol 58:453–488
6. Long JZ, Cravatt BF (2011) The metabolic serine hydrolases and their functions in mammalian physiology and disease. Chem Rev 111(10):6022–6063
7. Rauwerdink A, Kazlauskas RJ (2015) How the same core catalytic machinery catalyzes 17 different reactions: the serine-histidine-aspartate catalytic triad of α/β-hydrolase fold enzymes. ACS Catal 5(10):6153–6176
8. Hoyer KM, Mahlert C, Marahiel MA (2007) The iterative gramicidin S thioesterase catalyzes peptide ligation and cyclization. Chem Biol 14(1):13–22
9. Trauger JW, Kohli RM, Mootz HD, Mahariel MA, Walsh CT (2000) Peptide cyclization catalysed by the thioesterase domain of tyrocidine synthetase. Nature 407:215–218
10. Heberlig GW, Boddy CN (2020) Thioesterase from cereulide biosynthesis is responsible for

oligomerization and macrocyclization of linear tetradepsipeptide. J Nat Prod 83(6): 1990–1997

11. Huguenin-Dezot N, Alonzo DA, Heberlig GW, Mahesh M, Nguyen DP, Dornan MH et al (2019) Trapping biosynthetic acyl-enzyme intermediates with encoded 2,3-diaminopropionic acid. Nature 565(7737):112–117
12. Samel SA, Wagner B, Marahiel MA, Essen L-O (2006) The thioesterase domain of the fengycin biosynthesis cluster: a structural base for the macrocyclization of a non-ribosomal lipopeptide. J Mol Biol 359(4):876–879
13. Ray L, Yamanaka K, Moore BS (2016) A peptidyl-transesterifying type I thioesterase in salinamide biosynthesis. Angew Chem Int Ed Engl 55(1):364–367
14. Mulyani S, Egel E, Kittel C, Turkanovic S, Wohlleben W, Süssmuth RD et al (2010) The thioesterase Bhp is involved in the formation of beta-hydroxytyrosine during balhimycin biosynthesis in Amycoltopsis balhimycina. Chembiochem 11(2):266–271
15. Peschke M, Brieke C, Heimes M, Cryle MJ (2018) The thioesterase domain in glycopeptide antibiotic biosynthesis is selective for crosslinked aglycones. Chem Biol 13(1):110–120
16. Schmidt JJ, Khatri Y, Brody SI, Zhu C, Pietraszkiewicz H, Valeriote FA, Sherman DH (2020) A versatile chemoenzymatic synthesis for the discovery of potent cryptophycin analogs. Chem Biol 15(2):524–532
17. Wang M, Zhou H, Wirz M, Tang Y, Boddy CN (2009) A thioesterase from an iterative fungal polyketide synthase shows macrocyclization and cross coupling activity and may play a role in controlling iterative cycling through product offloading. Biochemistry 48(27):6288–6290
18. Heberlig GW, Wirz M, Wang M, Boddy CN (2014) Resorcylic acid lactone biosynthesis relies on a stereotolerant macrocyclizing thioesterase. Org Lett 16(22):5858–5861
19. Boddy CN, Schneider TL, Hotta K, Walsh CT, Khosla C (2003) Epothilone C macrolactonization and hydrolysis are catalyzed by the isolated thioesterase domain of epothilone polyketide synthase. J Am Chem Soc 125(12): 3428–3429
20. Zhou Y, Prediger P, Dias LC, Murphy AC, Leadlay PF (2015) Macrolide formation by the thioesterase of a modular polyketide synthase. Angew Chem Int Ed Engl 54(17): 5232–5235
21. Scaglione JB, Akey DL, Sullivan R, Kittendorf JD, Rath CM, Kim ES et al (2010) Biochemical and structural characterization of the tautomycetin thioesterase: analysis of a stereoselective polyketide hydrolase. Angew Chem Int Ed Engl 49(33):5726–5730
22. Kudo F, Kitayama T, Kakinuma K, Eguchi T (2006) Macrolactam formation catalyzed by the thioesterase domain of vicenistatin polyketide synthase. Tet Lett 47(10):1529–1532
23. Gamal AE, Agarwal V, Diethelm S, Rahman I, Schorn MA, Sneed JM et al (2016) Biosynthesis of coral settlement cue tetrabromopyrrole in marine bacteria by a uniquely adapted brominase-thioesterase enzyme pair. Proc Natl Acad Sci USA 113(14):3797–3802
24. Awakawa T, Yokota K, Funa N, Doi F, More N, Watanabe H, Horinouchi S (2009) Physically discrete beta-lactamase-type thioesterase catalyzes product release in atrochrysone synthesis by iterative type I polyketide synthase. Chem Biol 16(6):613–623
25. Tang MC, Fischer CR, Chari JV, Tan D, Suresh S, Chu A et al (2019) Thioesterase-catalyzed aminoacylation and thiolation of polyketides in fungi. J Am Chem Soc 141(20):8198–8206
26. Xu W, Chooi YH, Choi JW, Li S, Vederas JC, Da Silva NA et al (2013) LovG: the thioesterase required for dihydromonacolin L release and lovastatin nonaketide synthase turnover in lovastatin biosynthesis. Angew Chem Int Ed Engl 52(25):6472–6475
27. Zhou Y, Tao W, Qi Z, Wei J, Shi T, Kang Q et al (2022) Structural mechanistic insights into chain release of the polyene PKS thioesterase domain. ACS Catal 12(1):762–776
28. Cogan DP, Ly J, Nair SK (2020) Structural basis for enzymatic off-loading of hybrid polyketides by dieckmann condensation. ACS Chem Biol 15(10):2783–2791
29. Gokhale RS, Hunziker D, Cane DE, Khosla C (1999) Mechanism and specificity of the terminal thioesterase domain from the erythromycin polyketide synthase. Chem Biol 6(2):117125
30. Kim BS, Cropp TAC, Beck BJ, Sherman DH, Reynolds KA (2002) Biochemical evidence for an editing role thioesterase II in the biosynthesis of the polyketide pikromycin. J Biol Chem 277(50):48028–48034
31. Kohli RM, Trauger JW, Schwarzer D, Marahiel MA, Walsh CT (2001) Generality of peptide cyclization catalyzed by isolated thioesterase domains of nonribosomal peptide synthetases. Biochemistry 40(24):7099–7108
32. Bruner SD, Weber T, Kohli RM, Schwarzer D, Marahiel MA, Walsh CT, Stubbs MT (2002) Structural basis for the cyclization of the lipopeptide antibiotic surfactin by the thioesterase domain SrfTE. Structure 10(3):301–310

33. Galea CA, Roberts KD, Zhu Y, Thompson PE, Li J, Velkov T (2017) Functional characterization of the unique terminal thioesterase domain from polymyxin synthetase. Biochemistry 56(4):657–668
34. Grünewald J, Sieber SA, Marahiel MA (2004) Chemo- and regioselective peptide cyclization triggered by the N-terminal fatty acid chain length: the recombinant cyclase of the calcium-dependent antibiotic from Streptomyces coelicolor. Biochemistry 43(10): 2915–2925
35. Lou L, Chen H, Cerny RL, Li Y, Shen Y, Du L (2011) Unusual activities of the thioesterase domain for the biosynthesis of the polycyclic tetramate macrolactam HSAF in Lysobacter enzymogenes C3. Biochemistry 51(1):4–6
36. Argyropoulos P, Bergeret F, Pardin C, Reimer JM, Pinto A, Boddy CN et al (2016) Towards a characterization of the structural determinants of specificity in the macrocyclizing thioesterase for deoxyerythronolide B biosynthesis. Biochim Biophys Acta 1860(3):486–497
37. Koch AA, Hansen DA, Shende VV, Furan LR, Houk KN, Jiménez-Osés G (2017) A single active site mutation in the pikromycin thioesterase generates a more effective macrocyclization catalyst. J Am Chem Soc 139(38): 13456–13465
38. Sieber SA, Tao J, Walsh CT, Marahiel MA (2004) Peptidyl thiophenols as substrates for nonribosomal peptide cyclases. Angew Chem Int Ed Engl 43(4):493–498
39. Dawson PE, Muir TW, Clark-Lewis I, Kent SB (1994) Synthesis of proteins by native chemical ligation. Science 266(5186):776–779
40. Schnölzer M, Alewood P, Jones A, Alewood D, Kent SB (1992) In situ neutralization in Boc-chemistry solid phase peptide synthesis. Rapid, high yield assembly of difficult sequences. Int J Pept Protein Res 40(3–4): 180–193
41. Hackeng TM, Griffin JH, Dawson PE (1999) Protein synthesis by native chemical ligation: expanded scope by using straightforward methodology. Proc Natl Acad Sci USA 96(18):10068–10073
42. Camarero JA, Adeva A, Muir TW (2000) 3-Thiopropionic acid as a highly versatile multidetachable thioester resin linker. Lett Pept Sci 7:17–21
43. Crich D, Sana K (2009) Solid-phase synthesis of peptidyl thioacids employing a 9-fluorenylmethyl thioether-based linker in conjunction with Boc chemistry. J Org Chem 74(19):7383–7388
44. Ingenito R, Bianchi E, Fattori D, Pessi A (1999) Solid phase synthesis of peptide C-terminal thioesters by Fmoc/t-Bu chemistry. J Am Chem Soc 121(49):11369–11374
45. Camarero JA, Hackel BJ, deYoreo JJ, Mitchell AR (2004) Fmoc-based synthesis of peptide alpha-thioesters using an aryl hydrazine support. J Org Chem 69(12):4145–4151
46. Blanco-Canosa JB, Dawson PE (2008) An efficient Fmoc-SPPS approach for the generation of thioester peptide precursors for use in native chemical ligation. Angew Chem Int Ed Engl 47(36):6851–6855
47. Ollivier N, Dheur J, Mhidia R, Blanpain A, Melnyk O (2010) Bis(2-sulfnylethyl)amino native peptide ligation. Org Lett 12(22): 5238–5241
48. Duggan NM, Saez NJ, Clayton D, Budusan E, Watson EE, Tucker IJ et al (2021) Total synthesis of the spider-venom peptide Hi1a. Org Lett 23(21):8375–8379
49. Wolfender J-L (2009) HPLC in natural product analysis: the detection issue. Planta Med 75(7):719–734
50. Tseng CC, Bruner SD, Kohli RM, Marahiel MA, Walsh CT, Sieber SA (2002) Characterization of the surfactin synthetase C-terminal thioesterase domainas a cyclic depsipeptide synthase. Biochemistry 41(45):15550–13359
51. Yeh E, Lin H, Clugston SL, Kohli RM, Walsh CT (2004) Enhanced macrocyclization activity of the thioesterase from tyrocidine synthetase in presence of nonionic detergent. Chem Biol 11:1573–1582
52. Yan C, Wu Y, Weng Z, Gao Q, Yang G, Chen Z et al (2015) Development of an HPLC method for absolute quantification and QAMS of flavonoids components in psoralea corylifolia L. J Anal Methods Chem 2015
53. Michaelis L, Menten ML, Johnson KA, Goody RS (2011) The original Michaelis constant: translation of the 1913 Michaelis-Menten paper. Biochemistry 50(39):8264–8269
54. Cornish-Bowden A (2015) One hundred years of Michaelis-Menten kinetics. Perspect Sci 4:3–9
55. Bisswanger H (2014) Enzyme assays. Perspect Sci 1(6):41–55
56. Pfeifer BA, Admiraal SJ, Gramajo H, Cane DE, Khosla C (2001) Biosynthesis of complex polyketides in a metabolically engineered strain of E. coli. Science 291(5509):1790–1792
57. Kautsar SA, Lin K, Shaw S, Navarro-Munoz JC, Terlow BR, van der Hooft JJJ et al (2020) MIBiG 2.0: a repository for biosynthetic gene clusters of known function. Nucleic Acids Res 48(D1):454–458

58. Blin K, Shaw S, Kloosterman AM, Charlop-Powers Z, van Wezel GP, Medema MH et al (2021) antiSMASH 6.0: improving cluster detection and comparison capabilities. Nucleic Acids Res 49(W1):29–35
59. Finn RD, Bateman A, Clements J, Coggill P, Eberhardt RY, Eddy SR (2014) Pfam: the protein families database. Nucleic Acids Res 42: D222–D230
60. Finn RD, Clements J, Eddy SR (2011) HMMER web server: interactive sequence similarity searching. Nucleic Acid Res 39:W29–W37
61. Lu H, Tsai SC, Khosla C, Cane DE (2002) Expression, site-directed mutagenesis, and steady state kinetic analysis of the terminal thioesterase domain of the methymycin/picromycin polyketide synthase. Biochemistry 41(42):12590–12597
62. Wang M, Boddy CN (2008) Examining the role of hydrogen bonding interactions in the substrate specificity for the loading step of polyketide synthase thioesterase domains. Biochemistry 47(45):11793–11803
63. Pinto A, Wang M, Horsman M, Boddy CN (2012) 6-Deoxyerythronolide B synthase thioesterase-catalyzed macrocyclization is highly stereoselective. Org Lett 14(9): 2278–2281
64. Manns JM (2011) SDS-polyacrylamide gel electrophoresis (SDS-PAGE) of proteins. 22(1):A.3M.1–A.3M.13
65. Kim GJ, Li X, Kim SH, Yang I, Hahn D, Chin J et al (2018) Seongsanamides A-D: antiallergic bicyclic peptides from bacillus safensis KCTC 12796BP. Org Lett 20(23):7539–7543
66. Jahanshah G, Yan Q, Gerhardt H, Pataj Z, Lämmerhofer M, Pianet I et al (2019) Discovery of the cyclic lipopeptide Gacamide A by genome mining and repair of the defective GacA regulator in pseudomonas fluorescens Pf0-1. J Nat Prod 82(2):301–308
67. Gude M, Ryf J, White PD (2002) An accurate method for the quantitation of Fmoc-derivatized solid phase supports. Lett Pep Sci 6:203–206
68. Mali SM, Kumar MG, Katariya MM, Gopi HN (2014) HBTU mediated 1-hydroxybenzotriazole (HOBt) conjugate addition: synthesis and stereochemical analysis of β-benzotriazole N-oxide substituted γ-amino acids and hybrid peptides. Org Biomol Chem 12(42):8462–8472
69. Hughes AJ, Keatinge-Clay A (2011) Enzymatic extender unit generation for in vitro polyketide synthase reactions: structural and functional showcasing of Streptomyces coelicolor MatB. Chem Biol 18(2):165–176
70. Wilcoxen KM, Leman LJ, Weinberger DA, Huang ZZ, Ghadiri MR (2007) Biomimetic catalysis of intermodular aminoacyl transfer. J Am Chem Soc 129(4):748–749
71. Green MR, Sambrook J (2020) Cloning in plasmid vectors: directional cloning. Cold Spring Harb Protoc:485–487
72. Kurien BT, Danda D, Bachmann MP, Scofield RH (2015) SDS-PAGE to immunoblot in one hour. Methods Mol Biol 1312:449–454

Chapter 6

Chemo-Enzymatic Synthesis of Non-ribosomal Macrolactams by a Penicillin-Binding Protein-Type Thioesterase

Masakazu Kobayashi, Kei Fujita, Kenichi Matsuda, and Toshiyuki Wakimoto

Abstract

Penicillin-binding protein-type thioesterases (PBP-type TEs) are an emerging family of non-ribosomal peptide cyclases. PBP-type TEs exhibit distinct substrate scopes from the well-exploited ribosomal peptide cyclases and traditional non-ribosomal peptide cyclases. Their unique properties, as well as their stand-alone nature, highlight PBP-type TEs as valuable candidates for development as biocatalysts for peptide macrocyclization. Here in this chapter, we describe the scheme for the chemoenzymatic synthesis of non-ribosomal macrolactam by SurE, a representative member of PBP-type TEs.

Key words Cyclic peptides, Peptide cyclase, Penicillin-binding protein-type thioesterases, Chemoenzymatic synthesis

1 Introduction

Middle-sized molecules that fill the gap between small-molecule drugs (less than 500 Da) and large biologics (more than 5,000 Da) have emerged as a promising class of therapeutics to reach less explored pharmaceutical targets, such as intracellular protein–protein interactions and complex biological molecules on a cellular membrane. Among them, cyclic peptides are gaining increasing attention, with more than 40 cyclic peptides currently in clinical use [1]. Cyclic peptides generally exhibit enhanced drug-like properties, such as resistance to proteolytic degradation, cell membrane permeability, and increased target specificities, when compared to their linear counterparts. Therefore, the efficient synthesis of cyclic peptides is significant for developing and manufacturing cyclic

Masakazu Kobayashi and Kei Fujita contributed equally with all other contributors.

Michael Burkart and Fumihiro Ishikawa (eds.), *Non-Ribosomal Peptide Biosynthesis and Engineering: Methods and Protocols*, Methods in Molecular Biology, vol. 2670, https://doi.org/10.1007/978-1-0716-3214-7_6,

peptide-based pharmaceuticals. Although peptide macrocyclization methods have been intensively studied and various methodologies have been developed, macrocyclization remains a challenging task [2]. Regiospecific macrocyclization generally requires multiple protecting groups, which make the synthesis of cyclic peptides multistep. In addition, the use of classic coupling reagents potentially causes epimerization at the α-position of the carboxylic group, leading to the generation of a by product that is difficult to separate in downstream purification processes. Moreover, the macrocyclization must be performed under highly diluted conditions to suppress the undesired intermolecular oligomerization, thus requiring a large amount of anhydrous organic solvents.

The use of enzymes for peptide macrocyclization is one of the promising strategies, as their highly chemo-selective properties could potentially circumvent the needs for protecting groups and organic solvents. Several enzymes have been intensively examined as biocatalytic tools for peptide cyclization, including sortases [3, 4], trypsin [5], subtilisin-variants [6–8], asparaginyl endopeptidases (AEPs) like butelase-1 [9], OaAEP1 [10], and PatG-like cyanobactin macrocyclases [11, 12]. As their original functions are to catalyze digestion, ligation, or cyclization of ribosomally synthesized peptides in their native circumstances, here we refer to them as "ribosomal peptide cyclases." Although they are useful, and some of them are commercialized, the ribosomal peptide cyclases possess enzyme-specific drawbacks, such as low catalytic efficiency, incomplete cyclization, and retention of enzyme-recognition sequences in the cyclic products. Additionally, most of them are incapable of cyclizing short peptides (fewer than 10 residues), which are also difficult to cyclize by traditional synthetic methods [2]. An exception is found in the PatG-like macrocyclases, which are capable of cyclizing short peptides [11, 12], although they suffer from low catalytic efficiencies. Collectively, short peptides are currently out of the scope of the existing sets of ribosomal peptide cyclases.

Non-ribosomal peptides, typically with less than fifteen residues, are a prolific source of bioactive cyclic peptides with significant pharmaceutical importance. These are exemplified by the immunosuppressant cyclosporin and antibiotic lipopeptides such as daptomycins and polymyxins. The installations of head-to-tail or head-to-side chain cyclic scaffolds are typically mediated by *cis*-acting terminal domains that are fused to the *C*-terminus of the non-ribosomal peptide synthetases (NRPSs). These include type-I thioesterase (type-I TE) domains [13], terminal condensation (C_T) domains [14], and terminal reductase (R) domains [15–17]. Type-I TEs and C_Ts catalyze macrolactamization and macrolactonization, while Rs catalyze Dieckmann condensation [15] and the reductive release of aldehyde products, which undergo spontaneous cyclization to form macrocyclic imines [17].

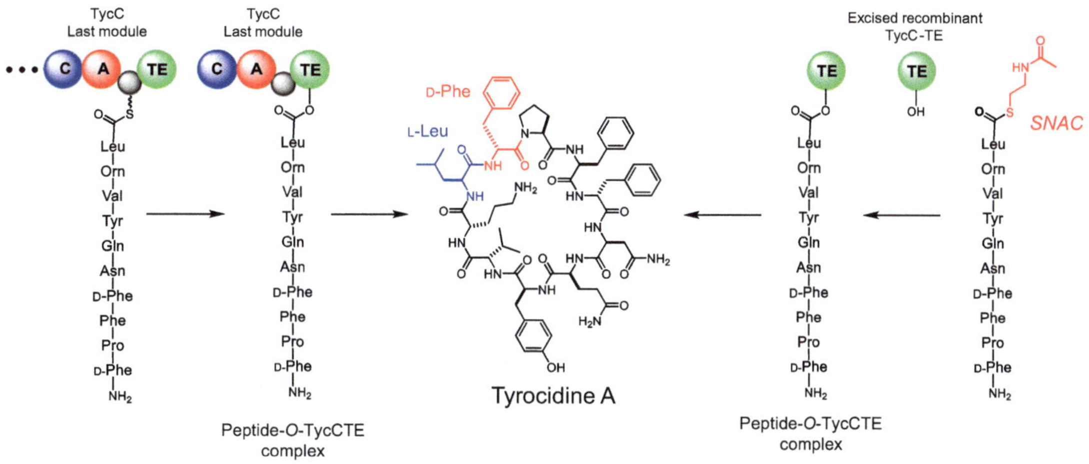

Fig. 1 Macrocyclization of non-ribosomal peptide mediated by type-I TE. Biosynthesis of tyrocidine is shown as an example. The grown peptide chain tethered on the last module of the assembly-line (TycC) is cyclized via two steps: The peptide is transferred to the catalytic serine residue of TycC-TE, and then the *N*-terminal amine intramolecularly attacks the ester linkage, which results in the cyclative release of a matured peptide (left half of figure). This process was reconstituted in vitro using the excised recombinant TycC-TE and the activated peptide with *N*-acetyl cysteamine at the *C*-terminus as a leaving group (right) [13]. TycC-TE catalyzes heterochiral coupling between *N*-terminal D-Phe (red) and *C*-terminal L-Leu (blue)

As the product size is considerably smaller, these non-ribosomal peptide cyclases could be powerful tools to complement the existing biocatalysts derived from ribosomal peptide cyclases. The utility of a non-ribosomal peptide cyclase as a biocatalyst was first demonstrated in the pioneering work by Walsh and colleagues in the early 2000s [13]. They overexpressed the excised version of the type-I TE domain (TycC-TE) from the NRPS (TycC) in the biosynthetic pathway of head-to-tail macrocyclic decapeptide tyrocidines (Fig. 1). The native substrates of TycC-TE are thioester *seco*-tyrocidines tethered to the peptidyl carrier protein (PCP) through a *C*-terminally linked phosphopantetheinyl group, and the substrate was successfully mimicked by the fragment of a pantetheinyl group referred to as SNAC, which stands for *N*-acetyl cysteamine. The excised TycC-TE efficiently cyclized SNAC-activated *seco*-tyrocidine in a regioselective manner (Fig. 1). In the follow-up study, they created a model of the minimal cyclization substrates [18] and combined TycC-TE with combinatorial solid phase peptide synthesis (SPPS) chemistry by performing the enzymatic cyclization of peptides tethered on a solid phase through a biomimetic linker [19]. This on-resin strategy enabled the synthesis of libraries of tyrocidine analogs and led to the identification of analogs with enhanced selectivity against bacterial membranes over eukaryotic ones. Since then, several non-ribosomal peptide cyclases have been experimentally characterized, including thioesterases in the biosyntheses of gramicidin S [20], surfactin [21], fengycin

[22], enterobactin [23], echinomycin [24], lysobactin [25], polymyxin [26], teixobactin [27], microcystin [28], cryptophycin [29], and others. Although promising, the exploitation of the *cis*-acting non-ribosomal peptide cyclases has not been fully realized due to their severe specificities for the native substrate sequences, which probably arose from their natural one-to-one relationship with a single assembly-line [21]. In addition, the excisions of *cis*-acting cyclases from assembly-lines sometimes cause the loss of enzymatic activities. In some cases, a phosphopantethenyl cofactor or covalently linked PCP is required to properly direct the peptide substrates into the active sites of cyclases [30].

Penicillin-binding protein-type thioesterases (PBP-type TEs) are an emerging family of non-ribosomal peptide cyclases that are distinct from others in many ways [31]. They lack sequence homology to *cis*-acting cyclases, but resemble penicillin-binding proteins (PBPs). In contrast to the *cis*-acting cyclases, PBP-type TEs are physically discrete from cognate NRPSs, thus acting in *trans*. Currently, at least seven PBP-type TEs have been described in the literature, including SurE in surugamide biosynthesis [32–35], PenA in pentaminomycin/BE-18257 biosynthesis [36], DsaJ in desotamide biosynthesis [37], Ulm16 in ulleungmycin biosynthesis [38], Nsm16 in noursamycin biosynthesis [39], Lon18 in longicatenamide biosynthesis [40], and MppK in mannnopeptimycin biosynthesis [41]. The products of the PBP-type TEs identified so far are exclusively head-to-tail macrolactams.

SurE, the first characterized example of a PBP-type TE, is employed in the biosynthesis of two groups of non-ribosomal macrolactams that are structurally distinct: octapeptidyl surugamides (surugamides A-E) and decapeptidyl surugamides (surugamide F and cyclosurugamide F) (Fig. 2) [32, 33]. The biosynthetic gene cluster (BGC) of surugamides (*sur*) contains four NRPS genes (*surA-D*) that all lack conventional cyclization domains at their *C*-terminus [42]. The pairs of NRPSs SurA/SurD and SurB/SurC form assembly-lines for octapeptides and decapeptides, respectively. The single stand-alone enzyme SurE acts in *trans* with both assembly-lines to mediate the cyclization of two peptide intermediates. The sizes and sequences of the two peptides are distinct, highlighting the broad substrate scope of SurE. Recombinant SurE quantitatively catalyzes the regioselective cyclization of a SNAC-based substrate in vitro [32]. The scope of SurE was examined thoroughly in a follow-up study [43]. Ala-substitution experiments revealed the broad tolerance of SurE for residues in the middle of the substrates. Conversely, SurE exhibits strict selectivity for the terminal residues: It requires a hydrophobic L-aa (amino acid) at the *N*-terminus and a hydrophobic D-aa at the *C*-terminus. The stereoselectivity of SurE against the *C*-terminal D-aa is consistent with the sequence homology of SurE to PBPs, whose substrates are a D-Ala-D-Ala motif in cell wall biosynthesis.

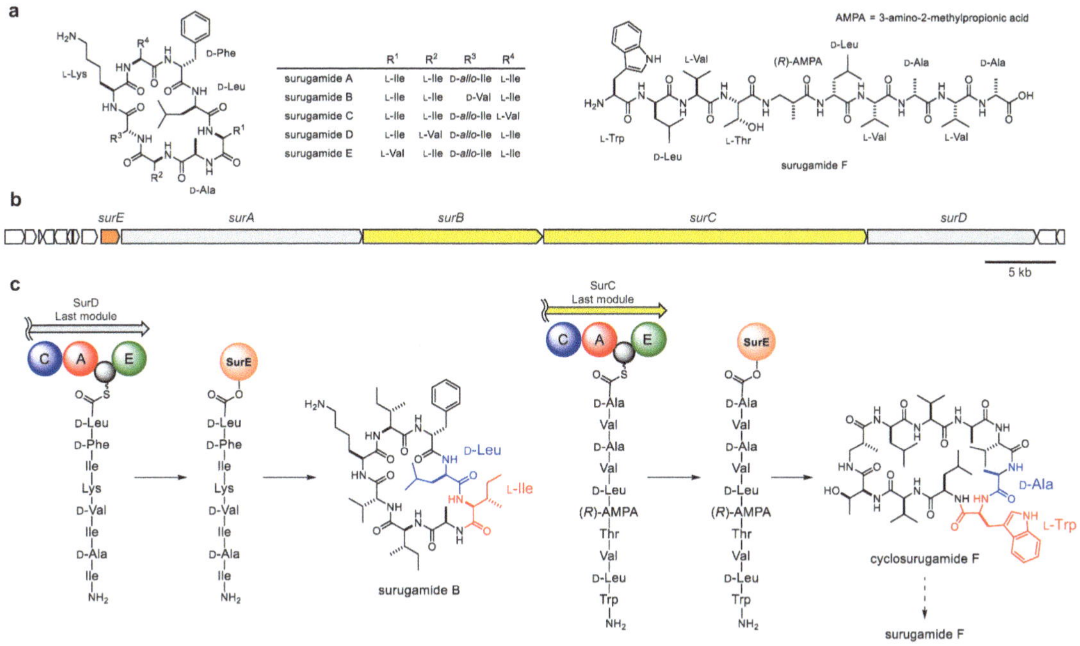

Fig. 2 Biosynthesis of surugamides. (**a**) Structures of surugamides. (**b**) *sur* BGC. Genes encoding the octapeptide assembly-line (SurA/SurD), decapeptide assembly line (SurB/SurC), and PBP-type TE SurE are colored gray, yellow, and orange, respectively. (**c**) Macrocyclization of surugamides mediated by SurE. The stand-alone enzyme SurE acts on both assembly-lines in *trans*. The reaction proceeds via a similar mechanism to conventional type-I TEs. SurE catalyzes heterochiral coupling between *N*-terminal L-Ile/L-Trp (red) and *C*-terminal D-Leu/D-Ala (blue)

The survey of non-ribosomal head-to-tail macrolactams shed light on a clear trend in the stereochemistry of the cyclization sites: First, most cyclization sites consist of heterochiral residues, and second, while traditional type-I TEs and C_Ts mediate coupling between an *N*-terminal D-aa and a *C*-terminal L-aa, PBP-type TEs mediate coupling between an *N*-terminal L-aa and a *C*-terminal D-aa [43]. The complementary stereoselectivity of *cis*- and *trans*-acting cyclases suggests that a suite of these cyclase family members can overcome the stereochemical constraints in the head-to-tail cyclization of non-ribosomal macrolactams and expand the variety of cyclization sites, in terms of stereochemistry. Overall, PBP-type TEs exhibit distinct substrate scopes from the well-exploited ribosomal peptide cyclases and traditional non-ribosomal peptide cyclases. Their unique properties, as well as their stand-alone nature, highlight PBP-type TEs as valuable candidates for further development as biocatalysts.

Here in this chapter, we describe the scheme for the chemoenzymatic synthesis of non-ribosomal macrolactam by SurE, a representative member of PBP-type TEs.

2 Materials

2.1 Synthesis of Seco-Surugamide B-SNAC

2.1.1 Solid-Phase Peptide Synthesis

1. CH_2Cl_2 (dichloromethane).
2. DMF (*N,N'*-dimethylformamide).
3. Methanol.
4. Ac_2O (acetic anhydrate).
5. 2-Chlorotrityl chloride resin.
6. Fmoc-D-Leu.
7. Fmoc-D-Phe.
8. Fmoc-L-Ile.
9. Fmoc-L-Lys(Boc).
10. Fmoc-D-Val.
11. Fmoc-D-Ala.
12. Boc-L-Ile.
13. *i*-Pr_2NEt (*N,N'*-diisopropylethylamine).
14. Oxyma.
15. DIC (*N,N'*-diisopropylcarbodiimide).
16. Piperidine.
17. HFIP (1,1,1,3,3,3-Hexafluoro-2-propanol).
18. Libra Tube.
19. Conventional lab shaker.
20. Vortex mixer.
21. Conventional centrifuge (20,630 × *g* for 1.5 mL tube and 3,500 × *g* for 50 mL tube).

2.1.2 Thioesterification, Deprotection, and Precipitation

1. CH_2Cl_2.
2. DMF.
3. EtOAc (ethyl acetate).
4. *N*-Acetylcysteamine.
5. 2,4,6-Collidine.
6. HCTU (1-[bis(dimethylamino)methylene]-5-chloro-1*H*-benzotriazolium 3-oxide hexafluorophosphate).
7. Deprotection solution: 95% TFA (trifluoroacetic acid), 2.5% H_2O, 2.5% *i*-Pr_3SiH (triisopropylsilane).
8. Et_2O (diethyl ether).

2.1.3 HPLC Purification

1. Conventional HPLC system.
2. Mobile phase of liquid chromatography: 40% MeCN (acetonitrile) supplemented with 0.05% TFA.
3. Preparative reversed phase column (COSMOSIL Cholester 10 × 250 mm).

2.2 Synthesis of Surugamide B

2.2.1 Solid-Phase Peptide Synthesis

1. CH_2Cl_2.
2. DMF.
3. 2-Chlorotrityl chloride resin.
4. Fmoc-D-Leu.
5. Fmoc-D-Phe.
6. Fmoc-L-Ile.
7. Fmoc-L-Lys(Boc).
8. Fmoc-D-Val.
9. Fmoc-D-Ala.
10. Oxyma.
11. DIC.
12. Piperidine.
13. HFIP.
14. Libra Tube.
15. Conventional lab shaker.
16. Vortex mixer.
17. Conventional centrifuge (20,630 × *g* for 1.5 mL tube and 3,500 × *g* for 50 mL tube).

2.2.2 Macrocyclization, Deprotection, and Precipitation

1. CH_2Cl_2.
2. DMF.
3. EtOAc.
4. 2,4,6-Collidine.
5. HOAt (1-hydroxy-7-azabenzotriazol).
6. PyBOP ((benzotriazol-1-yloxy)tripyrrolidinophosphonium hexafluorophosphate).
7. Deprotection solution: 95% TFA, 2.5% H_2O, 2.5% i-Pr_3SiH.
8. Et_2O.

2.2.3 HPLC Purification

1. Conventional HPLC system.
2. Mobile phase of liquid chromatography: 34% MeCN supplemented with 0.05% TFA.
3. Preparative reversed phase column (COSMOSIL Cholester 10 × 250 mm).

2.3 Preparation of a Recombinant SurE

2.3.1 Cloning of *surE* Gene

1. KOD One polymerase.
2. Oligonucleotide used as a primer, SurE_Fw: CCGGAATTCCATATGGGTGCCGAGGGGGCG.
3. Oligonucleotide used as a primer, SurE_Rv: CCCAAGCTTTCAGAGCCGGTGCATGGC.

4. The genomic DNA of *Streptomyces albidoflavus* NBRC12854.
5. Cloning vector pUC19.
6. Expression vector pET-28a(+) DNA.

2.3.2 Protein Purification

1. 2 × YT media: 1.6% Tryptone, 1.0% Bacto Yeast Extract, 0.5% NaCl.
2. 0.1 M isopropyl β-D-1-thiogalactopyranoside (IPTG): Weight 238.3 mg IPTG and dissolve in 10 mL of water, then filter through 0.22 μm syringe filter. Store at −20 °C.
3. Buffer A: 20 mM Tris–HCl pH 8.0, 150 mM NaCl.
4. Buffer B: 20 mM Tris–HCl pH 8.0, 150 mM NaCl, 20 mM imidazole pH 8.0.
5. Buffer C: 20 mM Tris–HCl pH 8.0, 150 mM NaCl, 500 mM imidazole pH 8.0.
6. Ni-NTA His Bind® Resin.
7. MWCO 10 kDa centrifugal ultrafiltration tubes (such as Amicon® Ultra 4 mL centrifugal filter).
8. Bio-Rad protein assay kit.

2.4 In Vitro Cyclization Mediated by SurE

1. 0.1% TFA.
2. Conventional HPLC system.
3. Mobile phase of liquid chromatography: 41% MeCN supplemented with 0.05% TFA.
4. Analytical reversed phase column (COSMOSIL 5C_{18}-MS-II 4.6 × 250 mm).

3 Methods

3.1 Synthesis of Seco-Surugamide B-SNAC (Fig. 3)

3.1.1 Solid-Phase Peptide Synthesis (SPPS)

1. Add CH_2Cl_2 (2 mL) to 2-chlorotrityl chloride resin (100 mg, 0.160 mmoL) in a Libra tube. Allow the resin to swell for 30 min and then remove the excess solvent by filtration.
2. Add a solution of Fmoc-D-Leu (113 mg, 0.318 mmoL) and *i*-Pr_2NEt (168 μL, 0.960 mmoL) in CH_2Cl_2 (0.5 mL) to the resin and shake for 30 min at 37 °C. Filter off the reaction mixture, wash the resin with DMF (×3), CH_2Cl_2 (×3), methanol, and Ac_2O, and then dry the resin under vacuum for 1 h to give Fmoc-D-Leu-2-chlorotrityl resin (**1**, Fig. 3) (*see* **Note 1**).
3. Add 20% piperidine in DMF (1 mL) to the dried Fmoc-D-Leu-2-chlorotrityl resin (**1**), and then shake the tube for 1 h. Dilute

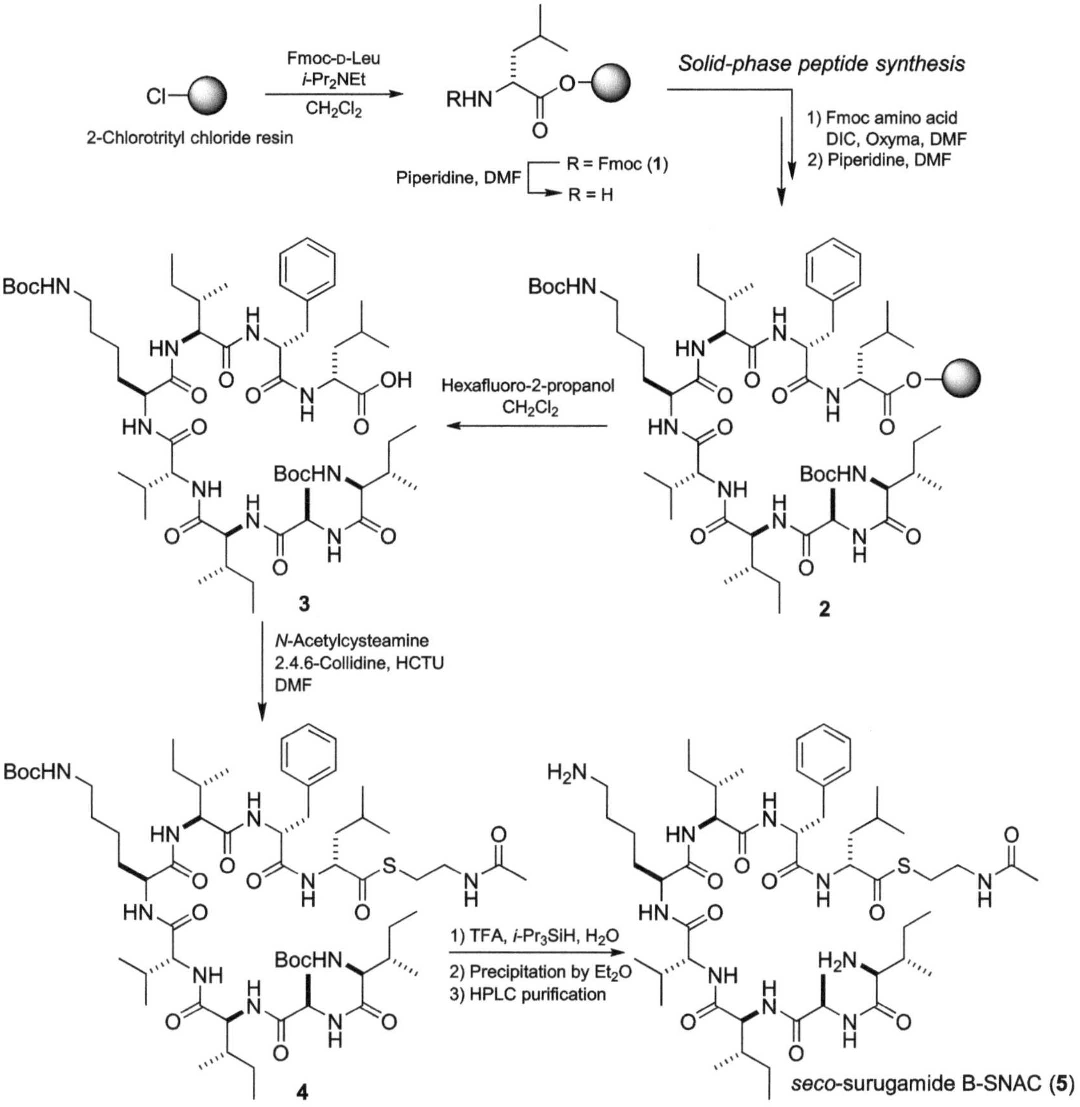

Fig. 3 Synthetic scheme of *seco*-surugamide B-SNAC (**5**)

the supernatant 100-fold with DMF and measure its UV absorption at 301 nm. The loading efficiency is calculated from the UV absorbance as follows (Eq. 1).

$$\text{Loading}\left(\text{mmolg}^{-1}\right) = \frac{(A_{301} \times 1000)}{M_{\text{resin}}[\text{mg}] \times 7800 \times \text{dilution rate}}$$

Where A_{301} = UV absorbance at 301 nm, M_{resin} = mass of the resin used for quantification (1)

4. The Fmoc group of the solid-supported peptide is removed by using a 20% piperidine/DMF solution (10 min, room temperature) (*see* **Note 2**). Skip this step in the first round of SPPS.
5. The resin in the reaction vessel is washed with DMF (×3) and CH_2Cl_2 (×3).
6. To the solution of Fmoc-D-Phe (4 eq), add DIC (4 eq, 0.5 M in DMF) and Oxyma (4 eq, 0.5 M in DMF). After 2–3 min of pre-activation, inject the mixture into the reaction vessel. Shake the resulting mixture for 30 min at 37 °C.
7. The resin in the reaction vessel is washed with DMF (×3) and CH_2Cl_2 (×3).
8. Condense amino acids [Fmoc-L-Ile, Fmoc-L-Lys(Boc), Fmoc-D-Val, Fmoc-L-Ile, Fmoc-D-Ala, Boc-L-Ile] onto solid-support by repeating the **Steps 4–7** for 6 cycles. This affords the resin-bound peptide (**2**, Fig. 3).
9. Add CH_2Cl_2/HFIP (=70:30) (0.5 mL) to **2** and then shake the tube for 30 min. Filter off the reaction mixture. Repeat this procedure twice. The removal of solvent from the filtrate affords the crude peptide (**3**, Fig. 3), which is used in the next reaction without further purification.

3.1.2 Thioesterification, Deprotection, and Precipitation

1. Dissolve the peptide (**3**) in DMF (1 mL), add *N*-acetylcysteamine (12 mg, 0.1 mmoL), 2,4,6-collidine (6.6 μL, 0.05 mmoL), and HCTU (22.8 mg, 0.055 mmoL), and then stir the mixture overnight at room temperature. Quench the mixture with saturated aqueous NH_4Cl, and then extract the solution with EtOAc (×3). Wash the organic layer with brine, dry it over $MgSO_4$, and then retain the filtrate. The solvent removal will afford peptide (**4**, Fig. 3).
2. Add slution of TFA/H_2O/*i*-Pr_3SiH = 95:2.5:2.5 (1 mL) to the residues to remove the protecting group and then stir for 1 h.
3. Add the reaction mixture dropwise into Et_2O (24 mL) and then cool it in a freezer for 10 min. Centrifuge at 3,500 × *g* for 5 min at 4 °C and then remove the Et_2O layer by decantation. This will afford the crude peptide (**5**, Fig. 3) (*see* **Note 3**).

3.1.3 HPLC Purification

Dissolve the obtained crude peptide into an appropriate volume of DMSO and then centrifuge it at 20,630 × *g* for 10 min. Fractionate the supernatant by reversed phase HPLC (COSMOSIL Cholester 10 × 250 mm) with MeCN:H_2O (=40:60) containing 0.05% TFA at a flow rate of 3.0 mL/min to purify the peptide (**5**), which will be obtained as a white solid (*see* **Note 4**). The column oven is set at 40 °C. From 0.05 mmoL of **1**, we typically obtain approximately 15 mg of **5** (30% for 17 steps).

3.2 Synthesis of Surugamide B (Fig. 4)

3.2.1 Solid-Phase Peptide Synthesis (SPPS)

1. Swell the Fmoc-D-Leu-2-chlorotrityl resin (**1**, 0.0500 mmoL) in a Libra tube with CH_2Cl_2 for 1 h, then subject it to 7 cycles [Fmoc-D-Phe, Fmoc-L-Ile, Fmoc-L-Lys(Boc), Fmoc-D-Val, Fmoc-L-Ile, Fmoc-D-Ala, Fmoc-L-Ile] of the solid-phase peptide synthesis protocol (see **steps 4–7** in Subheading 3.1.1), and then remove Fmoc group (see **steps 4** in Subheading 3.1.1) then wash the resin (see **steps 5** in Subheading 3.1.1). This will afford the resin-bound peptide (**6**, Fig. 4).
2. Add CH_2Cl_2/HFIP (=70:30) (0.5 mL) to the peptide (**6**), and then shake it for 30 min and filter off the reaction mixture. Repeat this procedure twice. Remove the solvent from the filtrate to obtain the crude peptide (**7**, Fig. 4), which is used in the next reaction without further purification.

3.2.2 Macrocyclization, Deprotection, and Precipitation

1. To a solution of the peptide (**7**) in CH_2Cl_2/DMF (=9:1) (25 mL), add 2,4,6-collidine (27 μL, 0.204 mmoL), HOAt (13.5 mg, 0.100 mmoL), and PyBOP (52.3 mg, 0.100 mmoL) and then stir overnight. Reduce the volume of the reaction mixture with an evaporator and then dilute the residue with EtOAc and saturated aqueous NH_4Cl. Extract the resulting mixture with EtOAc (×3), wash it with brine, dry it over $MgSO_4$, and concentrate the solution to afford the crude peptide (**8**, Fig. 4).
2. Add a cleaving cocktail of TFA/H_2O/i-Pr_3SiH = 95:2.5:2.5 (1 mL) to the residue and then stir the mixture for 1 h to remove the protecting group.
3. Add the reaction mixture dropwise into Et_2O (24 mL) and then cool it in a freezer for 10 min. Centrifuge at 3,500 × *g* for 5 min at 4 °C and then remove the Et_2O layer by decantation. This will afford the crude peptide (**9**, Fig. 4) (*see* **Note 3**).

3.2.3 HPLC Purification

Dissolve the obtained crude peptide into an appropriate volume of DMSO and then centrifuge it at 20,630 × *g* for 10 min. Fractionate the supernatant by reversed phase HPLC (COSMOSIL Cholester 10 × 250 mm) with MeCN:H_2O (=34:66) containing 0.05% TFA, at a flow rate of 3.0 mL/min to purify the peptide (**9**), which will be obtained as a white solid (*see* **Note 4**). From 0.05 mmoL of **1**, we typically obtain approximately 15 mg of **9** (34% for 18 steps). The column oven is set at 40 °C.

3.3 Preparation of Recombinant SurE

3.3.1 Cloning of the surE Gene

1. Amplify the DNA fragment encoding SurE by KOD One polymerase, using a single set of primers (SurE_Fw/SurE_Rv). Use genomic DNA from *Streptomyces albidoflavus* NBRC12854 as the template. PCR conditions are as follows: Initial denaturation is performed at 94 °C for 2 min. Next, 5 cycles of denaturation at 98 °C for 10 s and extension at 74 °C for 20 s are performed. Afterwards, perform 5 cycles of

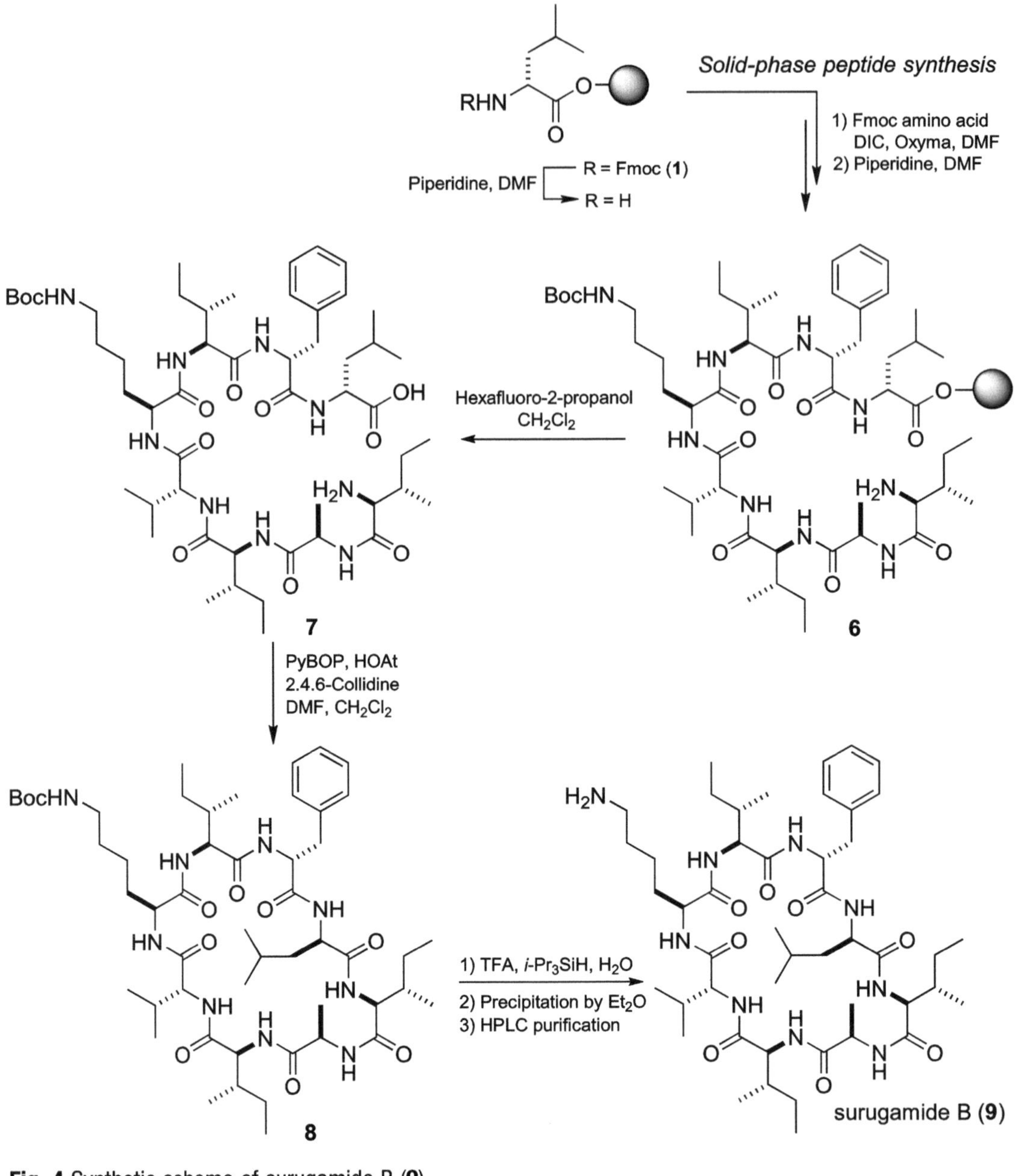

Fig. 4 Synthetic scheme of surugamide B (**9**)

denaturation at 98 °C for 10 s and elongation at 72 °C for 20 s, and then 5 cycles of denaturation at 98 °C for 10 s and elongation at 70 °C for 20 s. Finally, perform 30 cycles of denaturation at 98 °C for 10 s and extension at 68 °C for 20 s, followed by extension at 68 °C for 7 min (*see* **Notes 5** and **6**).

2. Digest the amplified fragment with *Eco*RI and *Hin*dIII (*see* **Note 7**) and insert it into the multicloning site of pUC19.

3. Transfer the fragment into the *NdeI/HindIII* sites in the multicloning site of the pET28a expression vector to generate SurE-pET28a.

3.3.2 Preparation of Recombinant SurE

1. Introduce SurE-pET28a into *E. coli* BL21(DE3) (*see* **Note 8**).
2. Inoculate a single colony of the transformant into 5 mL 2 × YT medium and then culture it overnight at 37 °C. Transfer 2 mL of the culture to 200 mL of fresh 2 × YT medium and incubate it at 37 °C for 3 h and then cool the culture on ice for 5 min. Add 200 μL of 0.1 M IPTG for induction and then continue the culture at 16 °C for 16 h (*see* **Note 9**).
3. Harvest the cells by centrifugation at 4 °C, 3,500 × *g*, for 10 min and wash them with buffer A (**Note 10**).
4. Homogenize the washed cells by using an ultrasonic homogenizer (*see* **Note 11**) and remove the debris by centrifugation at 4 °C, 20,630 × *g* for 20 min.
5. Apply the supernatant to a 1 mL Ni-NTA His Bind® Resin column (*see* **Notes 12** and **13**). Wash the column with buffer B and elute the bound protein with buffer C (*see* **Notes 14** and **15**). Subject the elution fraction to a conventional SDS-PAGE analysis to assess the quality of the protein solution (Fig. 5).
6. Subject the fraction containing the recombinant protein to ultrafiltration to remove the imidazole. Briefly, dilute the 1 mL fraction containing recombinant SurE with 4 mL of

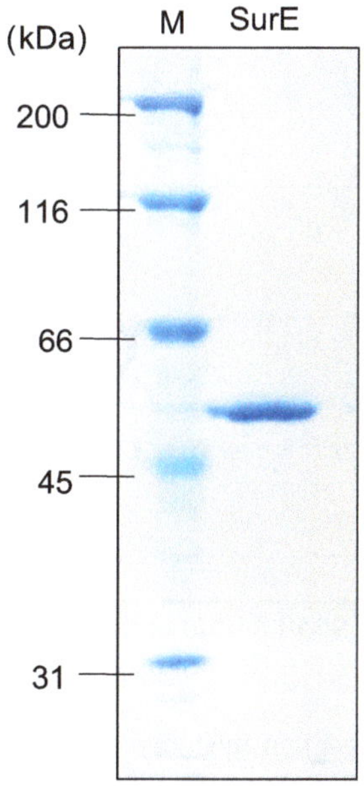

Fig. 5 SDS-PAGE of recombinant SurE. The theoretical molecular weight of the recombinant SurE is 49.6 kDa. This figure was regenerated from Fig. S13 in reference 32, with permission from Wiley

chilled buffer A, transfer the solution into MWCO 10 kDa centrifugal ultrafiltration tubes (such as Amicon® Ultra 4 mL centrifugal filter) and then centrifuge until 80% of the solution passes through the filter. Repeat these steps until the imidazole concentration becomes lower than 1 mM (*see* **Note 16**).

7. Measure protein concentrations with a Bio-Rad protein assay kit.

3.4 In Vitro Cyclization Mediated by SurE

3.4.1 In Vitro Reaction of SurE

1. Prepare the reaction mixture containing Tris–HCl (20 mM, pH 8.0) and the SNAC precursor peptide substrate (**5**) (0.5 mM) in a total volume of 100 μL and add recombinant SurE (2.0 μM) to initiate the reaction (*see* **Note 17**).
2. Incubate the reaction mixture at 30 °C for 2 h.
3. Quench the reaction by adding 100 μL of a 0.1% TFA solution.

3.4.2 Analysis of the Reaction Mixture

1. Centrifuge the resultant mixture at 20,630 × *g* for 10 min.
2. Analyze the supernatant by HPLC. The analytical HPLC conditions are as follows: COSMOSIL 5C_{18}-MS-II 4.6 × 250 mm column, using 41% MeCN supplemented with 0.05% TFA as a mobile phase at a flow rate of 0.8 mL/min. Monitor the column eluates by UV absorption at 215 nm. The substrate (**5**) and cyclic peptide (**9**) are typically eluted around the retention time of 7.5 min and 18.5 min, respectively, with this analytical condition (Fig. 6).

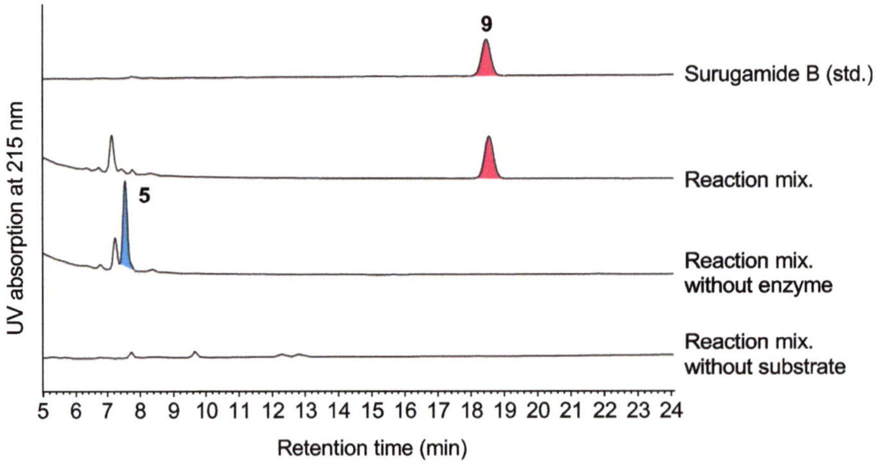

Fig. 6 HPLC analysis of reaction mixtures. The peaks corresponding to *seco*-surugamide B-SNAC (**5**) and surugamide B (**9**) are colored cyan and magenta, respectively. This figure was regenerated from Fig. 3b in reference 32, with permission from Wiley

4 Notes

1. The resin should be completely dried; otherwise, the accurate loading rate cannot be calculated.
2. If the deprotection of the Fmoc protecting group is incomplete, add 1,8-diazabicyclo[5.4.0]undec-7-ene and conduct the reaction for another 10 min [44].
3. The Et_2O should be placed in a 50 mL centrifugal tube and then cooled below − 20 °C prior to the addition of crude peptides. After the crude peptides are added into the centrifugal tube in a dropwise manner, the mixture should be cooled below − 20 °C for at least 10 min.
4. Avoid the injection of an excess amount of the DMSO solution, as this would cause severe tailing of peptide peaks and eventually make it extremely difficult to obtain peptides with good purity. We usually keep the volume of the injected sample below 50 μL for these separation conditions.
5. These primers are also responsible for adding restriction enzyme sites, which are underlined (*see* Subheading 2.3.1).
6. The addition of 1 μL of DMSO tends to improve the amplification of DNA fragments from high GC templates, such as the gDNA of *Streptomyces*.
7. Overnight incubation is recommended for complete digestion.
8. After mixing the solution of plasmid and cells, incubate it on ice for at least 20 min and then heat it at 42 °C for 30 s. Incubate the solution again on ice for 3 min, add 100 μL of 2 × YT medium, and then incubate it at 37 °C for at least 20 min.
9. It is convenient to prepare 5 M NaCl, 1 M Tris–HCl (pH 8.0), and 1 M imidazole (pH 8.0) solutions in advance. NaCl and Tris–HCl can be kept at room temperature, whereas imidazole should be stored at 4 °C.
10. Subsequent operations should always be performed at 4 °C.
11. After 30 s of sonication, the homogenizer and suspension must be cooled for 30 s. A total of 2 min of sonication should be sufficient for complete homogenization.
12. The column should be equilibrated with 10 mL of buffer A prior to the sample injection.
13. Apply the supernatant as gently as possible to keep from disturbing the surface of the resin.
14. By adding 1 mL at a time, for a total of 2 mL of buffer B, the proteins on the column wall can be washed away.

15. Add 1 mL of buffer C at a time, for a total of 5 mL. If the purification is successful, then the recombinant SurE will be eluted in the second fraction.
16. The imidazole removal by ultrafiltration is optional but recommended because we frequently observe a reduction of the SurE activity in the presence of imidazole.
17. Use DMSO solutions of the substrate peptides. If they are dissolved in other organic solvents, like methanol, the enzymatic activity will be severely reduced.

Acknowledgement

This work was partly supported by Hokkaido University, Global Facility Center (GFC), Pharma Science Open Unit (PSOU), funded by Grants-in-Aid from the Ministry of Education, Culture, Sports, Science and Technology (MEXT) under "Support Program for Implementation of New Equipment Sharing System", Global Station for Biosurfaces and Drug Discovery, a project of Global Institution for Collaborative Research and Education in Hokkaido University, the Asahi Glass Foundation, the Naito Foundation, the Uehara Memorial Foundation, the Sumitomo Foundation—Grant for Basic Science Research Projects, Daiichi Sankyo Foundation of Life Science, Japan Foundation for Applied Enzymology, TERUMO Life Science Foundation, the Japan Agency for Medical Research and Development JP19ae0101045, the Japan Science and Technology Agency (JST Grant Numbers ACT-X JPMJAX201F and A-STEP JPMJTR20US), Japan JP16H06448, JP18H02581, JP19K16390, JP21H02635, JP22K15302 and JP22J13818.

References

1. Zorzi A, Deyle K, Heinis C (2017) Cyclic peptide therapeutics: past, present and future. Curr Opin Chem Biol 38:24–29
2. White CJ, Yudin AK (2011) Contemporary strategies for peptide macrocyclization. Nat Chem 3:509–524
3. Jia X, Kwon S, Anderson Wang CI et al (2014) Semienzymatic cyclization of disulfide-rich peptides using Sortase A. J Biol Chem 289: 6627–6638
4. Stanger K, Maurer T, Kaluarachchi H et al (2014) Backbone cyclization of a recombinant cystine-knot peptide by engineered Sortase A. FEBS Lett 588:4487–4496
5. Thongyoo P, Roqué-Rosell N, Leatherbarrow RJ et al (2008) Chemical and biomimetic total syntheses of natural and engineered MCoTI cyclotides. Org Biomol Chem 6:1462–1470
6. Nuijens T, Toplak A, van de Meulenreek MBAC et al (2016) Chemo-enzymatic peptide synthesis (CEPS) using omniligases and selective peptiligases efficient biocatalysts for assembling linear and cyclic peptides and protein conjugates. Chem Today 34:16–19
7. Nuijens T, Toplak A, Quaedflieg PJML et al (2016) Engineering a diverse ligase toolbox for peptide segment condensation. Adv Synth Catal 358:4041–4048
8. Weeks AM, Wells JA (2020) Subtiligase-catalyzed peptide ligation. Chem Rev 120: 3127–3160
9. Nguyen GKT, Wang S, Qiu Y et al (2014) Butelase 1 is an Asx-specific ligase enabling

peptide macrocyclization and synthesis. Nat Chem Biol 10:732–738
10. Yang R, Wong YH, Nguyen GKT et al (2017) Engineering a catalytically efficient recombinant protein ligase. J Am Chem Soc 139: 5351–5358
11. Houssen WE (2019) Peptide cyclization catalyzed by cyanobactin macrocyclases. In: Nuijens T, Schmidt M (eds) Enzyme-mediated ligation methods. Springer, New York, pp 193–210
12. Sarkar S, Gu W, Schmidt EW (2020) Expanding the chemical space of synthetic cyclic peptides using a promiscuous macrocyclase from prenylagaramide biosynthesis. ACS Catal 10: 7146–7153
13. Trauger JW, Kohli RM, Mootz HD et al (2000) Peptide cyclization catalysed by the thioesterase domain of tyrocidine synthetase. Nature 407:215–218
14. Gao X et al (2012) Cyclization of fungal nonribosomal peptides by a terminal condensation-like domain. Nat Chem Biol 8:823–830
15. Sims JW, Schmidt EW (2008) Thioesterase-like role for fungal PKS-NRPS hybrid reductive domains. J Am Chem Soc 130:11149–11155
16. Becker JE, Moore RE, Moore BS (2004) Cloning, sequencing, and biochemical characterization of the nostocyclopeptide biosynthetic gene cluster: molecular basis for imine macrocyclization. Gene 325:35–42
17. Kopp F, Mahlert C, Grünewald J et al (2006) Peptide macrocyclization: the reductase of the nostocyclopeptide synthetase triggers the self-assembly of a macrocyclic imine. J Am Chem Soc 128:16478–16479
18. Trauger JW, Kohli RM, Walsh CT (2001) Cyclization of backbone-substituted peptides catalyzed by the thioesterase domain from the tyrocidine nonribosomal peptide synthetase. Biochemistry 40:7092–7098
19. Kohli RM, Walsh CT, Burkart MD (2002) Biomimetic synthesis and optimization of cyclic peptide antibiotics. Nature 418:658–661
20. Hoyer KM, Mahlert C, Marahiel MA (2007) The iterative gramicidin S thioesterase catalyzes peptide ligation and cyclization. Chem Biol 14: 13–22
21. Kohli RM, Trauger JW, Schwarzer D et al (2001) Generality of peptide cyclization catalyzed by isolated thioesterase domains of nonribosomal peptide synthetases. Biochemistry 40:7099–7108
22. Sieber SA, Walsh CT, Marahiel MA (2003) Loading peptidyl-coenzyme A onto peptidyl carrier proteins: a novel approach in characterizing macrocyclization by thioesterase domains. J Am Chem Soc 125:10862–10866
23. Frueh DP, Arthanari H, Koglin A et al (2008) Dynamic thiolation-thioesterase structure of a non-ribosomal peptide synthetase. Nature 454:903–906
24. Koketsu K, Oguri H, Watanabe K et al (2008) Enzymatic macrolactonization in the presence of DNA leading to triostin A analogs. Chem Biol 15:818–828
25. Hou J, Robbel L, Marahiel MA (2011) Identification and characterization of the lysobactin biosynthetic gene cluster reveals mechanistic insights into an unusual termination module architecture. Chem Biol 18:655–664
26. Galea CA, Roberts KD, Zhu Y et al (2017) Functional characterization of the unique terminal thioesterase domain from polymyxin synthetase. Biochemistry 56:657–668
27. Mandalapu D, Ji X, Chen J et al (2018) Thioesterase-mediated synthesis of teixobactin analogues: mechanism and substrate specificity. J Org Chem 83:7271–7275
28. Qiao L, Fang J, Zhu P et al (2019) A novel chemoenzymatic approach to produce cilengitide using the thioesterase domain from *Microcystis aeruginosa* microcystin synthetase C. Protein J 38:658–666
29. Schmidt JJ, Khatri Y, Brody SI et al (2020) A versatile chemoenzymatic synthesis for the discovery of potent cryptophycin analogs. ACS Chem Biol 15:524–532
30. Sieber SA, Marahiel MA (2003) Learning from nature's drug factories: nonribosomal synthesis of macrocyclic peptides. J Bacteriol 185:7036–7043
31. Matsuda K, Kuranaga T, Wakimoto T (2019) A new cyclase family catalyzing head-to-tail macrolactamization of non-ribosomal peptides. J Syn Org Chem Jpn 77:1106–1115
32. Kuranaga T, Matsuda K, Sano A et al (2018) Total synthesis of the nonribosomal peptide surugamide B and identification of a new off-loading cyclase family. Angew Chem Int Ed Engl 57:9447–9451
33. Matsuda K, Kobayashi M, Kuranaga T et al (2019) SurE is a *trans*-acting thioesterase cyclizing two distinct non-ribosomal peptides. Org Biomol Chem 17:1058–1061
34. Zhou Y, Lin X, Xu C et al (2019) Investigation of penicillin binding protein (PBP)-like peptide cyclase and hydrolase in surugamide non-ribosomal peptide biosynthesis. Cell Chem Biol 26:737–744
35. Thankachan D, Fazal A, Francis D et al (2019) A *trans*-acting cyclase offloading strategy for

nonribosomal peptide synthetases. ACS Chem Biol 14:845–849

36. Matsuda K, Fujita K, Wakimoto T (2021) PenA, a penicillin-binding protein-type thioesterase specialized for small peptide cyclization. J Ind Microbiol Biotechnol 48:kuab023
37. Li Q, Song Y, Qin X et al (2015) Identification of the biosynthetic gene cluster for the anti-infective desotamides and production of a new analogue in a heterologous host. J Nat Prod 78:944–948
38. Son S, Hong YS, Jang M et al (2017) Genomics-driven discovery of chlorinated cyclic hexapeptides ulleungmycins A and B from a *Streptomyces* Species. J Nat Prod 80: 3025–3031
39. Mudalungu CM, von Törne WJ, Voigt K et al (2019) Noursamycins, chlorinated cyclohexapeptides identified from molecular networking of *Streptomyces noursei* NTR-SR4. J Nat Prod 82:1478–1486
40. Liu C, Hashimoto J, Kudo K et al (2021) An atypical arginine dihydrolase involved in the biosynthesis of cyclic hexapeptide longicatenamides. Chem Asian J 16:1382–1387
41. Magarvey NA, Haltli B, He M et al (2006) Biosynthetic pathway for mannopeptimycins, lipoglycopeptide antibiotics active against drug-resistant gram-positive pathogens. Antimicrob Agents Chemother 50:2167–2177
42. Ninomiya A, Katsuyama Y, Kuranaga T et al (2016) Biosynthetic gene cluster for Surugamide A encompasses an unrelated decapeptide, Surugamide F. Chembiochem 17:1709–1712
43. Matsuda K, Zhai R, Mori T et al (2020) Heterochiral coupling in non-ribosomal peptide macrolactamization. Nat Catal 3:507–515
44. Ralhan K, KrishnaKumar VG, Gupta S (2015) Piperazine and DBU: a safer alternative for rapid and efficient Fmoc deprotection in solid phase peptide synthesis. RSC Adv 5:104417–104425

Chapter 7

Directed Evolution of the BpsA Carrier Protein Domain for Recognition by Non-cognate 4′-Phosphopantetheinyl Transferases to Enable Inhibitor Screening

Alistair S. Brown, Jeremy G. Owen, and David F. Ackerley

Abstract

4′-Phosphopantetheinyl transferases (PPTases) play an essential role in activating the carrier protein domains of mega-synthases involved in primary and secondary metabolism and have been validated as promising drug targets in multiple pathogens. Monitoring phosphopantetheinylation of the non-ribosomal peptidase synthetase BpsA, which produces blue indigoidine pigment upon activation, is a useful strategy to screen chemical collections for inhibitors of a target PPTase. However, PPTases can exhibit carrier protein specificity and some medically important PPTases do not activate BpsA. Here, we describe how to conduct a directed evolution campaign to evolve the BpsA carrier protein domain for improved recognition by a candidate PPTase, as exemplified for the human Sfp-like PPTase. This method can be applied to other non-cognate PPTases for discovery of new drug candidates or chemical probes, or to enable development of next-generation biosensors that utilize BpsA as a reporter.

Key words Directed evolution, Error-prone PCR, Indigoidine, PPTase, BpsA, Peptidyl carrier-protein domain, Human PPTase, Targeted screening

1 Introduction

4′-Phosphopantetheinyl transferases (PPTases) catalyze the attachment of a 4′-phosphopantetheinyl arm, derived from coenzyme A, to the carrier protein (CP) domains of large modular biosynthetic enzymes [1]. The attachment converts these enzymes from an inactive *apo* form to an active *holo* form and is essential for function. PPTases have been categorized into three groups [2]. Type I PPTases are small homo-trimers that typically possess a narrow substrate specificity and activate the carrier protein domains of fatty acid synthetases [3]. Type II PPTases are pseudo-dimers that usually have broader substrate specificities and can activate the carrier protein domains of a range of mega synthetases including fatty acid synthases, non-ribosomal peptide synthetases, and

Michael Burkart and Fumihiro Ishikawa (eds.), *Non-Ribosomal Peptide Biosynthesis and Engineering: Methods and Protocols*, Methods in Molecular Biology, vol. 2670, https://doi.org/10.1007/978-1-0716-3214-7_7,

polyketide synthetases [1]. Type III PPTases are less common and are fused to the C-termini of some fatty acid synthetases [4].

PPTases have attracted substantial interest as drug targets in a range of pathogenic microbes including *Mycobacterium tuberculosis* [5], *Pseudomonas aeruginosa* [6], and *Aspergillus fumigatus* [7]. We recently described an in vitro screen for PPTase inhibitors that utilize the non-ribosomal peptide synthetase BpsA as a colourimetric reporter [8]. BpsA synthesizes indigoidine, a blue pigment, from two molecules of L-glutamine. Before this can occur, *apo*-BpsA must be converted to *holo*-BpsA by a co-incubated PPTase; thus, the rate of indigoidine synthesis can be used to report on relative levels of PPTase inhibition in replicate reactions amended with candidate inhibitors [9]. Our primary goal in developing this screen was the discovery of novel PPTase inhibitors as potential antibiotic candidates, but we envisaged that it could also be adapted to counter-screen for undesirable inhibition of the human PPTase (hPPTase), as a likely source of toxicity. However, despite being a Type II PPTase, hPPTase exhibits a narrow substrate specificity (likely reflecting its primary role of activating the acyl carrier protein domains of fatty acid synthetases [10]) and cannot activate the native BpsA peptidyl carrier protein (PCP) domain [11]. We found that we could solve this problem by iteratively evolving the BpsA PCP domain to become an efficient substrate for hPPTase. Here, we describe how this evolution strategy might be applied more generically to evolve the BpsA PCP domain for activation by any target PPTase.

CP domains consist of a four helical bundle with an invariant serine (the site of 4′-phosphopantetheine attachment) located in the second alpha-helix [12]. The CP domain shuttles the substrate tethered to the 4′-phosphopantetheine between the different catalytic domains of the megasynthases. Recognition of an *apo* CP domain substrate by a PPTase is usually mediated by single residues or small patches of residues located on the CP domain [13]. A small number of substitutions can exert a profound effect on PPTase-CP interactions; coupled with their small size (approximately 80 residues), this makes CP domains excellent targets for improvement via directed evolution. As a model system to study PCP domain interactions, we previously developed a plasmid-borne *bpsA* variant (*slbpsA*) that has silent restriction sites flanking the PCP domain, to enable seamless insertion of alternative CP domains [14]. By amplifying the wild-type *bpsA* PCP domain in error-prone PCRs (epPCRs), we showed we could introduce mutations in an essentially random fashion, then ligate the amplicons into *slbpsA* to create a library of *bpsA* variants. This library can then be used to transform a strain of *E. coli* in which the endogenous Type II PPTase *entD* has been deleted, and the target PPTase is expressed from a compatible plasmid. It then becomes possible to screen for BpsA variants that are improved substrates for the target PPTase by

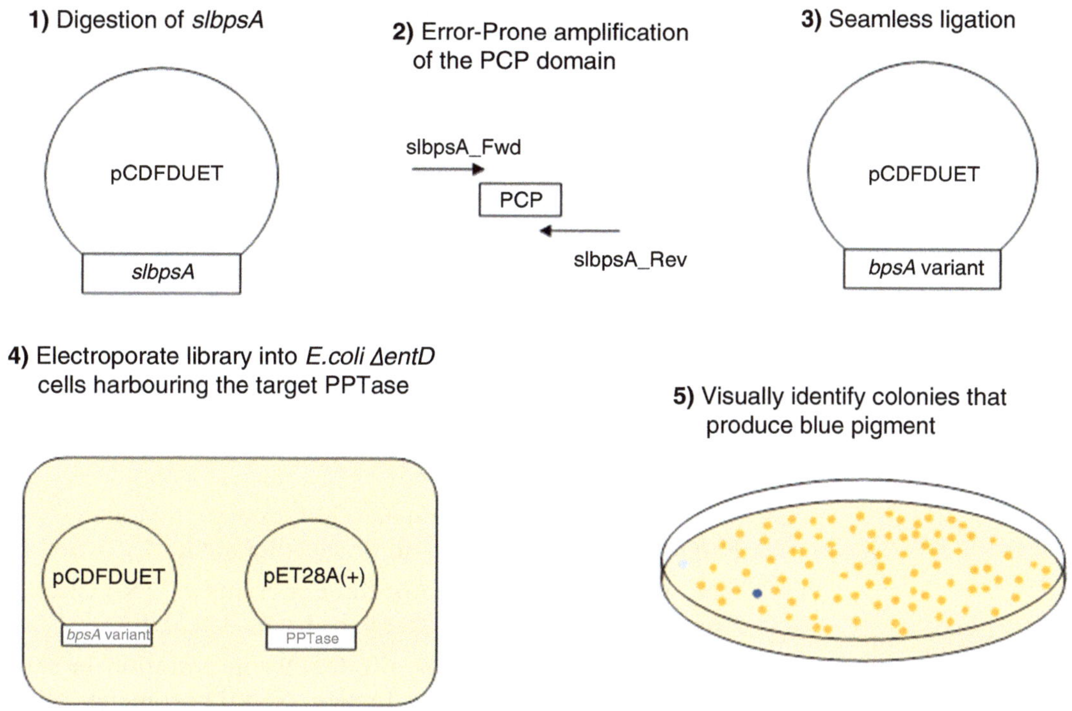

Fig. 1 Schematic diagram summarizing the library construction and screening process

detecting the production of indigoidine *in vivo* (Fig. 1). Mutations that confer improvement (i.e., that increase indigoidine production) can be rationally combined, or alternatively, additional rounds of directed evolution can be performed. While our directed evolution campaign focused on evolution of *bpsA* for recognition by the human PPTase, our methods can be broadly applied to other PPTases to enable screening for inhibitors of that PPTase, or alternatively, to modify BpsA for use as an *in vivo* reporter protein in a species where the endogenous PPTase does not recognize BpsA.

2 Materials

2.1 Library Preparation

1. pCDFDUET::*slbpsA* (a seamless swapping vector in which to stage the epPCR-amplified PCP domain), a compatible plasmid expressing the target PPTase (e.g., pET28A(+)::*hPPTase*), and a source of *bpsA* PCP domain for amplification (e.g., pCDFDUET::*bpsA*) (*see* **Note 1**).
2. 10 μM working stocks of oligonucleotide primers SL_Fwd and SL_Rev to amplify the *bpsA* PCP domain and sequence variant *bpsA* PCP domains (*see* **Note 2**).
3. High-fidelity DNA polymerase. For example, Q5 2× master mix (New England Biolabs; Ipswich, MA, USA).

4. Thermocycler with a heated lid. For example, Bio-Rad T100 Thermal Cycler (Bio-Rad; Hercules, CA, USA).
5. DNA loading dye. For example, 5× DNA loading buffer (Bioline, London, UK).
6. DNA ladder. For example, Hyperladder 1 kb (Bioline, London, UK).
7. Agarose gel (1% w/v); this can be stained with either SYBR safe (Invitrogen; Carlsbad, CA, USA) or ethidium bromide for standard DNA visualization, but SYBR safe should be used for gel extractions. A gel electrophoresis system, for example, Owl EasyCast B2 Mini Gel Electrophoresis Systems (ThermoFisher Scientific; Waltham, MA, USA) and a Gel UV visualization system. For example, UVITEC Gel Document System (UVITEC, Cambridge, UK).
8. Nucleic acid gel extraction kit. For example, Monarch Gel Extraction Kit (New England Biolabs; Ipswich, MA, USA).
9. DNA Clean and Concentrator Kit. For example, DNA Clean & Concentrator Kit (Zymo Research; Irvine, CA, USA).
10. Spectrophotometer to quantify DNA concentration in small volumes. For example, an Implen NanoPhotometer NP80 (Implen GmBH; Munich, Germany). Alternative methods to quantify DNA in small volumes such as the PicoGreen dsDNA Quantitation Reagent or a Qubit Fluorometer (ThermoFisher Scientific; Waltham, MA, USA) may also be suitable.
11. System for epPCR amplification of the *bpsA* PCP domain. We describe here use of the GeneMorph II Random Mutagenesis Kit (Agilent; Santa Clara, CA, USA), but alternative methods such as substituting Mg^{2+} by Mn^{2+} in *Taq* amplification reactions [15] may also be used.
12. Compatible restriction enzymes to cut *slbpsA* (*Nsi*I and *Spe*I) and the corresponding ends of PCP domain amplicons (*Pst*I and *Xba*I). We use NEB restriction enzymes (New England Biolabs; Ipswich, MA, USA), but alternative suppliers may be used.
13. Molecular biology grade water purchased from a supplier such as Sigma Aldrich (St. Louis, MI, USA) or produced in the laboratory with a Milli-Q water purification system.
14. Plasmid Miniprep Kit. For example, Presto Mini Plasmid Kit (Geneaid; New Taipei City; Taiwan).
15. T4 DNA ligase and ligation buffer. For example, Invitrogen T4 DNA Ligase kit (Thermo Fisher Scientific; Waltham, MA, USA).

2.2 Library Screening

1. Electrocompetent *E. coli* BL21 Δ*entD* (*see* **Note 3**) expressing the target PPTase gene (e.g., our study used pET28A(+)::*hPPTase*) and as a control, electrocompetent *E. coli* BL21 Δ*entD* expressing a broad-specificity PPTase (e.g., our study used pET28A(+)::*pcpS*, but any broad-specificity PPTase gene may be substituted).
2. Chilled electroporation cuvettes (0.2 cm). To perform the electroporation, we use a Gene Pulser Xcell Electroporation System (Bio-Rad; Hercules, CA, USA), using the pre-set *E. coli* Ec2 setting (2.5 kV, 200 Ω, and 25 μF, single pulse). Any electroporator capable of delivering similar conditions may be used.
3. Media and additives:
   (a) Lysogeny broth.
   (b) Agar.
   (c) Appropriate antibiotics for plasmid selection. For example, we use kanamycin (50 μg.mL^{-1} final concentration) for pET28A(+) plasmid selection and spectinomycin (100 μg.mL^{-1} final concentration) for *slbpsA* selection.
   (d) L-glutamine.
   (e) Glucose.
   (f) IPTG.

   We use Duchefa Biochemie (Haarlem, Netherlands) for lysogeny broth, micro-agar, kanamycin, spectinomycin and IPTG, and Sigma Aldrich (St. Louis, MI, USA) for L-glutamine and glucose, but any equivalent supplier may be used. Solutions of glucose (20% w/v), glutamine (100 mM), antibiotics (kanaymcyin 50 mg.mL^{-1} and spectinomycin 100 mg.mL^{-1}), and IPTG (100 mg.mL^{-1}) working stocks are filter-sterilized prior to use by passing them through a 0.22 μM syringe filter.
4. Sterile 80% (v/v) glycerol.
5. 90 mM Petri dishes (*see* **Note 4**).
6. Pigment-production-agar (1.5% agar (w/v), 100 mM L-glutamine, kanamycin 50 μg.mL^{-1}, and spectinomycin 100 μg.mL^{-1}).
7. Clear flat-bottom 96-well plates compatible with a standard plate reader. For example, 96 Well Tissue Culture Plate (Jet-Biofil; Guangzhou, China).
8. Plate reader capable of reading optical density at 590 nM and 800 nM. For example, PerkinElmer Enspire plate-reader (Waltham, MA, USA).

2.3 Protein Expression and Purification

1. Flasks that can provide adequate aeration to 400 mL of culture. For example, 2.5 L Ultra Yield Flasks (Thompsons; Oceanside, CA, USA).
2. BpsA purification buffers (buffers described here are for nickel-affinity purification of His_6-tagged proteins, but other protein purification systems can be used):
   (a) Bind buffer (50 mM Tris-Cl pH 8.0, 10 mM imidazole, 300 mM NaCl, 12.5% glycerol (v/v)).
   (b) Elution buffer (50 mM Tris-Cl pH 8.0, 400 mM imidazole, 300 mM NaCl, 12.5% glycerol (v/v)).
   (c) Desalting buffer (50 mM sodium phosphate, 12.5% glycerol (v/v) pH 8.0) (*see* **Note 5**).
3. PPTase purification buffers:
   (a) Bind buffer (50 mM Tris-Cl pH 8.0, 10 mM imidazole, 300 mM NaCl, 25% glycerol (v/v)) (*see* **Note 6**).
   (b) Wash buffer (50 mM Tris-Cl pH 8.0, 50 mM imidazole, 300 mM NaCl, 25% glycerol (v/v)).
   (c) Elute buffer (50 mM Tris-Cl pH 8.0, 400 mM imidazole, 300 mM NaCl, 25% glycerol (v/v)).
   (d) Desalting buffer (50 mM Tris-Cl, 25% glycerol (v/v) pH 8.0).
4. Sterile 80% glycerol (v/v).
5. French press (ThermoFisher Scientific; Waltham, MA, USA) (*see* **Note 7**).
6. Chilled centrifuge with rotors and adapters capable of holding 50 mL Falcon tubes and 15 mL Falcon tubes, and centrifuge bottles that can be used at 38,000 *g*. For example, Sorvall LYNX 6000 (ThermoFisher Scientific; Waltham, MA, USA).
7. Ni-NTA resin and purification system. For example, HisPur (ThermoFisher Scientific; Waltham, MA, USA) (*see* **Note 8**).
8. HiTrap desalting column (Cytiva; Marlorough, MA, USA). Alternative methods for buffer exchange may be used.
9. Protein concentration column. For example, Amicon® Ultra-15 Centrifugal Filter Unit 100 kDa cut off column.
10. A method to determine protein concentration. For example, A_{280} readings using a NanoPhotometer NP80 (Implen GmBH; Munich, Germany) ("Nanodrop"). Alternative methods such as the BCA assay may be used.

2.4 Kinetic Measurement of PPTase Activity

1. Stock reagents.
   (a) Sterile H_2O.
   (b) 1 M Tris-Cl pH 7.8.

(c) 1 M $MgCl_2$.

(d) 100 mM ATP.

(e) 100 mM L-glutamine.

(f) Purified BpsA.

(g) Purified target PPTase (for us this was hPPTase; subsequent references to hPPTase can be taken to refer generically to your target PPTase).

2. Clear flat-bottom 96-well plates compatible with a standard plate reader. For example, 96 Well Tissue Culture Plate (Jet-Biofil; Guangzhou, China).
3. Plate reader capable of reading a plate every 30 s at 590 nM. For example, PerkinElmer Enspire plate-reader (Waltham, MA, USA).
4. Plate shaker capable of 1000 rpm. For example, Eppendorf Mixmate (Hamburg, Germany).
5. Microsoft Excel (Redmond, WA, USA) or equivalent software capable of calculating slope values to determine kinetic rates.

3 Methods

3.1 PCP Domain Template Preparation and Error-Prone PCR

1. Amplify the wild type *bpsA* PCP domain from template DNA using the SL_Fwd and SL_Rev primers in a 25 μL high-fidelity PCR according to the manufacturer's protocol.
2. Confirm that appropriate amplification has occurred by electrophoresing 1–2 μL of the PCR alongside DNA loading dye on a 1% agarose gel (w/v) stained with ethidium bromide or SybrSafe and then visualize using a GelDoc or equivalent gel imaging system.
3. Remove any template DNA by electrophoresing the remainder of the PCR on a 1% agarose gel (w/v) stained with SYBR-safe. Extract and purify the PCR band using a nucleic acid gel extraction kit according to the manufacturer's instructions.
4. Quantify the purified PCR product using a Nanodrop spectrophotometer or equivalent method.
5. Set up the epPCR in a thin-walled, 0.2 mL PCR tube containing the following: 5 μL 10× Mutazyme II reaction buffer, 1 μL dNTP mix (for a final concentration of 200 μM of each dNTP), 2 μL SL_Fwd primer and 2 μL SL_Rev primer, 1 μL Mutazyme II DNA polymerase (2.5 units). Complete the reaction by adding the DNA template (Subheading 3.1, **step 3**) and molecular biology grade water to a final volume of 50 μL. The number of PCR cycles and the amount of template DNA used in the epPCR affects the mutation rate (*see* **Note 9**); a

mutation rate of approximately 2–4 substitutions across the PCP domain should be sufficient to confer detectable improvements without incorporating substantially deleterious mutations into a majority of amplicons. Mix the sample and then split the PCR mix into four separate tubes (12.5 μL each) (*see* **Note 10**). A negative control containing no template DNA should also be prepared to check for contamination.

6. Run the PCRs in a thermocycler using an appropriate PCR protocol. We use the following 2-step PCR protocol:
   (a) An initial denaturation step of 1 min at 95 °C.
   (b) 35 cycles of 20 s at 95 °C then 1.5 min at 72 °C
   (c) A final extension step of 10 min at 72 °C.
7. Pool the four completed PCRs and mix. Remove 2 μL of PCR product, mix with DNA loading dye, and load 5 μL onto a 1% agarose gel (w/v) alongside a DNA ladder to confirm the appropriate amplification of the PCP domain.
8. Purify the remaining epPCR product using the DNA Clean and Concentrate kit and elute in 25 μL of molecular biology grade water, pre-heated to 55 °C.
9. Quantify the purified epPCR product using a Nanodrop spectrophotometer.
10. Digest the epPCR product using the restriction enzymes *Pst*I and *Xba*I (30 U apiece), 1× CutSmart buffer, and sterile molecular biology grade water in a final volume of 60 μl for 5 h. Heat-inactivate the restriction enzymes by incubation at 80 °C for 20 min.
11. Purify the digested epPCR using the DNA Clean and Concentrate kit and elute in 25 μL of molecular biology grade water, pre-heated to 55 °C.
12. Quantify the purified, digested epPCR product using a Nanodrop spectrophotometer. At this point, the reaction can be stored at 4 °C or at −20 °C for longer-term storage.

3.2 Vector Preparation and Quality Assessment

1. Miniprep pCDFDUET::*slbpsA* using a miniprep kit. To obtain sufficient plasmid, we recommend use of a 12 mL overnight culture split into four separate minipreps. Elute each fraction in 30 μL of molecular biology grade water pre-heated to 55 °C and pool the eluted fractions.
2. Quantify the concentration of pCDFDUET::*slbpsA* using a Nanodrop spectrophotometer.
3. To ensure sufficient plasmid for library generation, we recommend digestion of approximately 16 μg of pCDFDUET::*slbpsA*. Incubate pCDFDUET::*slbpsA* at 70 °C for 20 min, to help relieve supercoiling. Then set up a restriction digest

containing 50 U of *Spe*I and *Nsi*I, CutSmart buffer and sterile molecular biology grade water in a final volume of 400 μL. Incubate the digest for 6 h at 37 °C. Then add a further 20 U of *Spe*I and *Nsi*I to the digest and incubate for an additional 16 h at 37 °C. Heat-inactivate the restriction enzymes by incubating at 80 °C for 20 min.

4. Split the restriction digest into three separate tubes and purify using a DNA Clean and Concentrate kit, then elute each sample in 20 μL of molecular biology grade water, pre-heated to 55 °C. Pool the three eluted fractions and quantify the DNA concentration using a Nanodrop spectrophotometer.
5. Assess the digested vector quality by electrophoresing a 200 ng sample mixed with loading dye and water on a 1% agarose gel (w/v) stained with ethidium bromide. Aliquot the remainder of the restriction digest into 5 μL samples and store them at − 20 °C until required.
6. We initially conduct a small test ligation to determine the quality of the digested vector prepared in Subheading 3.2, **step 3** and the purified PCR product from Subheading 3.1, **step 1**. Set up the ligation reaction with 50 ng of total DNA (using a ratio of 1:6 between vector and insert), 1× T4 ligase buffer, 1 U of ligase, and molecular biology grade water to a final volume of 50 μL. Incubate the ligation at 16 °C for 16 h. Add an additional 1 U of ligase and incubate for a further 6 h.
7. Transform the test ligation into electrocompetent *E. coli* BL21 Δ *entD* cells harboring a plasmid expressing a broad-specificity PPTase, for example pET28A(+)::*pcpS* (*see* **Note 11**). Thaw two 60 μL aliquots of *E. coli* BL21 pET28A(+)::*pcpS* on ice. Dilute 1–2 μL of purified ligation from Subheading 3.2, **step 6** in sterile molecular biology grade water to a final volume of 20 μL. Add the diluted ligation to the first aliquot of cells. Add 20 μL of molecular biology grade water to the second aliquot of cells as a negative control. Transfer the 80 μL aliquots to pre-chilled electroporation cuvettes and then electroporate (standard protocol 2.5 kV, 200 Ω and 25 μF using a GenePulser unit with Pulser Controller).
8. Immediately after electroporation, add 940 μL of sterile LB medium to the cuvette and gently transfer the cell suspension to a sterile 1.5 mL microcentrifuge tube. Let the cells recover at 37 °C, 200 rpm for 1 h and then plate out (*see* **Note 12**) 1 μL, 10 μL and 100 μL aliquots of the cell suspension on pigment-production plates (*see* **Note 13**) and incubate at 37 °C overnight.
9. Induce the expression of PcpS and BpsA by applying 200 μL of 3% IPTG (w/v) underneath the agar plate (Fig. 2a). To remove the agar plate from the Petri dish, gently lift the agar from the

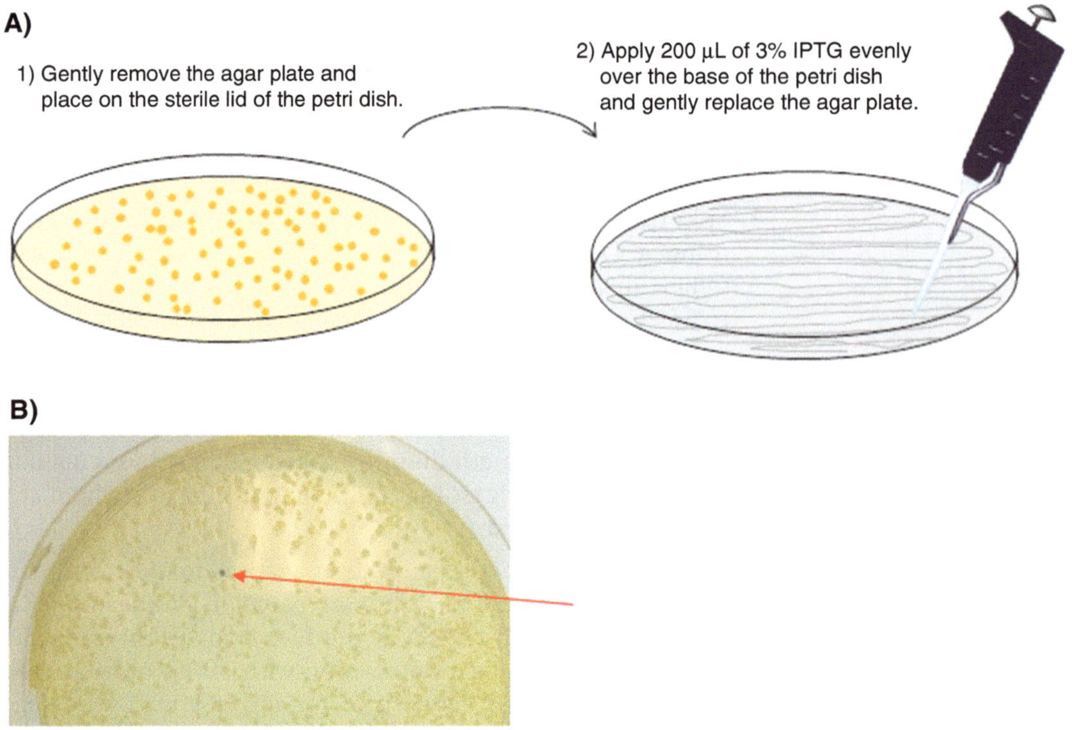

Fig. 2 Induction of pigment-production plates. (**a**) Schematic diagram summarizing the IPTG application process. Try to achieve a continuous film of IPTG across the bottom of the Petri dish, to ensure even distribution and diffusion of IPTG after the agar inset is replaced. (**b**) Photo of an improved variant producing indigoidine (red arrow) on a pigment-production plate. The photo was taken approximately 16 h post-induction

plate using a flat piece of metal or small spatula and place it on top of the inside of the sterile agar plate lid. Using a pipette, gently spread out the IPTG so it evenly covers the bottom of the agar plate and then replace the agar on top of the IPTG to let it diffuse through the agar (*see* **Note 14**). Incubate the plates at room temperature and within a few hours, the colonies containing a correctly ligated PCP-domain should begin to turn dark blue, indicating indigoidine production.

10. To determine the insert ratio, count a section of the plate (approximately 100 colonies) and determine the ratio of blue colonies to normal colonies. We aim for an insert ratio of at least 70% (*see* **Note 15**).

3.3 Library Construction and First-Tier Screening

1. The epPCR library ligation contains a total of 1–1.5 µg of DNA and is split into five separate 50 µL reactions, each containing a total DNA concentration of ~250 ng. Prepare the reactions as described in Subheading 3.2, **step 6**. After incubation, pool the ligations and purify them using a DNA Clean and Concentrate kit. Elute the reactions in 25 µL of molecular biology grade

water, pre-heated to 55 °C. Quantify the DNA concentration using a Nanodrop spectrophotometer. Aliquot the library into 2.5 μL samples and freeze at −20 °C for storage.

2. When screening the library, we are aiming for a plating density of between 2000 and 3000 colonies on each 90 mM Petri dish. This quantity enables us to pick individual colonies. Thaw two 60 μL aliquots of *E. coli* BL21 pET28A(+)::*hPPTase* on ice. Dilute 50 ng of ligation into a total volume of 20 μL sterile molecular biology grade water. Add the diluted ligation to the first aliquot of cells. Add 20 μL of sterile molecular biology grade water to the second aliquot of cells as a negative control.
3. Transfer the aliquots to a pre-chilled electroporation cuvette and electroporate (standard protocol 2.5 kV, 200 Ω, and 25 μF using a GenePulser unit with Pulser Controller). Immediately after electroporation, add 940 μL of sterile LB medium to the cuvette. Gently transfer the cell suspension to a sterile 1.5 mL microcentrifuge tube. Let the cells recover at 37 °C, 200 rpm for 1 h and then plate out 1 μL, 10 μL, and 100 μL aliquots of the cell suspension on pigment-production plates and incubate overnight at 37 °C.
4. To determine the cell density, count the number of colonies on the 10 μL plate and adjust the plating volume for subsequent library platings (*see* **Note 16**).
5. Induce the expression of the target PPTase and BpsA by adding 200 μL of 3% IPTG (w/v) underneath the plate as described in Subheading 3.2, **step 10**. Between 8- and 48-h post-induction, colonies bearing improved variants should begin to develop pigmentation (*see* **Note 17** and Fig. 2b).
6. If a "hit" colony can be accurately picked without contamination, transfer it into one of the inside 60 wells of 96-well plate containing 200 μL of LB supplemented with 50 $\mu g.mL^{-1}$ kanamycin, 100 $\mu g.mL^{-1}$ spectinomycin, and 0.4% glucose (w/v). If the colony cannot be accurately picked, streak it out onto a fresh pigment-production plate to get single colonies. In the 96-well plate establish the following controls: *E. coli* BL21 Δ*entD* harboring pCDFDUET::*bpsA* and pET28A(+)::*hPPTase, E.coli* BL21 Δ*entD* harboring wild-type BpsA and an empty pET28A(+) plasmid and a media only well. Thus, each 96-well plate can stage 57 non-control samples. We fill the outside wells with uninoculated media to minimize evaporation. Incubate the 96-well plate overnight at 37 °C, 200 rpm.
7. For long term storage, store the strains at −80 °C as glycerol stocks. To prepare each glycerol stock, aliquot 80 μL of 80% glycerol (v/v) into the inner wells of a 96-well plate. Then aliquot 80 μL of the cell suspension from Subheading 3.3, **step 6** into the corresponding well, mix thoroughly and freeze at −80 °C.

3.4 Second Tier Liquid Screening

Multiple factors can influence the rate at which pigment production occurs, including plating density, concentration of IPTG the colony is exposed to, and location on the agar plate. To gain a more quantitative measurement of activity, we conduct a medium-throughput liquid screen on initially selected variants, to identify the variants with the most improved activity. These can then be sequenced and purified for kinetic analysis.

1. Aliquot into each well of a 96-well plate 180 μL of LB supplemented with 50 $\mu g.mL^{-1}$ kanamycin, 100 $\mu g.mL^{-1}$ spectinomycin, and 0.4% glucose (v/v). Inoculate the inner wells with a sample of the previously glycerol stocked hits from the first-tier screening (we use a 96-well plate pinning tool, but samples can alternatively be individually toothpicked or transferred by scraping using a multiwell pipette) and incubate at 37 °C, 200 rpm for 16 h.
2. Aliquot into each well of a fresh 96-well plate 130 μL of LB amended with 100 mM L-glutamine, 0.6 mM IPTG, 50 $\mu g.mL^{-1}$ kanamycin, and 100 $\mu g.mL^{-1}$ spectinomycin (*see* **Note 18**). Transfer 20 μL of cell suspension from the plate prepared in Subheading 3.4, **step 1** into the corresponding inner wells of this fresh 96-well plate. Wrap the fresh plate in tinfoil and incubate at 18 °C, 200 rpm for 8–48 h. At appropriate time points based on visual inspection (*see* **Note 19**), the OD_{590} and OD_{800} values should be recorded using a PerkinElmer Enspire plate reader.
3. To quantify the normalized indigoidine value, we use the method described by Beer et al. [16] which was adapted from Myers et al. [17]. We record the optical density of the cell culture at two wavelengths: the indigoidine sensitive wavelength of 590 nM (OD^s) and the robust wavelength of 800 nM (OD^R) which accounts for the scattering contribution of the bacterial cells. The normalized indigoidine value can be calculated as:

$$\text{Indigoidine} = OD_{B,+I} - OD_{B,-I}$$

$OD_{B,+I}$ consists of absorption at OD^s from both bacterial cells and indigoidine, while $OD_{B,-I}$ accounts for scattering solely due to the bacterial cells at OD^s. $OD_{B,-I}$ is calculated as:

$$OD_{b,-I} = \delta.OD^R$$

where δ represents the correction factor and is calculated by measuring the ratio of pure cell culture at OD^s and OD^R (*see* **Note 20**). The normalized value for indigoidine is then calculated as:

$$\text{Normalized indigoidine} = OD^S - \delta.OD^R$$

To account for the contribution of the culture medium to the OD values, subtract the mean of pure culture medium replicates from each of the OD_{590} and OD_{800} measurements.

3.5 Expression and Purification of BpsA Variants

To purify each BpsA variant, we use the plasmid bearing it to transform an empty *E.coli* BL21 Δ*entD* strain, induce protein expression, and then use nickel affinity chromatography to isolate the protein.

1. Aliquot 3 mL of LB supplemented with 50 μg.mL^{-1} kanamycin, 100 μg.mL^{-1} spectinomycin, and 0.4% glucose (v/v) into a 15 mL Falcon tube and transfer the targeted variant from the previously glycerol stocked hits created in Subheading 3.3, **step** 7. Incubate at 37 °C, 200 rpm for 16 h.
2. Split the 3 mL overnight into two 1.5 mL microcentrifuge tubes and miniprep the variant plasmid using a Miniprep Kit then elute in 30 μL of molecular biology grade water pre-heated to 55 °C. Pool the two samples.
3. Quantify the DNA concentration of the miniprep using a Nanodrop spectrophotometer. At this stage, if desired, selected variants can be sent for sequencing to determine the mutations in each gene.
4. Thaw two 40 μL aliquots of electrocompetent *E. coli* BL21 Δ*entD* cells on ice. Dilute 1 ng of purified plasmid in a total volume of 10 μL sterile molecular biology grade water. Add the diluted plasmids to the first 40 μL aliquot of cells. Add 10 μL of sterile molecular biology grade water to the other 40 μL aliquot of *E. coli* BL21 Δ*entD* cells to act as a negative control. Transfer each aliquot to a separate pre-chilled electroporation cuvette and electroporate (standard protocol 2.5 kV, 200 Ω, and 25 μF using a GenePulser unit with Pulser Controller). Immediately add 950 μL of sterile LB medium to each cuvette and gently transfer each cell suspension to a separate sterile 1.5 mL microcentrifuge tube. In each case, let the cells recover at 37 °C, 200 rpm for 1 h and then plate out 100 μL of a 1/1000 and 1/100 dilution of the cell suspension on LB supplemented with spectinomycin 100 μg.mL^{-1} and incubate overnight at 37 °C.
5. Pick three colonies and transfer each colony to a 15 mL Falcon tube containing 3 mL LB supplemented with 50 μg.mL^{-1} kanamycin and a 15 mL Falcon tube containing 3 mL LB supplemented with 100 μg.mL^{-1} spectinomycin. Incubate overnight at 37 °C, 200 rpm. If you see growth in any of the tubes containing kanamycin, this indicates a double transformation event has occurred. Discard such samples.

6. Mix 400 μL of a non-double transformant cell suspension with 400 μL of sterile 80% glycerol (v/v) and freeze at −80 °C for long-term storage.
7. Set up a 10 mL overnight of *E. coli* BL21 Δ*entD* harboring each *bpsA* variant supplemented with 100 μg.mL^{-1} spectinomycin and 0.4% glucose (w/v).
8. Inoculate a 200 mL LB culture supplemented with 100 μg. mL^{-1} spectinomycin with the 10 mL overnight culture and incubate at 37 °C, 200 rpm until an OD_{600} of 0.6–0.8 is reached. Chill the culture on ice for 20 min with occasional swirling to mix. Induce the sample with IPTG to a final concentration of 500 μM. Incubate at 18 °C, 200 rpm for 16 h.
9. Pellet the cells using a centrifuge and discard the supernatant. The cell pellets can be frozen at −80 °C for long-term storage. Freezing the pellets will also improve the cell lysis step.
10. Resuspend the pellets in 15–30 mL of 1× BpsA bind buffer. Lyse the cells by passing them through a French Press three times. Collect the lysate and pellet by centrifuging at 38,000 *g* for 25 min. Transfer the supernatant to a fresh 50 mL Falcon tube.
11. BpsA binds relatively poorly to Ni-NTA agarose beads, so use a relatively large resin volume (approximately 6–7 mL) for purification. Apply the supernatant to the column. Wash the column with 50 mL of 1× BpsA bind buffer. Elute the BpsA with 15 mL of 1× BpsA elution buffer. Collect the elutant in a 50 mL Falcon tube.
12. Transfer the 15 mL of eluted BpsA to a 100 kDa molecular mass cut-off Millipore Amicon ultra-15 centrifuge filter unit. Centrifuge at 2600 *g* until approximately 750 μL of the BpsA solution remains. Resuspend the sample in 15 mL of desalting buffer and centrifuge again to approximately 500 μL. Adjust the final buffer composition to approximately 40% glycerol (v/v) and store aliquots at −20 °C.

3.6 Expression and Purification of PPTases

PPTases are relatively unstable and aggregate quickly. Thus, all reagents must be kept cold and the purification protocol must be completed as promptly as possible.

1. Set up a 10 mL overnight of *E. coli* BL21 Δ*entD* harboring pET28A(+)::*hPPTase* in LB supplemented with kanamycin 50 μg.mL^{-1} and 0.4% glucose (v/v).
2. Transfer 10 mL of the overnight to a 400 mL LB culture supplemented with 50 μg. mL^{-1} kanamycin. Grow until an OD_{600} of between 0.6 and 0.8 is reached. Chill the culture on ice for 20 min with occasional swirling to mix. Induce the sample with IPTG to a final concentration of 500 μM. Incubate at 18 °C, 200 rpm for 16 h.

3. Pellet the cells using a centrifuge and discard the supernatant. The cell pellets can be frozen at −80 °C for long-term storage. Freezing the pellets will also improve the cell lysis.
4. Resuspend the pellets in 15–30 mL of 1× PPTase bind buffer and lyse the cells by passing them through a French Press twice (alternative cell lysis methods may also be trialed here). Collect the lysate and pellet by centrifuging at 38,000 *g* for 25 min. Collect the supernatant and transfer it to a fresh 50 mL Falcon tube.
5. Apply the supernatant to the column (1 mL of resin). Wash the column with 10–15 mL 1× PPTase bind buffer. Wash the column with 10–15 mL 1× PPTase wash buffer. Elute the bound PPTase protein from the column with 2 mL of PPTase elution buffer.
6. Immediately desalt the PPTase using a Cytiva Healthcare HiTrap™ desalting columns with the PPTase desalting buffer. Adjust the glycerol concentration to 40% (v/v) and store at −80 °C in 100 μL aliquots.

3.7 In Vitro Quantification of Indigoidine Synthesis Rate

The method to determine the kinetic parameters of the BpsA variants is based on Owen et al. [9].

1. Measure the concentration of the purified BpsA variant and hPPTase by using a Nanodrop spectrophotometer or preferred quantification method.
2. In a standard 96-well plate, establish in triplicate a reaction mix containing 50 mM Tris-Cl pH 8.0, 10 mM $MgCl_2$, 5 mM L-glutamine, 5 mM ATP, 20 μM CoA, 4 μM *apo*-BpsA, and H_2O to a final volume of 80 μL.
3. To initiate the reaction, rapidly add 1.4 μM of hPPTase and H_2O in a final volume of 20 μL to each well.
4. Mix the 96-well plate in a plate mixer at 1000 rpm for 5 s.
5. Record the A_{590} values every 10–30 s for 1–2 h (*see* **Note 21**).
6. To determine the maximum velocity of target PPTase with each BpsA variant, graph the A_{590} values on the Y-axis and time in seconds on the X-axis. The Microsoft Excel slope function is then used to find the gradient every 3–12 points depending on the reaction rate (use fewer points if the reaction is fast). The resulting data points give the BpsA indigoidine synthesis velocity ($\Delta A_{590}/s$). The gradient between every 2–8 datapoints for the indigoidine synthesis velocity data set is then determined using the Microsoft Excel slope function. The Microsoft Excel Max function is used to identify the PPTase maximum velocity with the units $\Delta A_{590}/s^2$.

4 Notes

1. We use the seamless swapping vector pCDFDUET::*slbpsA* to stage the epPCR PCP domain. This vector was first described by Owen et al. [14] and is available on request. The PPTase that is being targeted can be staged in any vector that is compatible with the pCDFDUET backbone. The plasmid pCDFDUET::*bpsA* is available on request; alternatively, the PCP domain can be amplified from *S. lavendulae* genomic DNA.
2. The primers SL_Fwd (AGCTCTGCAGAGCTCGTCGAGC GCCCCTTCGTCGCCCCGCGCACG) and SL_Rev (AGCTTCTAGACTCCTGGGCGACCTCGCGCTCCAGG CGGCGGGCCA) are used to amplify the PCP domain. To sequence the BpsA PCP domain, we use the primer PCP_Seq (ACGAGCAGATCGGCCACGA) which is located in the A domain region immediately upstream of the PCP domain.
3. *E. coli* BL21 *ΔentD* is first described in Owen et al. [14] and has the endogenous Type II PPTase *entD* knocked out, as EntD has a limited ability to convert BpsA into the active *holo* form that can potentially be amplified during the evolutionary process. This strain cannot grow on iron-limited media but does not have a growth defect on rich laboratory media such as lysogeny broth.
4. An alternative screening method would be to use larger 135 mM Petri dishes to enable the screening of more colonies per plate. However, we use 90 mM Petri dishes, as it is easier to lift the agar without breaking it when applying the IPTG.
5. As BpsA appears to be sensitive to buffer composition, we exchange the buffer to sodium phosphate for long term storage [9]. We add 12.5% glycerol to help avoid aggregation and enhance stability.
6. For some PPTases (e.g., PptT from *M. tuberculosis*) the PPTase is highly unstable without magnesium and CoA co-factors. If this is the case for your target PPTase, you can supplement the lysis buffer with Co-enzyme A and $MgCl_2$ as described by Rottier et al. [18]. We add 25% glycerol to help avoid aggregation and enhance stability.
7. We have found the optimal cellular lysis method is physical disruption using a French Press. However, if this equipment is not available, it can be replaced with other cell disruption methods such as sonication or chemical lysis. Note that some chemical lysis solutions can promote precipitation of certain PPTases.

8. Alternative techniques such as FPLC are also suitable for purification. For purifying PPTases, it is imperative that the purification occurs rapidly. We utilize a peristaltic pump to increase the rate of flow relative to a gravity-drawn column.
9. Generally we have found that the mutation rate offered by this kit is lower than described in the manufacturer's instructions. We usually use 25–50% less template DNA than is advised in the protocol.
10. Splitting the reaction into four smaller reactions helps avoid a founder effect – i.e., reduces the likelihood that the library will be saturated by mutations introduced early in the PCR. This is especially important if a very small amount of template DNA and large number of PCR cycles are being used.
11. Any PPTase that efficiently activates BpsA can be used (e.g., the canonical PPTase Sfp from *B. subtilis*).
12. Because we subsequently need to lift the agar plate to add the IPTG (Fig. 2a), we prefer to use a "hockey stick" cell spreader rather than plating beads (which will spread colonies right to the edge of the plate).
13. We find it easiest to add the L-glutamine in powder form directly to the agar after it comes out of the autoclave. In our experience, the residual heat and the presence of two antibiotics are sufficient to avoid bacterial contamination. As a control we incubate a pigment-production plate without any culture alongside the library plates. When pouring the pigment production plates, it is important to make sure they are thick enough (approximately 30 mL per plate) and evenly poured to avoid breakage. If the agar used is of lower quality, the concentration can be increased to 1.6–1.7% to enhance stability.
14. We have investigated alternative induction methods such as auto-induction plates, or direct application of lactose to the colonies using a spray bottle with an attached vaporizer, but have found these methods to not be as effective. However, if a large-scale experiment is planned, or conditions occur where it is was not plausible to lift the agar plates, you may find that further optimization enables these methods to be used.
15. The higher the insert rate, the lower the number of colonies that have to be screened. We settled on 70% as a minimum benchmark, but it is more of a guideline than a rule.
16. We screened approximately 200,000 colonies over 100 agar plates to identify BpsA variants that were substantially improved in activation by hPPTase. To acquire sufficient cell suspension for plating, we conducted 10 individual

electroporations and then aliquoted out approximately 100 μL of cell suspension per plate. The plates were incubated overnight at 37 °C.

17. The rate of pigment production depends on several variables including cell density, temperature, agar thickness, and PPTase activity. If the laboratory is unusually warm or cool, we recommend storing the plates in an 18 °C incubator as this is an optimal temperature for pigment production.
18. We usually make 50 mL of LB supplemented with IPTG, spectinomycin, and kanamycin. We then add 0.73 g of L-glutamine (100 mM final concentration) and filter-sterilize through a 0.22 μM filter.
19. Depending on the PPTase and BpsA variant, it can take up to 48 h for the pigment to be produced. Selecting appropriate time points within the dynamic range is important, as indigoidine levels can saturate, resulting in artificially lower readings.
20. To obtain data for the pure cell culture we recommend measuring the well containing *E. coli* BL21 Δ*entD* pCDF::BpsA, pET28A(+) (i.e., the empty pET28A(+) control), as this strain does not produce detectable amounts of indigoidine.
21. Depending on the catalytic rate of the PPTase, you may need to measure in smaller or greater time intervals. For rapid PPTases (e.g., Sfp or PcpS with wtBpsA), we measure every 10–20 s and for slower PPTases (e.g., hPPTase with first round mutants) every 30–60 s. This helps to reduce signal noise when calculating the slope value.

Bibliography

1. Lambalot RH, Gehring AM, Flugel RS et al (1996) A new enzyme superfamily—the phosphopantetheinyl transferases. Chem Biol 3: 923–936
2. Beld J, Sonnenschein EC, Vickery CR et al (2014) The phosphopantetheinyl transferases: catalysis of a posttranslational modification crucial for life. Nat Prod Rep 31:61–108
3. Gehring AM, Lambalot RH, Vogel KW et al (1997) Ability of *Streptomyces* spp. aryl carrier proteins and coenzyme A analogs to serve as substrates *in vitro* for *E. coli* holo-ACP synthase. Chem Biol 4:17–24
4. Lomakin IB, Xiong Y, Steitz TA (2007) The crystal structure of yeast fatty acid synthase, a cellular machine with eight active sites working together. Cell 129:319–332
5. Leblanc C, Prudhomme T, Tabouret G et al (2012) 4′-Phosphopantetheinyl transferase PptT, a new drug target required for *Mycobacterium tuberculosis* growth and persistence *in vivo*. PLoS Pathog 8:e1003097
6. Barekzi N, Joshi S, Irwin S et al (2004) Genetic characterization of *pcpS*, encoding the multifunctional phosphopantetheinyl transferase of *Pseudomonas aeruginosa*. Microbiology 150: 795–803
7. Johns A, Scharf DH, Gsaller F et al (2017) A nonredundant phosphopantetheinyl transferase, PptA, Is a novel antifungal target that directs secondary metabolite, siderophore, and lysine biosynthesis in *Aspergillus fumigatus* and is critical for pathogenicity. MBio 8(4): e01504–e01516
8. Brown AS, Owen JG, Jung J et al (2021) Inhibition of indigoidine synthesis as a high-throughput colourimetric screen for antibiotics targeting the essential *Mycobacterium tuberculosis* phosphopantetheinyl transferase PptT. Pharmaceutics 13:1066

9. Owen JG, Copp JN, Ackerley DF (2011) Rapid and flexible biochemical assays for evaluating 4′-phosphopantetheinyl transferase activity. Biochem J 436:709–717
10. Joshi AK, Zhang L, Rangan VS et al (2003) Cloning, expression, and characterization of a human 4′-phosphopantetheinyl transferase with broad substrate specificity. J Biol Chem 278:33142–33149
11. Brown AS, Sissons JA, Owen JG et al (2020) Directed evolution of the nonribosomal peptide synthetase BpsA to enable recognition by the human phosphopantetheinyl transferase for counter-screening antibiotic candidates. ACS Infect Dis 6:2879–2886
12. Weber T, Baumgartner R, Renner C et al (2000) Solution structure of PCP, a prototype for the peptidyl carrier domains of modular peptide synthetases. Structure 8:407–418
13. Tufar P, Rahighi S, Kraas FI et al (2014) Crystal structure of a PCP/Sfp complex reveals the structural basis for carrier protein posttranslational modification. Chem Biol 21:552–562
14. Owen JG, Robins KJ, Parachin NS et al (2012) A functional screen for recovery of 4′-phosphopantetheinyl transferase and associated natural product biosynthesis genes from metagenome libraries. Environ Microbiol 14:1198–1209
15. Wilson DS, Keefe AD (2000) Random mutagenesis by PCR. Curr Protoc Mol Biol 51: 8.3.1–8.3.9
16. Beer R, Herbst K, Ignatiadis N et al (2014) Creating functional engineered variants of the single-module non-ribosomal peptide synthetase IndC by T domain exchange. Mol BioSyst 10:1709–1718
17. Myers JA, Curtis BS, Curtis WR (2013) Improving accuracy of cell and chromophore concentration measurements using optical density. BMC Biophys 6:4
18. Rottier K, Faille A, Prudhomme T et al (2013) Detection of soluble co-factor dependent protein expression in vivo: application to the 4′-phosphopantetheinyl transferase PptT from *Mycobacterium tuberculosis*. J Struct Biol 183: 320–328

Chapter 8

Unraveling Structural Information of Multi-Domain Nonribosomal Peptide Synthetases by Using Photo-Cross-Linking Analysis with Genetic Code Expansion

Julia Diecker, Wolfgang Dörner, Jennifer Rüschenbaum, and Henning D. Mootz

Abstract

Nonribosomal peptide synthetases (NRPSs) are large, multifunctional enzymes that facilitate the stepwise synthesis of modified peptides, many of which serve as important pharmaceutical products. Typically, NRPSs contain one module for the incorporation of one amino acid into the growing peptide chain. A module consists of the domains required for activation, covalent binding, condensation, termination, and optionally modification of the aminoacyl or peptidyl moiety. We here describe a protocol using genetically encoded photo-cross-linking amino acids to probe the 3D architecture of NRPSs by determining spatial proximity constraints. p-benzoyl-L-phenylalanine (BpF) is incorporated at positions of presumed contact interfaces between domains. The covalent cross-link products are visualized by SDS-PAGE-based methods and precisely mapped by tandem mass spectrometry. Originally intended to study the communication (COM) domains, a special pair of docking domains of unknown structure between two interacting subunits of one NRPS system, this cross-linking approach was also found to be useful to interrogate the spatial proximity of domains that are not connected on the level of the primary structure. The presented photo-cross-linking technique thus provides structural insights complementary to those obtained by protein crystallography and reports on the protein in solution.

Key words Genetic code expansion, p-Benzoyl-L-phenylalanine, MS-MS analysis, Protein–protein interaction, Docking domain, COM domain

1 Introduction

Nonribosomal peptide synthetases (NRPSs) are multifunctional enzymes involved in the biosynthesis of a large class of natural products, the nonribosomal peptides, many of which are of significant pharmaceutical importance [1–6]. This diverse class of biomolecules is characterized by short peptide backbones of typically about 7 to 15 amino acids that can be of linear, cyclic, or branched cyclic topology. These peptides typically contain various structural

Michael Burkart and Fumihiro Ishikawa (eds.), *Non-Ribosomal Peptide Biosynthesis and Engineering: Methods and Protocols*, Methods in Molecular Biology, vol. 2670, https://doi.org/10.1007/978-1-0716-3214-7_8,

elements not found in ribosomally produced polypeptides, such as nonproteinogenic amino acids, D-amino acids, and structural modifications like lipidation, *N*-methylation, glycosylation, and additional bond formation to restrain the conformational freedom. Some of these modifications are incorporated post-synthetically on the peptide backbone by dedicated, free-standing enzymes. However, several other modifications are incorporated during the NRPS-mediated synthesis by the protein template itself. Examples include the synthetases for the biosynthesis of cyclosporine, surfactin, tyrocidine, enterobactin, vancomycin, and teixobactin [7] from fungal and bacterial sources.

The amino acids and structural modifications for the NRPS-mediated backbone assembly are reflected by the domain composition of the enzymes [1–3]. Domains are the semiautonomous parts of the assembly line that work together in the integrated NRPS template. They carry out one of the catalytic reactions and serve to covalently bind one of the monomers or elongated peptidyl intermediates. A module contains several domains that are required to add an additional monomer in the biosynthesis. The integration of these functional parts as fused components to give the large, multifunctional enzymes holds the key to how the blueprint for a specific peptide product is programmed into the NRPS template. However, how exactly these instructions are encoded in the domain arrangement is not always clear [8]. In some cases, the correlation between the domain order and the peptide product is well understood and predictable from the protein composition. In other cases, the logic of domain interactions is not obvious. Understanding these rules is an important prerequisite for reprogramming the NRPS templates to obtain desired, novel products. Next to the arrangement of the domains on the polypeptide chain, other factors like affinities of domain–domain interactions and substrate specificities are likely to play important roles.

An adenylation (A) domain recognizes a free amino acid substrate and forms the activated aminoacyl-AMP under consumption of ATP [9]. It also catalyzes transfer of the aminoacyl moiety to the peptidyl-carrier-protein (PCP) or thiolation (T) domain, which binds the aminoacyl group as a thioester on its 4′-phosphopantetheine prosthetic group. A condensation domain binds two neighboring PCP units and catalyzes the elongation reaction, in which the aminoacyl or peptidyl group of the preceding PCP is transferred onto the aminoacyl group of the following PCP [10]. In a linear type of NRPS, the typical NRPS prototype (*see* below) [8], modules with the minimal domain composition of A, PCP, and C domains are arranged in a fashion that correlates with the sequence and length of the final peptide chain. A structural modification at one amino acid position can be programmed into this enzyme template by the occurrence of the respective domain in the responsible module, for example, an epimerization (E) or

N-methylation (M) domain. Termination of peptide assembly and cleavage of the peptide backbone in either linear or (branched) cyclic form from the NRPS template is achieved by a dedicated domain in the last module, typically a thioesterase (TE) domain, or a specialized condensation or reductive domain [1–3].

In general, NRPS templates can be composed according to three different biosynthesis concepts that reflect the utilization and interaction of the individual modules and domains [8]. The abovementioned linear NRPSs follow the colinearity rule, that is, they show a direct correlation between the domain arrangements and the synthesized peptide product. In iterative NRPS, some of the catalytic units are used more than once to synthesize substructures that appear in multimeric form in the product. Finally, nonlinear NRPSs contain one or more steps in the synthesis that are not reflected by the order of the domains in NRPS. Mixtures of these types are possible, for example, the head-to-tail dimerization of a linear precursor peptide by iterative usage of a colinear NRPS segment or a synthetic step based on a single nonlinear domain arrangement in an NRPS template that is otherwise based on the colinearity rule.

Thus, understanding domain interactions within NRPS templates can help decipher the underlying biosynthesis logic. Furthermore, physical contact or spatial proximity between domains can result from simple packing effects in 3D space even in cases where no functional interaction is required. Analyzing such proximities is thus also important to study the poorly understood supramolecular arrangement of the often large number of domains and modules [11]. Finally, many NRPSs, in particular those of bacterial origin, consist of more than one enzyme. In these cases, specific domain–domain interactions are required between two polypeptide chains. These intermolecular contacts are often brought about by so-called docking domains that are found at the C- and N-termini of the two enzymes in lieu of the genetic fusion [12]. While some of these docking domains have been structurally well characterized, others are still not completely understood, for example, the so-called communication (COM) domain at the interface between the C-terminal E domain and an N-terminal C domain in consecutive modules [13–16].

Given the enormous size of NRPSs, a complete description of intra- and intermolecular interactions between domains is extremely challenging. Although several high-resolution crystal structures of multi-domain and even multi-module fragments have been reported, these can only provide static snapshots and do not necessarily reflect the dynamic spatial domain organization in solution [6, 17–21]. To address these shortcomings, we have developed new experimental strategies to study NRPSs in solution and under catalytic conditions by exploring intramolecular Förster resonance energy transfer (FRET) [22, 23] and photoactivated chemical cross-linking, or photo-cross-linking, as reported here [14, 15].

Photo-cross-linking refers to using chemical moieties that are activated by light into a highly reactive, short-lived species and that can result in covalent bonds by insertion reactions with other groups in close spatial proximity [24–26]. Such photo-induced cross-links thus capture structural information that is typically analyzed on the molecular level by using tandem mass spectrometry (MS-MS). The widely used benzophenone group forms a diradical species following irradiation with light of 350 to 365 nm that inserts in C–H bonds [27]. The reactivity of the C–H bonds can be increased by neighboring heteroatoms. In particular, methionine side chains show efficient reactions [28]. Unreacted benzophenone can be repeatedly activated, in contrast to other photo-cross-linking groups like aromatic azides and diazirines that irreversibly lose a dinitrogen and form a reactive nitrene or carbene species, respectively. The reactive carbonyl diradical of the benzophenone group has a reactive range of 3.1 Å [29]. However, the spatial resolution can be compromized to up to 20 Å by rotations by conformational flexibility around the attachment point of the photo-cross-linking moiety [30, 31].

A key development for probing large proteins with photo-cross-linkers has been the genetic code expansion technology, which enables site-specific incorporation of noncanonical amino acids (ncAA) into recombinant proteins [32]. Examples for such residues with photo-cross-linking moieties in their side chains are p-azido-L-phenylalanine (AzF) [33], p-benzoyl-L-phenylalanine (BpF, also termed BzF or BpA) [34, 35], and diazirine derivatives such as 4′-[3-(trifluoromethyl)-3H-diazin-3-yl]-L-phenylalanine (TfmdPhe) [36].

We have studied structural aspects of NRPSs with photo-cross-linking and subsequent SDS-PAGE gel and MS-MS analyses to reveal proximity information within the first two modules of the gramicidin S and tyrocidine synthetases, namely, GrsA/TycA and GrsB1/TycB1 [14, 15]. These two modules are functionally exchangeable between the two systems. Figure 1 shows the domain composition of the GrsA and TycB1 modules described as exemplary proteins in the chapter, with TycB1 carrying the photo-cross-linking amino acid BpF at position 5. Specifically, we studied COM domains and collected evidence for their architecture according to the helix-hand model. Mapping of cross-links by MS-MS led us to conclude on the orientation of the C-terminal donor-COM helix and to propose a revised upside-down, helix-hand model [14]. Incorporation of BpF into the acceptor-COM of TycB1 also resulted in a surprising cross-link into the small subdomain of TycA/GrsA [15]. This finding led us to propose an unexpected spatial proximity between these two functionally unrelated and

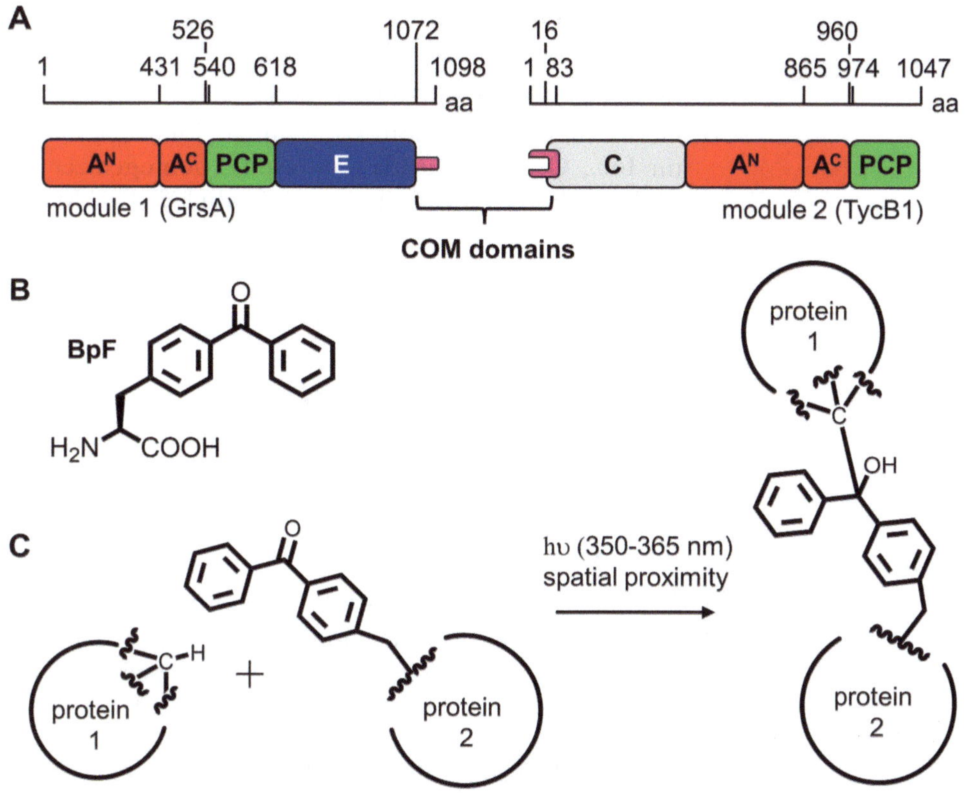

Fig. 1 Model NRPSs and photo-cross-linking with BpF. (**a**) The initiation module gramicidin S synthetase I (GrsA) and the first elongation module of the tyrocidin synthetase II (TycB1) are used as sample proteins in this chapter. GrsA/TycA and GrsB1/TycB1 are functionally interchangeable [14, 47]. Shown are the primary sequences of the two modules with the domain organization. (**b**) Structure of p-benzoyl-L-phenylalanine (BpF, also termed BzF or Bpa). (**c**) Photo-cross-linking reaction of BpF following UV irradiation. A short-lived diradical is formed that can insert into C–H bonds in close proximity

non-neighboring domains that could be explained with a model for the 3D arrangement of these domains as part of the largely unexplored supermolecular structure of NRPS modules in solution.

In this chapter, we detail the workflow for photo-cross-linking with the genetically incorporated BpF and analysis of the cross-links by SDS-PAGE gel analysis and MS-MS. While the protocols should be applicable to various types of proteins and protein–protein interactions when making adjustments to some specific experimental conditions, we have provided the specific parameters for our analysis of TycB1(BpF) with its partner protein GrsA.

2 Materials

2.1 Equipment

1. Temperature-controlled shaker for culture flasks up to 2 L.
2. Temperature-controlled shaker (25–100 °C) for 1.5 mL tubes.
3. Microcentrifuge for 1.5 mL tubes.

4. Cooled centrifuge for 50 mL tubes and 500 mL to 1 L centrifuge bottles, force ≥25,000 × *g*.
5. SDS-PAGE and Western blot unit.
6. Device for cell disruption: For example, EmulsiFlex-C5 (Avestin Inc., Ottawa, Canada), ultrasonic homogenizer SONOPULS 3100 (Bandelin electronic GmbH & Co. KG, Berlin, Germany).
7. Gravity flow column.
8. Dialysis tube (MWCO = 14 kDa).
9. Thin-walled polypropylene tube 0.2 mL.
10. UV lamp (UV-16 L): λ = 365 nm; 8 W, 1900 μW/cm^2 (Herolab GmbH Laborgeräte, Wiesloch, Germany).
11. Scalpel.
12. 0.2 mL flat-bottom micro glass insert or 1.5 mL low-binding microcentrifuge tube.
13. Rotary vacuum concentrator or lyophilizer.
14. UltiMate™ 3000 RS LC nano system (Thermo Fisher Scientific GmbH, Dreieich, Germany) equipped with a C18 trapping column (Acclaim™ PepMap™ 100, 5 μm, 100 Å, ID 300 μm × L 5 mm, Thermo Scientific) and a C18 nano column (Acclaim™ PepMap™ 100 C18, 2 μm, 100 Å, ID 0.075 mm × L 250 mm, Thermo Fisher Scientific GmbH, Dreieich, Germany) connected to a maXis II UHR-TOF mass spectrometer with a CaptiveSpray nano-ESI source (Bruker Daltonics GmbH & Co. KG, Bremen, Germany) equipped with a nanoBooster device.
15. PC with Windows 10 and DataAnalysis 4.4 (Bruker Daltonics GmbH & Co. KG, Bremen, Germany) and StavroX 3.6.6 installed.

2.2 Chemicals and Solutions

1. Plasmid pGV172 (encoding SBP-[GrsA]A-PCP-E) [14], pED45 (encoding [TycB1]C-A-PCP(S5X)-H_6) [15], and pEVOL-BpF (encoding the orthogonal BpF-specific tRNA-synthetase/$tRNA_{CUA}$ pair) [34, 37] (*see* **Note 1**).
2. Competent *E. coli* BL21 Gold (DE3) cells, genotype: *E. coli* B F^- *ompT hsdS*(r_B^- m_B^-) *dcm*$^+$ Tet^r *gal*λ(DE3) *endA* Hte.
3. LB medium: 5 g/L NaCl, 10 g/L tryptone, 5 g/L yeast extract, pH 7.
4. Agar-agar, Kobe 1.
5. Antibiotics:
   (a) Kanamycin stock solution: 50 mg/mL kanamycin sulfate.
   (b) Ampicillin stock solution: 100 mg/mL ampicillin sodium salt.
   (c) Chloramphenicol stock solution: 34 mg/mL chloramphenicol.

6. Inducing agents:
   (a) IPTG stock solution: 400 mM Isopropyl-ß-thiogalactoside, sterile.
   (b) L-(+)-arabinose.
7. p-Benzoyl-L-phenylalanine (BpF).
8. 1 M NaOH.
9. 4× SDS sample buffer: 250 mM Tris–HCl (pH 6.8), 8% (w/v) sodium dodecyl sulfate, 40% (v/v) glycerol, 20% (v/v) ß-mercaptoethanol, 0.2% (w/v) bromophenol blue.
10. Ni^{2+}-NTA affinity chromatography:
    (a) Ni^{2+}-NTA agarose.
    (b) HEPES buffer A: 50 mM HEPES, 100 mM NaCl, pH 8.
    (c) Imidazole stock solution: 5 M imidazole, pH 8.
11. SDS-PAGE:
    (a) Aqueous acrylamide, bisacrylamide stock solution in the ratio 37.5:1.
    (b) 1.5 M Tris, adjusted to pH 8.8 with HCl.
    (c) 0.5 M Tris, adjusted to pH 6.8 with HCl.
    (d) 10% (w/v) (sodium dodecyl sulfate).
    (e) 10% APS (w/v) (ammonium peroxodisulfate).
    (f) TEMED (*N*,*N*,*N*′,*N*′-tetramethyl ethylenediamine).
12. Coomassie staining solution: 2.5 g/L Coomassie brilliant blue R250, 10% (v/v) acetic acid, 40% (v/v) ethanol.
13. Coomassie destaining solution: 10% (v/v) acetic acid, 40% (v/v) ethanol.
14. NRPS assay buffer: 50 mM HEPES, 100 mM NaCl, 1 mM EDTA, 10 mM $MgCl_2$, pH 7.
15. DTT (DL-dithiothreitol).
16. Glycerol.
17. Destaining solution (for SDS-PAGE gel slice): 50% (v/v) methanol, 50% (v/v) pure water + 0.1% trifluoroacetic acid.
18. Ammonium bicarbonate buffers:
    (a) 50 mM NH_4HCO_3.
    (b) 100 mM NH_4HCO_3.
19. 2-Iodacetamide.
20. Sequencing Grade Modified Trypsin (Promega GmbH, Walldorf, Germany).
21. ProteaseMAX™ surfactant, trypsin enhancer (Promega GmbH, Walldorf, Germany).
22. Formic acid (pure).

2.3 Material for MS Analysis

Prepare all solutions using ultrapure water (resistivity of 18.2 MΩ × cm at 25 °C).

1. Eluent A: 0.1% (v/v) analytical grade formic acid.
2. Eluent B: 0.1% (v/v) analytical grade formic acid in LC-MS grade acetonitrile.
3. Calibrant solution: 0.5 mM sodium formate in 50% (v/v) 2-propanol.

3 Methods

3.1 Design of the Expression Plasmid for the NRPS Protein with the Unnatural Photo-Cross-Linking Amino Acid

The unnatural amino acid with the photo-cross-linking moiety is incorporated by the amber stop codon suppression technology [38], more recently also referred to as genetic code expansion [39, 40]. To incorporate the unnatural amino acid, which is provided as a chemical to the bacterial growth medium, the expression plasmid for the NRPS protein of interest requires an amber stop codon within the encoding gene at the desired position. Co-expression of the tRNA-synthetase/$tRNA_{CUA}$ pair specific for the unnatural amino acid from the pEVOL-plasmid as a second plasmid in the bacterial strain provides the required components for decoding the amber stop codon [37]. To introduce the amber stop codon (TAG) typically the standard plasmid used for expressing the unaltered NRPS protein can be used as long as the antibiotic resistance cassette and the origin of replication are compatible with the second pEVOL-plasmid. We do not provide an experimental protocol for the site-directed mutagenesis as this is usually common laboratory practice or done by ordering a synthetic gene. In the following, we detail only a few general steps in the choice of expression plasmid and amber stop codon mutagenesis to avoid potential pitfalls:

1. For the choice of the expression plasmid to encode the NRPS protein with the unnatural photo-cross-linking amino acid, a few parameters have to be considered. Select a vector backbone that is compatible to the second required plasmid encoding the tRNA-synthetase/$tRNA_{CUA}$ pair. In case of the pEVOL-series, this plasmid contains a p15A replicon and a chloramphenicol resistance cassette (Cm^r) [37], such that plasmids with a compatible replicon, for example the most widely used ColE1 replicon, and a compatible antibiotic resistance gene, such as the beta-lactamase (providing ampicillin resistance, Amp^r) or the aminoglycoside 3′-phosphotransferase (providing kanamycin resistance, Kan^r), are best suited. Different promotors can be used. In our example, the TycB1-encoding plasmid is based on the pTrc99a backbone (pBR322 origin, Amp^r) that contains an IPTG-inducible Trc promotor.

2. Check the stop codon at the end of the gene encoding for the NRPS protein of interest. As an amber stop codon (TAG) is required for incorporation of the unnatural amino acid and hence this codon will be suppressed during translation, one of the other two stop codons should be used, that is, either the opal codon (TGA) or the ochre stop codon (TAA). If your gene contains an amber stop codon to terminate translation, change this codon to one of the two other codons by a separate site-directed mutagenesis experiment.
3. Choose the plasmid providing the aminoacyl-tRNA-synthetase/tRNA pair for the unnatural amino acid, in this case pEVOL-BpF [37] (*see* **Note 2**).
4. Choose suitable positions for the incorporation of the photo-cross-linking amino acid in the NRPS protein of interest. When protein–protein interactions are to be investigated, the unnatural amino acid should be positioned on the surface of protein where interaction with other proteins is expected. In the present example of TycB1(S5BpF), a position in the N-terminal tail region of the C domain of TycB1 was used that is located in the COM acceptor domain to investigate its supposed role in mediating protein–protein interaction with the preceding TycA module (or the homologous GrsA module, as specified in this chapter) (*see* **Note 3**).

3.2 Production of NRPS Protein with p-benzoyl-L-phenylalanine (BpF) Incorporated by Using Nonsense Suppression

For site-specific incorporation of the photo-cross-linking amino acid into the NRPS protein of interest (here the TycB1 protein), the encoding expression plasmid (here pED45: pTrc99a-TycB1 (S5X)-His$_6$ [15]), as well as the plasmid encoding the tRNA-synthetase/tRNA$_{CUA}$ pair specific for the unnatural amino acid (here pEVOL-BpF [37]) have to be in the same *E. coli* expression strain (here *E. coli* BL21 Gold (DE3) cells). Other proteins that will be used as interaction partners in cross-linking experiments and do not require the unnatural amino acid can be prepared by standard protocols.

1. Co-transform competent *E. coli* BL21 Gold (DE3) cells with the pEVOL-BpF and pTrc99a-TycB1(S5X) plasmids. Spread the transformed bacteria onto an LB agar plate containing ampicillin (100 μg/mL) for the selection of the latter and chloramphenicol (34 μg/mL) for selection of the former plasmid. Incubate the plate overnight at 37 °C, until visible colonies have grown.
2. Transfer a single colony to liquid LB medium (4 mL) containing 100 μg/mL ampicillin and 34 μg/mL chloramphenicol and incubate at 37 °C overnight in a shaker or roller incubator.
3. Use the overnight culture to inoculate the main culture (300 mL LB medium with the respective antibiotics as

described before) in a ratio of 1:100 (v/v). Incubate the culture in a shaker at 37 °C under aerobic conditions until it has reached an optical density at 600 nm of 0.6 to 0.7 (*see* **Note 4**).

4. Remove 500 μL sample as preinduction sample and keep incubating at the same conditions as the main culture.
5. Before inducing the protein production, add 330 μL BpF solution (81 mg BpF in 330 μL 1 M NaOH) dropwise to the main culture while swirling to give a final concentration of 1 mM BpF. After an additional incubation time of 10 min under shaking, induce the TycB1 production by adding IPTG to a final concentration of 0.4 mM (add 300 μL of a 400 mM sterile-filtered stock solution) and induce the production of the tRNA-synthetase/$tRNA_{CUA}$ pair by adding 0.2% L-arabinose (w/v, i.e., 600 mg for 300 mL of main culture). Reduce the temperature to 28 °C and incubate while still shaking the culture for another 4 h (*see* **Note 5**).
6. Take 500 μL postinduction sample and pellet these cells as well as the cells of the preinduction sample by centrifugation (1 min at 20,000 × *g*) and discard the supernatant. Resuspend the cells in 45 μL H_2O and add 15 μL 4× SDS sample buffer.
7. Harvest the cells of the main culture by centrifugation (20 min at 3000 × *g* and 4 °C) in a precooled centrifuge rotor. Discard the supernatant. The pellet can be stored at −20 °C for a prolonged period of time, either before or after resuspension in the buffer for cell lysis.

3.3 Cell Lysis and Protein Purification

Here we describe the purification of the His_6-tagged TycB1 module containing BpF using Ni^{2+}-NTA affinity chromatography. Other methods for protein purification can be used as alternatives.

1. If a frozen cell pellet is used then thaw the cells on ice.
2. Resuspend the cell pellet in 10–15 mL ice-cold HEPES buffer A and rupture the cells using your preferred method.
3. Separate the soluble fraction from insoluble material and cell debris by centrifugation (30 min at 25,000 × *g* and 4 °C) in a precooled centrifuge rotor.
4. In the meantime, fill a disposable gravity-flow column (diameter of 1.5 cm) with 500 μL Ni^{2+}-NTA agarose (*see* also **Note 3**). Alternatively, an FPLC chromatography system with a buffer gradient can be used. Equilibrate the Ni^{2+}-NTA column with 5 mL HEPES buffer A including 20 mM imidazole (*see* **Note 6**).
5. After centrifugation, transfer the cell lysate in a 50 mL tube and add imidazole (from a 5 M stock solution) to give a final concentration of 20 mM. For later analysis, take a cell lysate

sample (9 μL) and mix it with 3 μL of 4× SDS sample buffer. Take also a sample of the cell debris with a 10 μL pipette tip. The amount of the pellet should be nearly the same amount of the preinduction and postinduction pellet. Resuspend the cells in 45 μL H_2O and add 15 μL 4× SDS sample buffer.

6. The following steps should be done at ≤4 °C and using precooled buffers.
7. Load the cell lysate twice onto the Ni^{2+}-NTA column and collect the flow-through. Take a sample of the flow-through fraction (9 μL) and mix it with 3 μL 4× SDS sample buffer as described before.
8. Wash the column with the bound protein using 5 mL HEPES buffer A with 20 mM imidazole (10 column volumes (cv)) and in a second step with 3 mL HEPES buffer A with 35 mM imidazole (6 cv). Again, remove 9 μL sample from each wash fraction and mix it with 3 μL of 4× SDS buffer as described before for further analysis.
9. Subsequently, the protein can be eluted from the column by addition of 3 mL HEPES buffer B (6 cv) containing 250 mM imidazole. Collect 6 elution fractions each with a volume of 0.5 mL. From each elution fraction, take a 9 μL sample and mix it with 3 μL of 4× SDS buffer for further analysis as described before.
10. Boil all fraction samples at 95 °C for 10 min. Spin the samples down by brief centrifugation.
11. Analyze all fraction samples by an 8% SDS-PAGE. Load approximately 2 μL of pre-, postinduction and pellet fractions, 1 μL of lysate, flow-through and wash fraction samples and 3 μL of the elution samples onto the gel. After conducting the electrophoresis, stain the gel with Coomassie brilliant blue and pool the elution fractions with the highest concentration of purified protein. Analyze the gel for protein expression, solubility, and purity of the protein of interest in the elution fractions (*see* **Notes** 7 and **8**).
12. Dialyze the pooled fractions against a 100-fold excess volume of precooled NRPS assay buffer (i.e., 100 mL buffer for 1 mL eluate). Add 2 mM DTT during the first dialysis step. Exchange buffer twice and add NRPS assay buffer with 10% glycerol to the last dialysis step to protect the purified protein upon freezing for storage. Each dialysis step should take at least 60 min. Perform all these steps in the cold room to keep the buffer temperatures ≤4 °C.
13. Determine the protein concentration of the dialyzed samples by measuring the absorbance at 280 nm and using the calculated molecular extinction coefficient (for TycB1(S5BpF): 104,060 M^{-1} cm^{-1}) at this wavelength (*see* **Note 9**).

14. Optionally, at this stage the NRPS protein can be 4′-phosphopantetheinylated at its PCP domain(s), if applicable. To this end, use, for example, the 4′-phosphopantetheine (Ppant) transferase Sfp and coenzyme A as widely described in the literature (e.g., Ref. [22]). Of note, only the Ppantylated holo-protein can be considered to be fully functional, which may affect the protein–protein interactions (*see* **Note 10**).
15. Aliquot the protein solution in suitable volumes for subsequent assays, for example, 50 μL. Flash freeze them in liquid nitrogen and store the protein at −80 °C.

3.4 Photo-Cross-Linking Assay

To monitor the cross-linking ability of the constructs, each BpF-containing protein is UV-irradiated either in the absence or in the presence of its partner protein. Ideally, the protein with the incorporated photo-cross-linking amino acid should be tested for biochemical activity beforehand to evaluate the potential effect of the amino acid substitution on protein folding and protein–protein interactions (*see* also **Note 10**).

1. Thaw the proteins on ice. Set up a mixture of the BpF-containing protein (here TycB1(S5BpF)) and the native partner protein (here GrsA) in NRPS assay buffer. We use for this assay each protein at a concentration of 5 μM in a final volume of 30 μL in a 0.2 mL thin-walled polypropylene PCR tube (*see* **Note 11**).
2. Also prepare a separate assay that contains only the BpF-containing protein at the same concentration (5 μM TycB1). This sample serves as a control for various potential forms of intra- or intermolecular self-cross-links.
3. Preincubate for 45 min at 25 °C.
4. Remove a sample of 12 μL from each assay as nonirradiated controls (−UV samples) and mix with 4 μL of 4× SDS sample buffer.
5. For irradiation of the remaining 18 μL assay mixture (+UV samples), use a UV lamp ($\lambda = 365$ nm, 8 W) with a fan cooling included. Place the tubes in a rack in a distance of 1.5 cm between the lamp and the tube. Irradiate the samples for 45 min (*see* **Note 12**).
6. Add 6 μL of 4× SDS sample buffer to each sample and boil the samples at 95 °C for 10 min. Spin down after boiling by centrifugation. The samples can now be immediately analyzed (next section) or can be stored at −20 °C.

3.5 Analysis of Photo-Cross-Linked Proteins by Coomassie-Stained SDS-PAGE and by Western Blotting

Potential cross-linked products are first analyzed by SDS-PAGE as these typically reveal themselves by a different migration behavior (*see* **Note 13**). Western blotting of the SDS-PAGE gel can be useful to distinguish between different and potentially overlapping protein bands in cases when a specific antibody exists that can be used to selectively probe one of the proteins, for example, because only this protein carries a short epitope tag (such as using an anti-His IgG for His_6-tagged proteins). In case of the proteins used here, TycB1 carries a C-terminal His_6-tag and GrsA contains an N-terminal SBP-tag that can be probed with streptavidin conjugates. As shown in Fig. 2, Western blotting helped identify a TycB1-GrsA cross-link product (m band) that overlapped with a cross-link product arising from the TycB1(S5BpF) protein alone and thus was not clearly identifiable from the Coomassie-stained SDS-PAGE gel. Western blotting can also be helpful to detect weak cross-linked products because it is typically more sensitive than SDS-PAGE analysis by Coomassie brilliant blue staining. We here only give the protocol for the SDS-PAGE procedure.

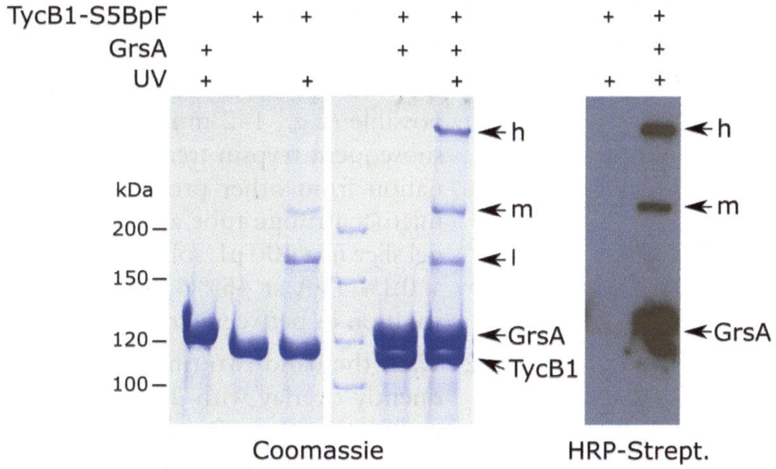

Fig. 2 Gel analysis of photo-cross-linking. Shown is the analysis for the sample proteins TycB1(S5BpF) (125.2 kDa) and GrsA (131.1 kDa). GrsA carries an N-terminal streptavidin-binding peptide (SBP) and can therefore be selectively probed using a horseradish-peroxidase-streptavidin conjugate (HRP-Strept.). Note that TycB1(S5BpF) forms two prominent bands (low, l; medium, m) upon irradiation (365 nm) that likely represent an intramolecular cross-link and an intermolecular cross-link between two TycB1 monomers, respectively. The cross-linking experiment between TycB1(S5BpF) and GrsA reveals the appearance of an additional band (high, h). Probing with HRP-Strept. revealed that both the m and h bands contain GrsA. Further analysis by MS-MS revealed two different cross-link isoforms in these bands (*see* Fig. 3)

1. Run all samples by SDS-PAGE using a 6% separation SDS-PAGE gel. To this end, load 6 μL of each sample onto the gel. After conducting the electrophoresis, stain the gel with Coomassie brilliant blue (*see* **Notes 14** and **15**).
2. Analyze the destained gel for differences between the (−UV) and (+UV) samples as these are most likely to stem from the photo-cross-linking. Of note, also compare the control samples lacking the partner protein. New protein bands might either result from self-cross-linking of the BpF-containing protein in intra- or intermolecular fashion or they might result from cross-linking between the BpF-containing protein and the partner protein. Importantly, the migration behavior of the cross-linked species not only depends on the associated molecular weights, but also on the topology of the cross-links and can therefore result in quite unpredictable outcomes (*see* also **Note 13**). *See* Fig. 2, for example, using TycB1(S5BpF) with its partner protein GrsA.

3.6 In-Gel Tryptic Digest of Photo-Cross-Linked Proteins

For the following steps, wear gloves to protect the samples from contaminations:

1. Analyze the gel and excise a gel slice containing the band of the cross-link product with a scalpel. Keep the slice as thin as possible (e.g., 1–2 mm) to minimize the gel material for the subsequent trypsin treatment and to avoid potential contamination from other protein species. Transfer the band into a microcentrifuge tube and destain the slice by submerging the gel slice in a 400 μL solution of 50% methanol/50% pure water + 0.1% TFA at 45 °C in a shaker. If necessary, exchange the solution in between (*see* **Note 16**).
2. Wash the band two times with 200 μL pure water and subsequently overlay with 200 μL acetonitrile until the gel slice is completely white. Remove the supernatant and dry the gel slice using a rotary vacuum concentrator or a lyophilizer.
3. Incubate the dried slice for 20 min at 56 °C in 40 μL solution of 10 mM DTT/100 mM NH_4HCO_3 (must be freshly made).
4. Remove the supernatant completely and wash the gel slice using 400 μL 100 mM NH_4HCO_3. Remove the supernatant and add 40 μL of 55 mM 2-Iodacetamide/100 mM NH_4HCO_3 (must be freshly made and kept in the dark). Incubate 30 min at room temperature in the dark.
5. Remove the supernatant completely and wash the gel slice twice with 200 μL pure water and subsequently overlay with 200 μL acetonitrile until the gel slice is completely white. Remove the supernatant and dry the gel slice using a rotary vacuum concentrator or a lyophilizer. Transfer the gel slice into a 0.2 mL flat-bottom micro glass vial or into a low-binding tube.

6. To increase the activity of trypsin, the manufacturer suggests heating the trypsin 15 min at 30 °C. Afterward, dissolve 4 µL of 400 ng trypsin in 16 µL 50 mM NH_4HCO_3, pH 8. Add the solution to your gel slice and allow to soak in for 10 min at room temperature.
7. Add 30 µL of 0.01% (w/v) ProteaseMAX™ (Promega) in 50 mM NH_4HCO_3. Incubate the sample at 37 °C for 2 h. To avoid evaporation of the reaction solution, a consistent heat distribution (like in an incubator) is necessary (*see* **Note 17**).
8. Take the supernatant and acidify with 5% formic acid to a final concentration of 0.5% (v/v). Check the pH with a pH-indicator paper (pH 2). Centrifuge the sample 5 min at 15,000 rpm and transfer the supernatant to an HPLC vial for subsequent MS analysis (*see* below).

3.7 MS-MS Analysis

The quality of cross-link mapping strongly depends on the sensitivity and mass accuracy of the MS setup. Access to a high-resolution mass spectrometer is thus essential for XL-MSMS. According to Iacobucci et al. [41], by far most cross-linking studies have thus been carried out with Orbitrap mass spectrometers (Thermo Scientific™), some with high-resolution ToF instruments like maXis II [14] (Bruker Daltonics GmbH & Co. KG, Bremen, Germany) and more recently with a timsTOF Pro setup [42] (Bruker Daltonics GmbH & Co. KG, Bremen, Germany). The following procedure is based on measurements with a maXis II instrument:

1. For desalting and to protect the analytical column, load 10 µL of the peptide solution on the C18 trapping column and wash at a flow rate of 20 µL/min in 2% eluent B.
2. To improve mass accuracy, a calibrant solution can be injected during the desalting phase (e.g., 10 min washing at 2% B) to enable post-run calibration of MS and MS/MS spectra.
3. Apply a 50 min gradient (2% to 60% B, flow rate 300 nL/min) for the separation of peptides on the C18 nano column. MS settings for the maXis II instrument are capillary voltage 1200 V or higher, depending on the state of the capillary, mass range: m/z 150–2200. MS survey scans are performed with a cycle time of 3 s. After each survey scan, the 10 to 20 most abundant precursor ions with $z > 1$ are selected for fragmentation using collision-induced dissociation. MS/MS summation time is adjusted depending on the precursor intensity. The applied precursor isolation window and the collision energy are chosen depending on the precursor m/z and charge.

3.8 Data Processing

Cross-link analysis is very demanding for bifunctional chemical cross-linkers, leading to large numbers of possible cross-link peptides. In contrast, photo-cross-linking with one known and well-defined partner, as is the case for proteins containing recombinantly introduced unnatural amino acids, greatly reduces the complexity of the task for search engines.

There are several tools available for cross-link peptide analysis. StavroX [43], pLink [44], and MeroX [45] have been the most commonly used tools [41], but new tools keep occurring, with MS Annika being the most recent search engine [46].

In general, MS data have to be converted into fragment spectra collections accommodated in Mascot generic format (mgf), mzXML or mzML files prior to the analysis.

1. Open the MSMS data folder with DataAnalysis, carry out an internal calibration, and generate an mgf file using the tool ProteinAnalysis.
2. Generate a fasta file containing the protein sequences of interest.
3. Load fasta and mgf files in StavroX 3.6.6.
4. By default, p-benzoyl-L-phenylalanine is included as BPA in the list on the Amino acids tab of the StavroX 3.6.6 settings with the code x and the composition C16H13NO2.
5. Select BPA as cross-linker by double-clicking at the respective line on the Cross-linker tab of Settings.
6. For all other settings, use the defaults as a starting point of the analysis.
7. Click Start to run the analysis.
8. Check the decoy analysis. For small numbers of proteins, the target decoy analysis may not bear statistical significance.
9. Inspect the cross-link table. Scores above 100 can be considered reliable.
10. Click on details of cross-link to view the annotated spectrum and peak list. For examples of fragmentations patterns derived from the experimental data, *see* Fig. 3.

4 Notes

1. This plasmid is available from Addgene under the name pEVOL-pBpF with plasmid number #31190.
2. The same NRPS expression plasmid can be combined with tRNA-synthetase/$tRNA_{CUA}$ pairs for different photo-cross-linking amino acids [14].

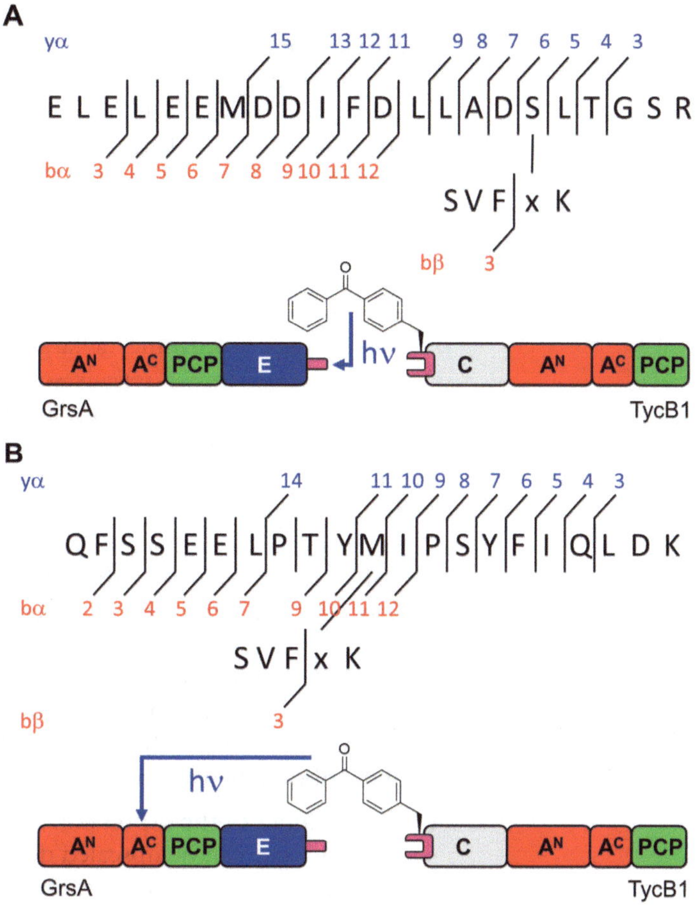

Fig. 3 MS-MS mapping of photoinduced cross-links. Shown is the analysis for the sample proteins TycB1(S5BpF) and GrsA. (**a**) The fragmentation pattern of the m band sample (*see* Fig. 2) revealed a cross-link from BpF in TycB1(S5BpF) (indicated as "x") into the C-terminal COM domain of GrsA. Specifically, the score and site probability were highest for S1096 with the COM donor domain as a hit, resulting in a nearly linear connectivity of the two polypeptide chains (L-form, *see* **Note 13**). (**b**) The fragmentation pattern of the h band sample (*see* Fig. 2) revealed a cross-link from BpF in TycB1(S5BpF) (shown as "x") into the A^C subdomain of GrsA. Specifically, the score and site probability were highest for M499 within the A^C subdomain, resulting in a branched connectivity of the two polypeptide chains (T-form, *see* **Note 13**). This surprising finding suggested the spatial proximity of the functionally unrelated and in the primary sequence non-neighboring COM and A^C domains [15]

3. Similar to all amino acid substitutions, alteration of the side chain may have undesired and unexpected structural changes. Ideally, the mutated protein should be checked for functional integrity [14, 15]. Due to the structure of BpF, the most similar canonical amino acids are the aromatic residues tyrosine, tryptophan, and phenylalanine, of which tyrosine is often

found on the surface of domains and in regions of protein–protein interactions. However, also other residues on the surface should be probed. Due to the large side chain of BpF, larger than those of all proteinogenic amino acids including the aromatic ones, a substitution with BpF brings about a significant increase in steric bulk. While this fact must be taken into consideration as it might affect the local structural integrity of the protein or a protein–protein interaction, it can also be used to an experimental advantage. It can be regarded as the equivalent of an alanine scan with opposite qualities. While in a substitution to alanine, a larger side chain is stripped away and the experiment reveals something of the importance of this side chain now missing, a substitution to BpF adds steric bulk at the respective position that may or may not be tolerated, also depending on structural packing of the side chains and sufficient room to accommodate the large BpF side chain. For example, our structural analysis of the COM domain with a BpF scan revealed that three hydrophobic amino acids in the donor-COM helix did not tolerate the incorporation of BpF, that is, the respective GrsA mutants showed clearly diminished interaction with their downstream module partner and a decreased amount of dipeptide formation. This finding was attributed to the tight hydrophobic packing in the proposed hydrophobic helix-hand interaction [14].

4. The volumes of the expression cultures and the volume of Ni^{2+}-NTA agarose can be upscaled depending on how much protein is needed. We recommend pilot experiments using smaller volumes due to the cost of BpF.
5. Suppression times and temperatures may be adapted according to what is beneficial to the protein of interest in terms of expression yields and solubility. While a lower-expression temperature like 20 °C can be beneficial for the solubility of large NRPS proteins, in our experience the best suppression yields with the BpF aminoacyl-tRNA-synthetase are obtained at 37 °C. Therefore, a compromise between NRPS protein solubility and yield (temperature lower than 37 °C) and efficient suppression yields (temperature as close as possible to 37 °C) should be identified.
6. Imidazole at a certain amount will help suppress unspecific binding of untagged proteins to the affinity column. Use an imidazole concentration here under which the protein of interest is not eluted. Depending on the protein and on the length of the histidine tag (e.g., His_6 versus His_{10}), the imidazole concentration can be further optimized.
7. Suppression of the amber stop codon usually results in lower expression yields because the amber stop codon still partially results in termination of translation.

8. The other fractions serve to check for protein expression and protein solubility and are useful for troubleshooting. For example, if the expression levels are fine (compare postinduction sample vs. preinduction sample), however, the protein does not appear to be of good solubility (much or all of the protein in the pellet fraction and only little or nothing in the soluble fraction) then repeat the expression with a lower temperature as this is often helpful for better solubility. When the expression temperature is in the range of only 20–25 °C, then longer expression times are typically required, for example, overnight.
9. Use a standard protein analysis program or use an online bioinformatic tool (i.e., "ExPASy," *see* https://web.expasy.org/protparam/) to determine the extinction coefficient of proteins. A suitable substitution for the unnatural amino acid BpF is phenylalanine.
10. Conversion of the PCP domain(s) from their apo into the holo form can affect domain–domain interactions [17] and therefore may change the outcome of a photo-cross-linking analysis [15].
11. Depending on the affinity between the proteins, also other concentrations should be tested. The affinity between GrsA and TycB1 was determined to be $K_d = 5.0 \pm 0.9$ μM [15].
12. Also other setups of UV lamps can be used. Critical parameters are the power of the lamp, the distance between lamp and sample as well as the wavelength distribution of the lamp. Cooling can be important to avoid overheating of the samples that might lead to loss of native protein–protein interactions and ultimately to protein unfolding and denaturation.
13. Both intra- and intermolecular cross-links can result in altered migration behavior of the proteins in SDS-PAGE gels, sometimes with unpredictable consequences on the migration behavior. As a rule of thumb, expect the added molecular weights, when two proteins are cross-linked are connected in a rather linear fashion (L-form). However, when a more branched cross-link pattern arises (T-form), then significantly slower migration behavior can be observed as exemplified in the m and h bands of the TycB1(S5BpF) cross-links with GrsA [15] (*see* Fig. 2).
14. We recommend loading (−UV) and (+UV) sample next to one another. Depending on the sizes of the proteins and the cross-linked products, also higher percentage SDS-PAGE gels might be used.
15. Use sufficiently low-percentage SDS gels to resolve the potentially higher molecular weight bands. The gel should be adjusted according to sizes of proteins and observed cross-link products.

16. We recommend exchanging the destaining solution if it starts to turn blue. By increasing the temperature up to 50 °C, it is possible to accelerate the destaining process. Higher temperatures should, however, be avoided because they might result in the artefactual *O*-methylation of acidic side chains.
17. Incubation overnight is also possible.

Acknowledgments

This work was supported by the Deutsche Forschungsgemeinschaft (DFG; grant number MO1073/8-1 to H. D. M.). We thank Peter Schultz for kindly providing the pEVOL-BpF plasmid.

References

1. Marahiel MA, Stachelhaus T, Mootz HD (1997) Modular peptide synthetases involved in nonribosomal peptide synthesis. Chem Rev 97:2651–2674
2. Cane DE, Walsh CT, Khosla C (1998) Harnessing the biosynthetic code: combinations, permutations, and mutations. Science 282:63–68
3. Sussmuth RD, Mainz A (2017) Nonribosomal peptide synthesis-principles and prospects. Angew Chem Int Ed Engl 56:3770–3821
4. Reimer JM, Haque AS, Tarry MJ et al (2018) Piecing together nonribosomal peptide synthesis. Curr Opin Struct Biol 49:104–113
5. Meier JL, Burkart MD (2009) The chemical biology of modular biosynthetic enzymes. Chem Soc Rev 38:2012–2045
6. Gulick AM (2016) Structural insight into the necessary conformational changes of modular nonribosomal peptide synthetases. Curr Opin Chem Biol 35:89–96
7. Ling LL, Schneider T, Peoples AJ et al (2015) A new antibiotic kills pathogens without detectable resistance. Nature 517:455–459
8. Mootz HD, Schwarzer D, Marahiel MA (2002) Ways of assembling complex natural products on modular nonribosomal peptide synthetases. Chembiochem 3:490–504
9. Conti E, Stachelhaus T, Marahiel MA et al (1997) Structural basis for the activation of phenylalanine in the non-ribosomal biosynthesis of gramicidin S. EMBO J 16:4174–4183
10. Bloudoff K, Schmeing TM (2017) Structural and functional aspects of the nonribosomal peptide synthetase condensation domain superfamily: discovery, dissection and diversity. Biochim Biophys Acta Proteins Proteom 1865: 1587–1604
11. Tarry MJ, Haque AS, Bui KH et al (2017) X-ray crystallography and electron microscopy of cross- and multi-module nonribosomal peptide synthetase proteins reveal a flexible architecture. Structure 25(783–793):e784
12. Richter CD, Nietlispach D, Broadhurst RW et al (2008) Multienzyme docking in hybrid megasynthetases. Nat Chem Biol 4:75–81
13. Hahn M, Stachelhaus T (2004) Selective interaction between nonribosomal peptide synthetases is facilitated by short communication-mediating domains. Proc Natl Acad Sci U S A 101:15585–15590
14. Dehling E, Volkmann G, Matern JC et al (2016) Mapping of the communication-mediating Interface in nonribosomal peptide synthetases using a genetically encoded photocrosslinker supports an upside-down helix-hand motif. J Mol Biol 428:4345–4360
15. Dehling E, Ruschenbaum J, Diecker J et al (2020) Photo-crosslink analysis in nonribosomal peptide synthetases reveals aberrant gel migration of branched crosslink isomers and spatial proximity between non-neighboring domains. Chem Sci 11:8945–8954
16. Fage CD, Kosol S, Jenner M et al (2021) Communication breakdown: dissecting the COM interfaces between the subunits of nonribosomal peptide synthetases. ACS Catal 11:10802–10813
17. Drake EJ, Miller BR, Shi C et al (2016) Structures of two distinct conformations of holo-non-ribosomal peptide synthetases. Nature 529:235–238
18. Tanovic A, Samel SA, Essen LO et al (2008) Crystal structure of the termination module of a nonribosomal peptide synthetase. Science 321:659–663

19. Reimer JM, Aloise MN, Harrison PM et al (2016) Synthetic cycle of the initiation module of a formylating nonribosomal peptide synthetase. Nature 529:239–242
20. Reimer JM, Eivaskhani M, Harb I et al (2019) Structures of a dimodular nonribosomal peptide synthetase reveal conformational flexibility. Science 366:eaaw4388
21. Izore T, Cryle MJ (2018) The many faces and important roles of protein-protein interactions during non-ribosomal peptide synthesis. Nat Prod Rep 35:1120–1139
22. Alfermann J, Sun X, Mayerthaler F et al (2017) FRET monitoring of a nonribosomal peptide synthetase. Nat Chem Biol 13:1009–1015
23. Mayerthaler F, Feldberg AL, Alfermann J et al (2021) Intermediary conformations linked to the directionality of the aminoacylation pathway of nonribosomal peptide synthetases. RSC Chem Biol 2:843–854
24. Suchanek M, Radzikowska A, Thiele C (2005) Photo-leucine and photo-methionine allow identification of protein-protein interactions in living cells. Nat Methods 2:261–267
25. Sinz A (2006) Chemical cross-linking and mass spectrometry to map three-dimensional protein structures and protein-protein interactions. Mass Spectrom Rev 25:663–682
26. Ptackova R, Jecmen T, Novak P et al (2014) The application of an emerging technique for protein-protein interaction interface mapping: the combination of photo-initiated cross-linking protein nanoprobes with mass spectrometry. Int J Mol Sci 15:9224–9241
27. Dorman G, Nakamura H, Pulsipher A et al (2016) The life of Pi star: exploring the exciting and forbidden worlds of the benzophenone photophore. Chem Rev 116:15284–15398
28. Wittelsberger A, Thomas BE, Mierke DF et al (2006) Methionine acts as a "magnet" in photoaffinity crosslinking experiments. FEBS Lett 580:1872–1876
29. Winnik MA (1981) Cyclization and the conformation of hydrocarbon chains. Chem Rev 81:491–524
30. Belsom A, Mudd G, Giese S et al (2017) Complementary benzophenone cross-linking/mass spectrometry photochemistry. Anal Chem 89: 5319–5324
31. Wittelsberger A, Mierke DF, Rosenblatt M (2008) Mapping ligand-receptor interfaces: approaching the resolution limit of benzophenone-based photoaffinity scanning. Chem Biol Drug Des 71:380–383
32. Liu CC, Schultz PG (2010) Adding new chemistries to the genetic code. Annu Rev Biochem 79:413–444
33. Chin JW, Santoro SW, Martin AB et al (2002) Addition of p-azido-L-phenylalanine to the genetic code of *Escherichia coli*. J Am Chem Soc 124:9026–9027
34. Chin JW, Martin AB, King DS et al (2002) Addition of a photocrosslinking amino acid to the genetic code of *Escherichia coli*. Proc Natl Acad Sci U S A 99:11020–11024
35. Kauer JC, Erickson-Viitanen S, Wolfe HR Jr et al (1986) p-Benzoyl-L-phenylalanine, a new photoreactive amino acid. Photolabeling of calmodulin with a synthetic calmodulin-binding peptide. J Biol Chem 261:10695–10700
36. Tippmann EM, Liu W, Summerer D et al (2007) A genetically encoded diazirine photocrosslinker in *Escherichia coli*. Chembiochem 8: 2210–2214
37. Young TS, Ahmad I, Yin JA et al (2010) An enhanced system for unnatural amino acid mutagenesis in *E. coli*. J Mol Biol 395:361–374
38. Wang L, Brock A, Herberich B et al (2001) Expanding the genetic code of *Escherichia coli*. Science 292:498–500
39. Young DD, Schultz PG (2018) Playing with the molecules of life. ACS Chem Biol 13:854–870
40. de la Torre D, Chin JW (2021) Reprogramming the genetic code. Nat Rev Genet 22:169–184
41. Iacobucci C, Piotrowski C, Aebersold R et al (2019) First community-wide, comparative cross-linking mass spectrometry study. Anal Chem 91:6953–6961
42. Ihling CH, Piersimoni L, Kipping M et al (2021) Cross-linking/mass spectrometry combined with ion mobility on a timsTOF pro instrument for structural proteomics. Anal Chem 93:11442–11450
43. Gotze M, Pettelkau J, Schaks S et al (2012) StavroX--a software for analyzing crosslinked products in protein interaction studies. J Am Soc Mass Spectrom 23:76–87
44. Yang B, Wu YJ, Zhu M et al (2012) Identification of cross-linked peptides from complex samples. Nat Methods 9:904–906
45. Gotze M, Pettelkau J, Fritzsche R et al (2015) Automated assignment of MS/MS cleavable cross-links in protein 3D-structure analysis. J Am Soc Mass Spectrom 26:83–97
46. Pirklbauer GJ, Stieger CE, Matzinger M et al (2021) MS Annika: a new cross-linking search engine. J Proteome Res 20:2560–2569
47. Stachelhaus T, Mootz HD, Bergendahl V et al (1998) Peptide bond formation in nonribosomal peptide biosynthesis. Catalytic role of the condensation domain. J Biol Chem 273: 22773–22781

Chapter 9

A Chemoenzymatic Approach to Investigate Cytochrome P450 Cross-Linking in Glycopeptide Antibiotic Biosynthesis

Y. T. Candace Ho, Yongwei Zhao, Julien Tailhades, and Max J. Cryle

Abstract

Glycopeptide antibiotics (GPAs) are important and medically relevant peptide natural products. In the context of antimicrobial resistance (AMR), understanding and manipulating GPA biosynthesis is essential to discover new bioactive derivatives of these peptides. Among all the enzymatic steps in GPA biosynthesis, the most complex occurs during the maturation (cross-linking) of the peptide aglycone. This is achieved—while the peptide remains attached to the nonribosomal peptide synthetase (NRPS) machinery—through the action of a cytochrome P450 (CYP450 or Oxy)-mediated cyclization cascade. There is great interest in understanding the formation of the cross-links between the aromatic side chains in GPAs as this process leads to the cup-shaped aglycone, which is itself a requirement for antibiotic activity. In this regard, the use of in vitro experiments is crucial to study this process. To address the process of peptide cyclization during GPA biosynthesis, a series of peptide substrates and different Oxy enzymes are required. In this chapter, we describe a practical and efficient route for the synthesis of peptidyl-CoAs, the expression of proteins/enzymes involved in the in vitro cyclization assay, the loading of the PCP with peptidyl-CoAs, an optimized CYP450-mediated cyclization cascade and assay workup followed by mass spectrometry (MS) characterization. This in vitro assay affords high conversion to cyclic peptides and demonstrates the tolerance of the P450s for novel GPA precursor peptide substrates.

Key words Glycopeptide antibiotics, Solid-phase peptide synthesis, Coenzyme A, Bioconjugation, Nonribosomal peptide synthetase, Cytochrome P450

1 Introduction

Glycopeptide antibiotics (GPAs) are an important class of nonribosomal peptide synthetase (NRPS) metabolites; specifically, vancomycin (type I) and teicoplanin (type IV), members of the GPA class, are both used in the treatment of life-threatening infections by gram-positive bacteria (such as *Staphylococcus aureus*) [1]. These peptide natural products are assembled from amino acid precursors by an NRPS assembly line and are modified by the action of post-assembly line tailoring enzymes [2]. These complex molecules are typically heptapeptide aglycones comprising a combination of

Michael Burkart and Fumihiro Ishikawa (eds.), *Non-Ribosomal Peptide Biosynthesis and Engineering: Methods and Protocols*, Methods in Molecular Biology, vol. 2670, https://doi.org/10.1007/978-1-0716-3214-7_9,

proteinogenic and nonproteinogenic amino acids [3]. The peptide then undergoes oxidative cross-linking of aromatic side chains mediated by multiple cytochrome P450-catalyzed reactions, resulting in the insertion of aryl and phenolic links to achieve the rigid, active GPA conformation [4]. These peptide aglycones are then further modified after cleavage from the NRPS through a variety of post-assembly line modifications, including glycosylation and methylation [5].

The vancomycin peptide contains nonproteinogenic amino acids at residues 4, 5 (4-D-hydroxyphenylglycine, Hpg) and 7 (3,5-L-dihydroxyphenylglycine, Dpg) and contains three oxidative cross-links between aryl side chains at residues 2–4, 4–6, and 5–7 (*see* Fig. 1). These cross-links are introduced through the activity of cytochrome P450 (Oxy) enzymes to form the active GPA aglycone in the following order [6]: (1) C-*O*-D ring (linking residues 4 and 6 performed by OxyB), (2) D-*O*-E ring (linking residues 2 and 4 performed by OxyA), and (3) A-B ring (linking residues 5 and 7 performed by OxyC). The F-*O*-G ring (linking residues 1 and 3 performed by OxyE) is achieved in teicoplanin biosynthesis after the insertion of the C-*O*-D ring and before the insertion of the D-*O*-E ring. These cross-links are installed when the corresponding peptide is linked via a thioester to the peptidyl carrier protein (PCP) domain of the final NRPS module (C-A-PCP-X-TE), with the activity of the cross-linking cascade depending on the presence of a specialized neighboring Oxy recruitment domain, a condensation-like domain known as the X-domain [7]. X-domains provide a docking platform for these P450s to sequentially bind and cross-link their peptide substrates [8].

The Oxy enzymes are members of the cytochrome P450 superfamily of heme-containing monooxygenases. P450 enzymes display a high degree of substrate diversity throughout secondary metabolism and show flexibility in the oxidation chemistry that they catalyze [9]. The active site of P450s contains a heme cofactor that is essential for their function as monooxygenases, with the iron within the heme moiety tethered to the protein via a cysteine thiolate ligand [10]. The cysteine ligand bound to the heme iron needs to be deprotonated (i.e., a thiolate) for P450s to be active; in the protonated (thiol) form these enzymes are inactive, a state that generates a 420 nm soret maximum on reduction and CO complexation [11]. P450s generate a highly reactive intermediate in their active cycle by binding an oxygen molecule to the heme iron and then sequentially delivering two electrons and two protons; electron delivery to P450s is mediated by specific redox partner proteins from a cofactor such as NAD(P)H [12].

Due to the increasing threat of antibiotic resistance, such as methicillin-resistant *Staphylococcus aureus* (MRSA) and vancomycin-intermediate *S. aureus* (VISA), it is important to

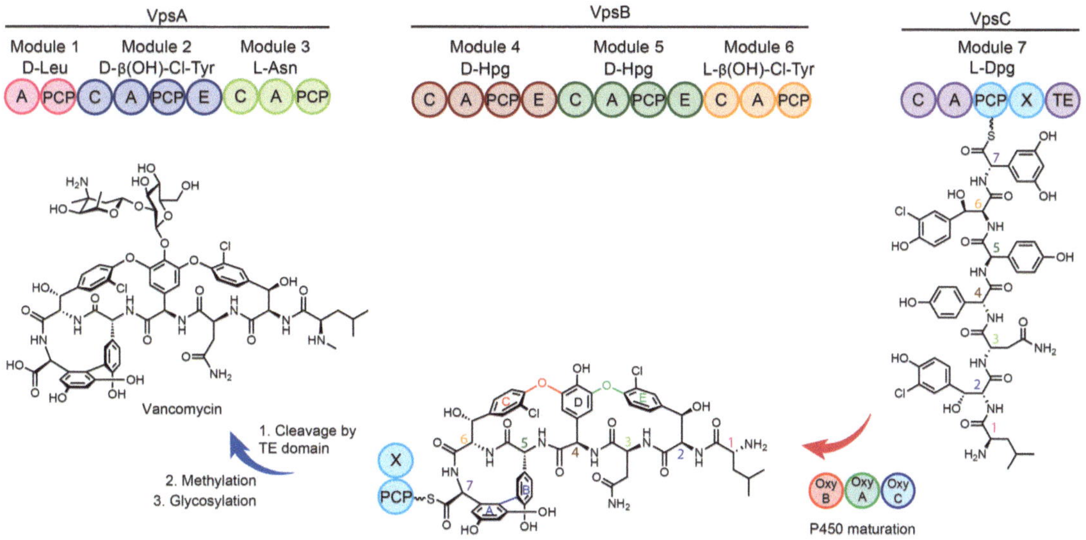

Fig. 1 Schematic description of vancomycin biosynthesis as performed by nonribosomal peptide synthesis, cytochrome-P450-mediated cross-linking, and peptide modification post-cleavage from the nonribosomal peptide synthetase. The domains (PCP-X) as discussed in this chapter for in vitro P450 cyclization cascade are indicated in cyan. Domain key—A: adenylation; C: condensation; E: epimerization; PCP: peptidyl carrier protein; X: oxygenase recruitment domain; TE: thioesterase; Oxy: Cytochrome P450

identify GPA modifications that could be useful for biosynthetic redesign of this antibiotic class. As the conversion of linear peptides into GPA aglycones is the source of the antibiotic activity of these compounds, it is important to understand the selectivity and flexibility of the P450 enzymes that perform this chemically challenging transformation [13, 14]. With a better understanding of the selectivity of these enzymes, it would then be possible to engage in redesign of the NRPS and resultant peptide in order to generate new compounds together with the confidence that the peptide products of the NRPS would be able to be appropriately modified by the P450 cross-linking enzymes.

The central domains in NRPS-mediated biosynthesis are peptidyl carrier proteins (PCPs), as they are involved in the delivery of amino acid building blocks and peptide intermediates to the various catalytic NRPS domains during peptide assembly [15]. Substrates are tethered to PCP domains as a thioester via the phosphopantetheine (PPant) posttranslational modification, which is originally derived from coenzyme A (CoA) [16]. In GPA biosynthesis, the oxidative cross-linking of the linear peptides occurs while the peptide is tethered to the PCP domain in the final module of the NRPS.

To explore the selectivity and flexibility of the P450 monooxygenase enzymes toward their PCP-bound peptide substrates, a straightforward method is required; in order to aid discussion, a general workflow of the chemical synthesis of peptide substrates

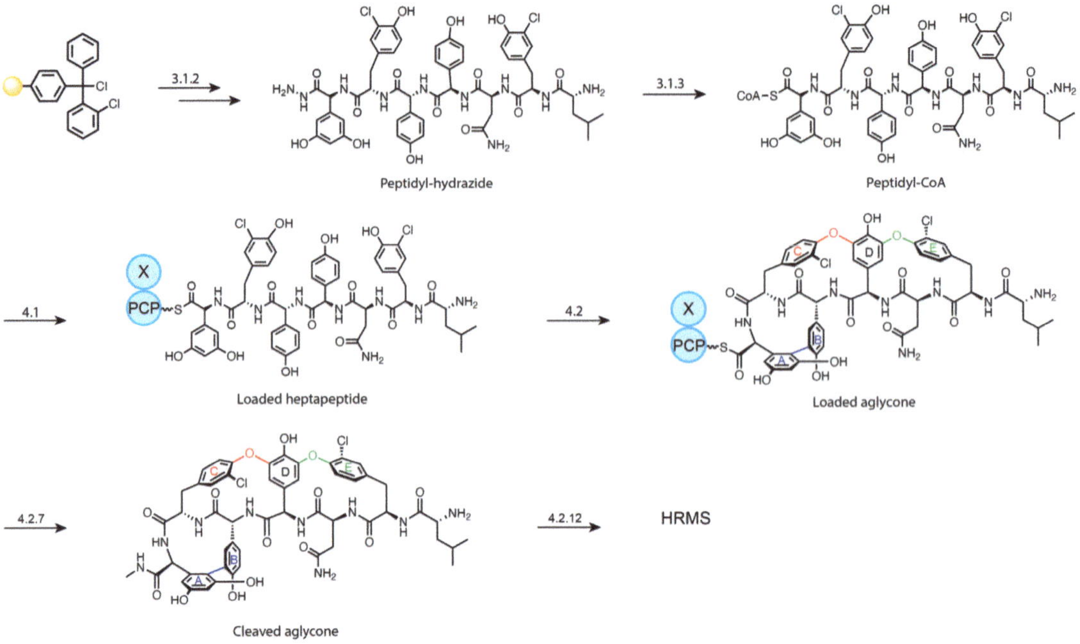

Fig. 2 The general workflow described in this chapter. The numbering refers to the corresponding numbering found in the text. Key—yellow ball: 2-chlorotrityl chloride resin; PCP: peptidyl carrier protein; X: oxygenase recruitment domain; Oxy: Cytochrome P450

and subsequent enzymatic cyclization with Oxy enzymes in vitro is shown in Fig. 2.

First, a linear peptidyl-hydrazide is synthesized by a solid-phase peptide synthesis (SPPS) based on Fmoc chemistry. With optimized conditions for amino acid coupling and cleavage of the Fmoc-protecting group on the amino-functionality, racemization (especially of Hpg and Dpg) can be dramatically reduced during SPPS [17]. This also significantly reduces workload and time (coupling reaction and purification) for the synthesis as well as producing less waste. Furthermore, incorrect peptide stereochemistry can be deleterious for P450-mediated cross-linking. We use commercially available 2-chlorotrityl chloride resin to produce a linear peptidyl hydrazide (*see* Subheading 3.1.2) before subsequently activating the hydrazide and exchanging it with CoA to generate the desired peptidyl-CoA (*see* Subheading 3.1.3).

Second, all the proteins that are required for the in vitro assay, such as PCP-X didomain, Sfp, Oxy (B_{van}, A_{ris}, C_{cep}, and E_{ris}), and PuR/PuxB must be expressed and purified (*see* Subheading 3.2). All proteins that contain PCP domains should be expressed and purified in their *apo* form, which can be converted into their *holo* form using the phosphopantetheinyl transferase Sfp (R4-4 mutant) and desired peptidyl-CoAs [18].

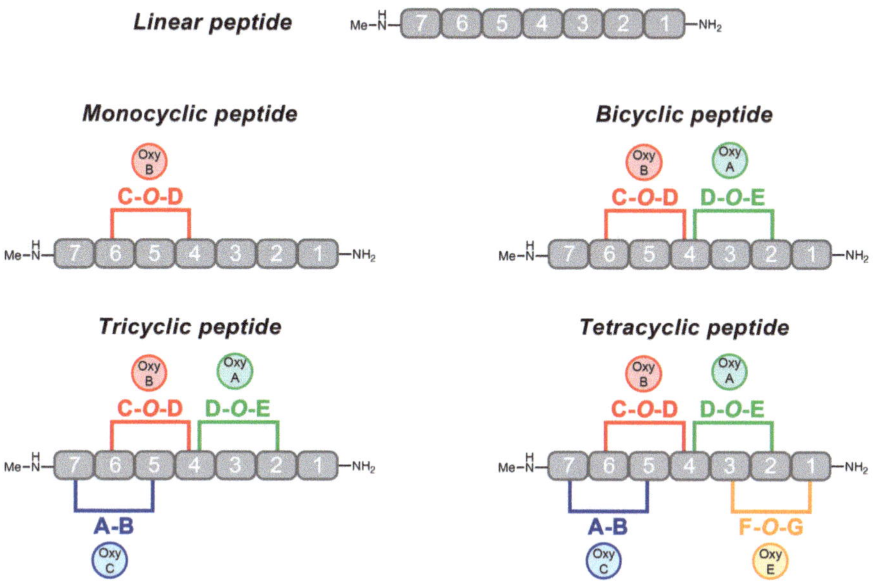

Fig. 3 Examples of cross-linking patterns reconstituted using different combinations of Oxy enzymes in the in vitro P450 cyclization assay

Third, the chemoenzymatic cyclization is performed in three steps: (1) peptidyl-CoA loading onto the PCP-X protein (*see* Subheading 4.1)—the PCP-X didomain from module 7 of teicoplanin $PCP\text{-}X_{tei}$ biosynthesis can be exploited to study cytochrome P450-catalyzed oxidative cross-linking reactions for type I-IV GPAs [17]. As an alternative, the PCP-X didomains from complestatin ($PCP\text{-}X_{com}$) or kistamicin ($PCP\text{-}X_{kis}$) biosynthesis can be used to study the cross-linking reactions for type V GPAs [19]. (2) P450-catalyzed cross-linking reactions (*see* Fig. 3 and Subheading 4.2). (3) Cleavage of the cross-linked peptides from the PCP-X construct (*see* Subheading 4.2, **step** 7). Methylamine is used to chemically cleave the peptide from the protein to terminate the reaction. This results in the formation of peptide methylamides, including unreacted substrates (linear peptidyl-methylamide) and the cyclized products (mono-/bi-/tri-cyclized peptidyl-methylamide). The methylamide peptides are purified from the reaction mixture via solid-phase extraction (SPE) before analysis by liquid chromatography mass spectrometry (LC-MS). From the analysis of the LC-MS results, the conversion of methylamide peptides performed by individual Oxy enzymes can be calculated. This can have major implications for biosynthetic engineering of the peptide substrate, with an example being the major differences seen in P450 conversion of GPA peptides containing positively charged residues in either position 1 or 3; at position 1 these were shown to be well accepted, while at position 3 these are largely incompatible. We believe that this methodology could be of general relevance for the exploration of other NRPS-derived natural products containing P450-mediated aromatic cross-linking such as arylomycin [20] and the newly discovered type V GPA corbomycin [21].

2 Materials

2.1 Synthesis of Protected Amino Acids

1. Unprotected amino acids: (S)-3,5-dihydroxyphenylglycine (H-L-Dpg-OH, Netchem Inc), (S)/(R) 4-hydroxyphenylglycine (H-L/D-Hpg-OH, Oakwood chemical), H-L/D-Tyr(3Cl)-OH (Mimotopes).
2. Fmoc N-hydroxysuccinimide ester (Fmoc-OSu, GL Biochem).
3. Sodium carbonate solution: 10% Na_2CO_3 solution in Milli-Q water (Sigma-Aldrich).
4. Acetonitrile (ACN, Thermo Fisher).
5. Brine: saturated aqueous sodium chloride.
6. Diethyl ether (Sigma-Aldrich).
7. 2 M hydrochloric acid (Sigma-Aldrich).
8. Ethyl acetate (Sigma-Aldrich).
9. Hexane (Sigma-Aldrich).
10. Sodium sulfate (Na_2SO_4, Sigma-Aldrich).

2.2 Solid-Phase Peptide Synthesis (SPPS)

1. Commercially available amino acids: Boc-Gly-OH (Sigma-Aldrich), Fmoc-L-Asn(Trt)-OH (Sigma-Aldrich), and Boc-D-Leu-OH (Sigma-Aldrich).
2. 2-chlorotrityl chloride resin (GL Biochem).
3. Dichloromethane (DCM, Thermo Fisher).
4. *N*,*N*-dimethylformamide (DMF, Thermo Fisher).
5. 5% hydrazine solution in DMF.
6. A capping solution: 70% DMF, 20% triethylamine (TEA), and 10% methanol (MeOH).
7. COMU (Oakwood chemical).
8. 0.12 M of 2,6-lutidine in DMF.
9. 1% 1,8-diazabicyclo[5.4.0]undec-7-ene (DBU) in DMF.
10. TFA cleavage cocktail: 95% trifluoroacetic acid (TFA), 2.5% triisopropylsilane (TIS), and 2.5% water.
11. Ice-cold diethyl ether (Sigma-Aldrich).

2.3 Peptidyl Hydrazide Displacement with CoA

1. Urea buffer: 6 M urea, 0.2 M $NaHPO_3$ at pH 3.
2. 0.5 M aqueous sodium nitrate (Sigma-Aldrich).
3. Coenzyme A (CoA) trilithium salt (Chem-Impex).
4. 1 M potassium phosphate buffer at pH 8: 6% KH_2POH_4 and 94% K_2HPO_4 buffer.
5. Salt/ice bath at −15 °C.

2.4 Protein Expression and Purification

1. Plasmids of interest: PCP-X_{tei} didomain, Sfp mutant R4-4 (Sfp), Oxy (B_{van}, A_{ris}, C_{cep}, and E_{ris}), palustrisredoxin reductase (PuR), and palustrisredoxin B A105V (PuxB) (*see* **Notes 1** and **2**).
2. Competent *E. coli* BL21(DE3) cells.
3. Competent *E. coli* BL21(DE3) (*entD-*) cells.
4. Competent *E. coli* ArcticExpress (DE3) cells.
5. 50 mg/mL kanamycin stock solution (Sigma-Aldrich).
6. 20 mg/mL gentamicin stock solution (Sigma-Aldrich).
7. Lysogeny broth (LB) media.
8. Terrific broth (TB) media.
9. 5-aminolevulinic acid hydrochloride (Sigma-Aldrich).
10. Isopropyl-β-D-thiogalactopyranoside (IPTG, Sigma-Aldrich).
11. Protease inhibitor cocktail tablets (SIGMAFAST Protease Inhibitor Cocktail Tablets, EDTA-Free; Sigma-Aldrich).
12. Benzonase® Nuclease (Sigma-Aldrich).
13. Ni-NTA agarose (Macherey-Nagel).
14. Sodium dodecyl sulfate (SDS) gel.
15. Precision Plus Protein™ Dual Colour Standards (Bio-Rad).
16. 0.2 mL PCR tubes (Pacific Laboratory Products).

2.5 Loading of Peptidyl-CoA onto apo PCP-X Didomain

1. *Apo* PCP-X didomain.
2. 4′-Phosphopantetheinyl-transferase Sfp R4-4 mutant (Sfp).
3. Peptidyl-CoA.
4. 1 M 4-(2-hydroxyethyl)-1-piperazineethanesulfonic acid (HEPES), pH 7.0 (Sigma-Aldrich).
5. 5 M sodium chloride (NaCl, Sigma-Aldrich).
6. 1 M magnesium chloride ($MgCl_2$, Sigma-Aldrich).
7. 1.5 mL microtubes.
8. Centrifugal concentrators (Amicon® Ultra-0.5 mL centrifugal filter units (30 kDa MWCO)).
9. PCP-loading buffer: 50 mM HEPES pH 7.0, 50 mM NaCl, and 10 mM $MgCl_2$.

2.6 CYP450-Catalyzed Cross-Linking Reaction

1. 50 mM NADH stock solution in Milli-Q water, stored at −80 °C (Sigma-Aldrich).
2. 2 mg/mL glucose dehydrogenase stock solution in Milli-Q water, stored at −80 °C (Sigma-Aldrich).
3. 20% glucose stock solution in Milli-Q water, stored at −80 °C (Sigma-Aldrich).
4. From PCP-loading reaction: 30 mM *holo* peptidyl-PCPX in PCP-loading buffer (*see* Subheading 4.1).
5. Redox system: PuR and PuxB.
6. Cytochrome P450s (Oxy B_{van}, A_{ris}, C_{cep}, and E_{ris}).
7. 60 mM cysteamine (Sigma-Aldrich) (*see* **Note 3**).
8. 60 mM tris(2-carboxyethyl) phosphine (TCEP, Sigma-Aldrich) (*see* **Note 4**).
9. Methylamine (40% in water, Sigma-Aldrich).
10. Solid-phase extraction (SPE) columns: Bond Elut Plexa 30 mg/mL (Agilent Technologies).
11. SPE column activation solvent: 0.1% formic acid (FA) in MeOH.
12. SPE column equilibration buffer: 0.1% FA in water.
13. SPE column elution solvent: 50% acetonitrile in Milli-Q water with 0.1% formic acid (FA).
14. 1.5 mL microtubes.

2.7 Equipment

1. Tribute—UV peptide Synthesizer with UV monitoring (Protein Technologies).
2. Rotary evaporator (Vaccubrand PC 3001 Vario Pro).
3. Freeze dryer (Christ Alpha 1–2 LD plus).
4. Spark-proof floor centrifuge (Spintron GT-175).
5. High-performance liquid chromatography (HPLC, Shimadzu LC-20AP) system equipped with analytical C18 columns (Agilent ZORBAX 300SB-C18, 7 μm, 212 × 250 mm).
6. Liquid chromatography mass spectrometry (LC-MS, Shimadzu LCMS-2020 LC/MS) system equipped with analytical C18 column (Agilent ZORBAX 300SB-C18, 5 μm, 4.6 × 250 mm).
7. Thermomixer (Eppendorf).
8. Floor incubator (Infors HT Multitron standard).
9. Oven (Steridium).
10. Floor centrifuge (Beckman Coulter).
11. Cell disruptor (Avestin EmulsiFlex, ATA scientific).
12. Thermomixer (Eppendorf).

13. Amicon® Ultra Centrifugal Filters (Merck Millipore).
14. Benchtop centrifuge (Eppendorf).
15. HiTrap Q HP anion-exchange chromatography column (Cytiva).
16. Gel-filtration chromatography SRT 10 SEC 300 (105 mL) column (Sepax Technologies).
17. ÄKTA PURE system (Cytiva).
18. Electrophoresis chamber (Bio-rad).
19. PowerPac™ Basic Power Supply (Bio-Rad).
20. Mini-PROTEAN Tetra Vertical Electrophoresis Cell (Bio-Rad).
21. NanoDrop One microvolume UV-vis spectrophotometer (Thermo Scientific).
22. High-resolution mass spectrometry (HRMS, Thermo Scientific, Orbitrap Fusion mass spectrometer) coupled online to a nano-LC (Ultimate 3000 RSLCnano; Thermo Scientific), equipped with a 50 cm reverse-phase column (Acclaim PepMap RSLC, 75 μm × 50 cm, nanoViper, C18, 2 μm, 100 Å; Thermo Scientific) after binding to a trap column (Acclaim PepMap 100, 100 μm × 2 cm, nanoViper, C18, 5 μm, 100 Å; Thermo Scientific).

3 Methods

3.1 Peptidyl-CoA Synthesis

Synthesis of peptidyl-hydrazides on solid phase at 0.05 mmol scale using chemically synthesized as well as commercially acquired Fmoc-amino acids (*see* **Notes 5** and **6**). Subsequent hydrazide activation and displacement with coenzyme A generates the desired peptidyl-CoA thioesters [22].

3.1.1 Amino Acid Synthesis

Protect all the unprotected amino acid with Fmoc-OSu for SPPS as follows:

1. Dissolve an unprotected amino acid (30 mmol) in a solution of ACN/10% Na_2CO_3 (v/v′, 1: 1, 100 mL).
2. Stir the mixture and cool it to 4 °C.
3. Add Fmoc-OSu (33 mmol) portion wise over 30 min.
4. Leave the reaction to stir overnight.
5. Dilute the reaction with water (100 mL) and extract with diethyl ether (2 × 100 mL).
6. Acidify the aqueous phase to pH 3 with aqueous hydrochloric acid (2 M).
7. Extract with ethyl acetate (3 × 100 mL).

8. Wash the combined organic layers with brine (2 × 100 mL), dry over Na_2SO_4, and remove the solvent in vacuo.
9. Purification:
   (a) Fmoc-L-Dpg-OH—sonicate the powder into a 1:9 mixture of diethyl ether/hexane and remove the solvent by filtration (2 × 30 min).
   (b) Fmoc-L/D-Hpg-OH—recrystallize the compound into a mixture of ethyl acetate and hexane.
   (c) Fmoc-L/D-Tyr(3Cl)-OH—flash column chromatography using a gradient of 0–100% ethyl acetate in hexane (+0.1% acetic acid).

3.1.2 Solid-Phase Peptide Synthesis

1. Resin swelling: shake the 2-chlorotrityl chloride resin (200 mg) in DCM (8 mL) for 30 min, followed by washing with DMF (3×) (*see* **Notes** 7 and **8**).
2. Incubate the resin with a 5% hydrazine solution in DMF (6 mL, 2 × 30 min), followed by washing with DMF (3×).
3. Add a solution of DMF/TEA/MeOH (v/v′/ v″, 7:2:1, 4 mL) to cap unreacted 2-chlorotrityl groups of the resin, shake for 15 min, and subsequently wash with DMF (3×).
4. Coupling the first Fmoc-protected amino acid: activate the first Fmoc-protected amino acid (0.06 mmol) with COMU (0.06 mmol) and 2,6-lutidine (0.12 M in DMF, 3 mL) for 1 min. Add the solution to the resin and shake overnight, followed by washing with DMF (3×).
5. Cap the unreacted hydrazine group with Boc-Gly-OH (0.15 mmol) that has been activated using COMU (0.15 mmol) and 2,6-lutidine (0.15 mmol, 0.12 M in DMF, 3 mL) and shaken for 1 h, followed by washing with DMF (3×).
6. Fmoc deprotection: treat the resin with 3 mL of a 1% DBU solution in DMF and shake for 30 s (3×), following by washing with DMF (4×) (*see* **Note 9**).
7. Amino acid activation and coupling: activate Fmoc-protected amino acids (0.15 mmol) with COMU (0.15 mmol) and 2,6-lutidine (0.12 M in DMF, 3 mL) for 1 min, then add to the resin. Shake the solution for 40 min, followed by washing with DMF (3×) (*see* **Note 10**).
8. Repeat **steps 6** and 7 until completion of the peptide sequence (*see* **Notes 11** and **12**).
9. Cleavage: cleave the hydrazide peptide from the resin and remove protecting groups (Trt and Boc) using a solution of TFA/TIS/H_2O (v/v′/v″, 95:2.5:2.5, 5 mL), and shake the reaction for 1 h (*see* **Note 13**).

10. Remove the resin by filtration and wash it with TFA (2×).
11. Concentrate the filtrate under a N_2 stream to ~1 mL.
12. Precipitate the peptide with ice-cold diethyl ether (~8 mL).
13. Collect the peptide by centrifugation in a flame-resistant centrifuge at 2470 *g*.
14. Wash the peptide with ice-cold diethyl ether (3×).
15. Purify the crude peptide using preparative RP-HPLC (gradient in ACN; 10–40% or 15–45% over 30 min).

3.1.3 Peptidyl Hydrazide Displacement with CoA

1. Dissolve purified hydrazide peptides in buffer A (6 M urea and 0.2 M NaH_2PO_4, pH 3) to a final concentration of 5 mM (*see* **Note 14**).
2. Cool the solution to −15 °C using salt/ice bath.
3. Add 0.5 M aqueous $NaNO_2$ (0.95 equiv.) to the mixture and stir the reaction for 10 min (*see* **Note 15**).
4. Add CoA (1.2 equiv., dissolved in buffer A) to the reaction and adjust the pH to 6.5 using 1 M potassium phosphate buffer at pH 8.
5. Stir the reaction mixture for an additional 2 h (warming from −15 to 4 °C).
6. Directly purify the crude peptidyl-CoA using preparative RP-HPLC (gradient in ACN; 10–40% or 15–45% over 30 min).
7. Aliquot the desired peptidyl-CoA into microtubes in the amount required for an enzymatic reaction, freeze-dry and store at −20 °C (*see* **Note 16**).

3.2 Protein Expression and Purification

Express and purify proteins PCP-X didomain, Oxy (B_{van}, A_{ris}, C_{cep}, and E_{ris}), and Sfp for the enzymatic assay as follows:

3.2.1 PCP-X Didomain Expression and Purification

1. Transform the PCP-X expression plasmid pET-GB1-1d into chemically competent *E. coli* BL21(DE3) (*entD*-) cells (*see* **Note 17**).
2. Incubate the mixture on ice for 30 min.
3. Heat-shock the bacterial cells at 42 °C for 30 s.
4. Place the cells on ice for 2 min.
5. Recover the bacterial cells by adding LB media (0.5 mL) and incubating at 37 °C for 45 min.
6. Plate the cells on a LB agar plate supplemented with 50 μg/mL kanamycin and incubate the plate at 37 °C overnight.

7. Pick a single colony from the agar plate and inoculate into a 200 mL LB pre-culture supplemented with 50 μg/mL kanamycin and incubate at 37 °C overnight with shaking at 180 rpm.
8. Use the pre-culture to inoculate 12 × 1 L of TB medium supplemented with 50 μg/mL kanamycin.
9. Incubate the culture at 37 °C with shaking at 180 rpm until the $OD_{600\ nm}$ reaches 0.4–0.6.
10. Add IPTG to the TB culture to a final concentration of 0.1 mM and cool the culture to 18 °C.
11. Continue to shake at 180 rpm at 18 °C overnight.
12. Harvest the cells by centrifugation at 3064 *g* for 20 min at 4 °C.
13. Resuspend the cells in 200 mL of Ni-NTA buffer A (50 mM Tris–HCl pH 8.0, 300 mM NaCl, 10 mM imidazole) supplemented with a protease inhibitor cocktail tablet and Benzonase® Nuclease (10 μL).
14. Lyse the cells by a cell disruptor operating at 14,000–19,000 psi, and clarify the lysate by centrifugation at 22,680 *g* for 45 min at 4 °C.
15. Equilibrate 2 mL of Ni-NTA agarose with Ni-NTA buffer A.
16. Incubate the supernatant at 4 °C for 1 h with the Ni-NTA agarose with gentle stirring.
17. After incubation, wash the beads with 20 bed volumes of Ni-NTA buffer A.
18. Elute bound protein with 5 bed volumes of Ni-NTA buffer B (50 mM Tris–HCl pH 8.0, 300 mM NaCl, 1 M imidazole).
19. Dialyze the protein against a buffer of 50 mM Tris–HCl pH 7.4 and 300 mM NaCl at 4 °C for 2 h.
20. Further purify the protein by gel-filtration chromatography using an SRT 10 SEC 300 (105 mL) column connected to an ÄKTA PURE system.
21. Equilibrate the column with 1.2 column volumes of gel-filtration buffer (50 mM Tris–HCl, pH 7.4; 300 mM NaCl).
22. Concentrate the protein to a desired volume and inject onto the column.
23. Fractionate the eluate into 1.5 mL fractions.
24. Analyze the elution fractions containing monomeric protein by SDS-PAGE.
25. Combine and concentrate appropriate fractions using centrifugal filter units (30 kDa MWCO).

26. Determine the protein concentration by measuring protein absorbance at 280 nm using a NanoDrop One microvolume UV-vis spectrophotometer.
27. Concentrate and aliquot the protein into chilled 0.2 mL PCR tubes, flash-freeze in liquid nitrogen and store at −80 °C.

3.2.2 Cytochrome P450 Oxy Expression and Purification

1. Transform the plasmid encoding the protein of Oxy (cloned in the pET 28a(+)) into chemically competent *E. coli* ArcticExpress (DE3) cells.
2. Incubate the mixture on ice for 30 min.
3. Heat-shock the bacterial cells at 42 °C for 30 s.
4. Place the cells on ice for 2 min.
5. Recover the bacterial cells by adding LB media (0.5 mL) and incubating at 37 °C for 45 min.
6. Plate the cells on an LB agar plate supplemented with 50 μg/mL kanamycin and 20 μg/mL gentamicin and incubate the plate at 37 °C overnight.
7. Pick a single colony from the agar plate and inoculate into a 200 mL LB pre-culture supplemented with 50 μg/mL kanamycin and 20 μg/mL gentamicin and incubate at 37 °C overnight with shaking at 180 rpm.
8. Use the pre-culture to inoculate 12 × 1 L of TB medium supplemented with 50 μg/mL kanamycin.
9. Add 5-aminolevulinic acid (heme precursor) to the final concentration of 200 μg/mL into the culture (*see* **Note 18**).
10. Incubate the culture at 37 °C with shaking at 180 rpm until the $OD_{600\ nm}$ reaches 0.6–1.0.
11. Perform cold shock by incubating the culture at 4 °C for 15 min (*see* **Note 19**).
12. Add IPTG to the TB culture to a final concentration of 0.05 mM.
13. Continue to shake at 120 rpm at 18 °C overnight.
14. Harvest the cells by centrifugation at 3064 *g* for 20 min at 4 °C.
15. Resuspend the cells in 200 mL of Ni-NTA buffer A (50 mM Tris–HCl pH 8.0, 300 mM NaCl, 10 mM imidazole) supplemented with a protease inhibitor cocktail tablet and Benzonase® Nuclease (10 μL).
16. Lyse the cells by a cell disruptor operating at 14,000–19,000 psi and clarify the lysate by centrifugation at 22,680 *g* for 45 min at 4 °C.
17. Equilibrate 4 mL of Ni-NTA agarose with Ni-NTA buffer A.

18. Incubate the supernatant at 4 °C for 1 h with the Ni-NTA agarose with gentle stirring.
19. After incubation, wash the beads with 20 bed volumes of Ni-NTA buffer A.
20. Elute bound protein with 5 bed volumes of Ni-NTA buffer B (50 mM Tris–HCl pH 8.0, 300 mM NaCl, 1 M imidazole).
21. Dialyze the protein against a buffer of 50 mM Tris–HCl pH 7.4 and 10 mM NaCl at 4 °C for 2 h.
22. Purify the protein by HiTrap Q HP anion-exchange chromatography column connected to an ÄKTA PURE system (*see* **Note 20**).
23. Equilibrate the column with 5 column volumes of anion-exchange buffer A (50 mM Tris–HCl, pH 7.4; 10 mM NaCl).
24. Concentrate the protein to a desired volume and load onto the column.
25. Fractionate the eluate into 1.5 mL fractions by eluting the column with 5 column volumes using a gradient of 0–50% of anion-exchange buffer B (50 mM Tris–HCl, pH 7.4; 1 M NaCl) (*see* **Note 21**).
26. Analyze the elution fractions containing monomeric protein by SDS-PAGE.
27. Combine and concentrate appropriate fractions using centrifugal filter units (30 kDa MWCO).
28. Further purify the protein by gel-filtration chromatography using an SRT 10 SEC 300 (105 mL) column connected to an ÄKTA PURE system.
29. Equilibrate the column with 1.2 column volumes of gel-filtration buffer (50 mM Tris–HCl, pH 7.4; 100 mM NaCl).
30. Concentrate the protein to a desired volume and inject onto the column.
31. Fractionate the eluate into 1.5 mL fractions.
32. Analyze the elution fractions containing monomeric protein by SDS-PAGE.
33. Combine and concentrate appropriate fractions using centrifugal filter units (30 kDa MWCO).
34. Determine the protein concentration by measuring protein absorbance at 280 nm using a NanoDrop One microvolume UV-vis spectrophotometer.
35. Concentrate and aliquot the protein into chilled 0.2 mL PCR tubes, flash-freeze in liquid nitrogen and store at −80 °C.

3.2.3 Sfp Mutant R4-4 Expression and Purification

1. Transform the Sfp expression plasmid pET28a(+) into chemically competent *E. coli* BL21(DE3) cells.
2. Incubate the mixture on ice for 30 min.
3. Heat-shock the bacterial cells at 42 °C for 30 s.
4. Place the cells on ice for 2 min.
5. Recover the bacterial cells by adding LB media (0.5 mL) and incubating at 37 °C for 45 min.
6. Plate the cells on a LB agar plate supplemented with 50 μg/mL kanamycin and incubate the plate at 37 °C overnight.
7. Pick a single colony from the agar plate and inoculate into a 100 mL LB pre-culture supplemented with 50 μg/mL kanamycin and incubate at 37 °C overnight with shaking at 180 rpm.
8. Use the pre-culture to inoculate 6 × 1 L of LB medium supplemented with 50 μg/mL kanamycin.
9. Incubate the culture at 37 °C with shaking at 180 rpm until the $OD_{600\ nm}$ reaches 0.4–0.6.
10. Add IPTG to the LB culture to a final concentration of 0.1 mM and cool the culture to 18 °C.
11. Continue to shake at 180 rpm at 18 °C overnight.
12. Harvest the cells by centrifugation at 3064 *g* for 20 min at 4 °C.
13. Resuspend the cells in 200 mL of Ni-NTA buffer A (20 mM Tris–HCl pH 7.9, 500 mM NaCl, 5 mM imidazole) supplemented with a protease inhibitor cocktail tablet and Benzonase® Nuclease (5 μL).
14. Lyse the cells by a cell disruptor operating at 14,000–19,000 psi, and clarify the lysate by centrifugation at 22,680 *g* for 45 min at 4 °C.
15. Equilibrate 2 mL of Ni-NTA agarose with Ni-NTA buffer A.
16. Incubate the supernatant at 4 °C for 1 h with the Ni-NTA agarose with gentle stirring.
17. After incubation, wash the beads with 20 bed volumes of Ni-NTA buffer A.
18. Elute bound protein with 5 bed volumes of Ni-NTA buffer B (20 mM Tris–HCl pH 7.9, 500 mM NaCl, 300 M imidazole).
19. Analyze the elution fractions containing monomeric protein by SDS-PAGE.
20. Dialyze the protein twice against a buffer of 10 mM Tris–HCl pH 7.5, 1 mM EDTA, and 10% glycerol at 4 °C for 2 h.
21. Concentrate and aliquot the protein solution into 0.2 mL PCR tubes, flash-freeze in liquid nitrogen, and store at −80 °C.

3.2.4 PuR/PuxB Expression and Purification [23]

1. Transform the PuR or PuxB expression plasmid pET26 into chemically competent *E. coli* BL21(DE3) cells.
2. Follow **steps 2** through **12** in Subheading 3.2.3.
3. Resuspend the cells in 200 mL of anion-exchange buffer A (20 mM Tris–HCl pH 7.4, 50 mM NaCl) supplemented with a protease inhibitor cocktail tablet and Benzonase® Nuclease (5 μL).
4. Lyse the cells by a cell disruptor operating at 14,000–19,000 psi, and clarify the lysate by centrifugation at 22,680 *g* for 45 min at 4 °C.
5. Follow **steps 22** through **35** in Subheading 3.2.2.

4 Enzymatic Cyclization

To perform these assays, first load the peptidyl-CoA onto an *apo* PCP-X didomain followed by P450-catalyzed cross-linking. The cross-linked peptides are then cleaved from the PCP-X construct and analyzed by HRMS and MSMS.

4.1 Loading of Peptidyl-CoA onto an Apo *PCP-X Didomain*

Each enzymatic assay is carried out at a volume of 87.5 μL per reaction (*see* **Note 22**), using the protocol below (*see* **Note 23**):

1. Take an aliquot of peptidyl-CoA (7.3 μg) and dilute it with Milli-Q water (63 μL) (*see* **Notes 24** and **25**).
2. Add the peptidyl-CoA (45 μM) to a microtube tube containing PCP-X (30 μM) in PCP-loading buffer (50 mM HEPES, pH 7.0; 50 mM NaCl; 10 mM $MgCl_2$) on ice (*see* **Note 26**).
3. Add Sfp (3 μM) into the tube.
4. Mix the reaction and incubate for 1 h at 30 °C.
5. Remove the unloaded peptidyl-CoA and Sfp mutant R4-4 by sequential concentration/dilution steps using centrifugal concentrator (30 kDa MWCO) with PCP-loading buffer at 4 °C.
6. Repeat **step 5** (×3).
7. Bring the volume of the reaction to 87.5 μL with PCP-loading buffer.
8. Keep the reaction on ice until further use.

4.2 CYP450-Catalyzed Cross-Linking Reaction on Type I–IV GPAs

1. Set up the turnover reaction in a total volume of 105 μL in reaction buffer in a 1.5 mL microtube (*see* **Note 27**).
2. Into the tube containing peptidyl-PCP-X_{tei} (30 μM), add Oxy-B_{van} (0.5 μM), PuR (0.66 μM), PuxB A105V (2.5 μM), glucose (0.33%), glucose dehydrogenase (0.033 mg/mL), cysteamine (250 μM), TCEP (2500 μM), and NADH (2 mM).

3. Incubate the reaction for 20 min at 30 °C with 300 rpm shaking.
4. Optional: for teicoplanin-like peptidyl substrate (F-*O*-G ring), add $OxyE_{ris}$ (1.0 μM) and incubate the reaction for 20 min at 30 °C at 300 rpm [24].
5. Add $OxyA_{ris}$ (1.0 μM (2.0 μM if OxyE has been added)) and incubate the reaction for 20 min at 30 °C with 300 rpm shaking.
6. Add $OxyC_{cep}$ (2.0 μM (4.0 μM if OxyE has been added)) and incubate overnight at 30 °C with 300 rpm shaking.
7. Cleave the cross-linked peptides from the PCP-X didomain by addition of 40% methylamine solution in water (0.5 M) and incubate for 15 min at room temperature.
8. Meanwhile, prepare the solid-phase extraction columns to purify the peptides; activate the cartridge with 0.1% FA in MeOH (1 mL) and equilibrate with 0.1% FA in water (1 mL).
9. Add water with 0.1% FA (850 μL) to each reaction and load the solution onto the equilibrated SPE-column via gravity flow (*see* **Note 28**).
10. Wash the column with 0.1% FA in water (1 mL) and elute with 1 mL of 0.1% FA in ACN/H_2O (50/50) (*see* **Note 29**).
11. Flash-freeze the samples with liquid nitrogen and dry them using a freeze dryer.
12. For RP-HPLC-MS analysis, dilute the elution fraction to a final volume of 50 μL using 10% ACN in water. Analyze the samples by analytical reversed-phase HPLC-MS.

5 Notes

1. $PCP\text{-}X_{tei}$ (tcp12, UniProt protein ID: Q70AZ6), $OxyB_{van}$ (UniProt protein ID: G4V4R5), $OxyA_{ris}$ (UniProt protein ID: A0A075V230), $OxyC_{cep}$ (UniProt protein ID: O52825 and O52816), $OxyE_{ris}$ (UniProt protein ID: A0A075UYD8), PuR (UniProt protein ID: Q6N3B2), PuxB (UniProt protein ID: Q6N2U2), Sfp mutant R4-4 was generously provided by Professor Jun Yin [25].
2. PuR/PuxB is an NADH-dependent redox cascade that is compatible with GPA Oxy enzymes.
3. Always prepare the stock solution of cysteamine freshly.
4. Always prepare the stock solution of TCEP freshly.
5. All reactions are carried out at room temperature.

6. The solid-phase peptide synthesis can be performed on a Tribute UV peptide synthesizer or synthesized manually.
7. Do not use a magnetic stir bar to mix the resin as this destroys the beads.
8. It is essential that the resin is well swollen to guarantee efficient solvent access to the resin.
9. Best results of Fmoc cleaved after DBU treatment are obtained by incubating the resin with several batches of fresh DBU solution rather than incubating it once over a longer period.
10. Effective mixing of the resin is important to ensure proper coupling.
11. The final amino acid to be added to the peptide is preferably Boc-protected.
12. Resin can be dried by diethyl ether and stored in a closed tube until further use.
13. Shake for 2 h for arginine-containing peptides.
14. Make sure the urea buffer is pH 3 before use.
15. Make sure the solution is not frozen for the entire duration of the reaction.
16. Each enzymatic assay is carried at a volume of 87.5 μL and the molarity of the peptidyl-CoA is 45 μM.
17. To avoid producing *holo* PCP, as the phosphopantetheine arm can be installed by the phosphopantetheinyl transferase EntD that phosphopantetheinylates some PCP domains when they are expressed in *E. coli* under low iron conditions [26].
18. 5-aminolevulinic acid is a precursor of heme [27].
19. Cold-shock chaperones are considered to help to increase the efficiency of translation.
20. To remove non-heme bound proteins.
21. Add 420 nm wavelength channel to detect the heme soret of the P450 enzyme.
22. The loading can be achieved at once for the same peptidyl-CoA in function of the number of reaction n leading to $n \times 87.5$ μL.
23. The loading reaction utilizes a 1:1.5:0.1 molar ratio of the PCP domain to peptidyl-CoA to Sfp.
24. Always freshly dissolve the peptide-CoA conjugate prior to each experiment due to potential instability of the peptide-CoA conjugate once dissolved in water.
25. Dissolve peptide with 0.25% ACN in water to aid with the solubility of hydrophobic peptide sequences.

26. The peptide-loading reaction is best carried out at pH 7.0 as a higher pH leads to more rapid hydrolysis of peptide-CoA thioesters, which reduces the yield of the loading reaction.
27. Keep the mixture on ice at all times before incubation.
28. For proper product adsorption onto the solid phase use gravity flow in the loading step.
29. For washing and elution steps, some pressure can be generated by use of a syringe.

Acknowledgments

We would like to thank the Monash-Warwick Alliance for scholarship support for Y. T. C. Ho and J. Yin (University of Chicago) for the R4-4 Sfp expression plasmid. This work was supported by Monash University and EMBL Australia. This research was conducted by the Australian Research Council Centre of Excellence for Innovations in Peptide and Protein Science (CE200100012) and funded by the Australian Government.

References

1. Asseray N, Jacqueline C, Le Mabecque V, Batard E, Bugnon D, Potel G et al (2005) Activity of glycopeptides against *Staphylococcus aureus* infection in a rabbit endocarditis model: MICs do not predict in vivo efficacy. Antimicrob Agents Chemother 49(2):857–859
2. Hubbard BK, Walsh CT (2003) Vancomycin assembly: nature's way. Angew Chem Int Ed Engl 42(7):730–765
3. Sussmuth RD, Mainz A (2017) Nonribosomal peptide synthesis-principles and prospects. Angew Chem Int Ed Engl 56(14):3770–3821
4. Peschke M, Gonsior M, Sussmuth RD, Cryle MJ (2016) Understanding the crucial interactions between cytochrome P450s and non-ribosomal peptide synthetases during glycopeptide antibiotic biosynthesis. Curr Opin Struct Biol 41:46–53
5. Yim G, Thaker MN, Koteva K, Wright G (2014) Glycopeptide antibiotic biosynthesis. J Antibiot (Tokyo) 67(1):31–41
6. Bischoff D, Pelzer S, Bister B, Nicholson GJ, Stockert S, Schirle M et al (2001) The biosynthesis of vancomycin-type glycopeptide antibiotics-the order of the cyclization steps. Angew Chem Int Ed 40(24):4688–4691
7. Bischoff D, Bister B, Bertazzo M, Pfeifer V, Stegmann E, Nicholson GJ et al (2005) The biosynthesis of vancomycin-type glycopeptide antibiotics--a model for oxidative side-chain cross-linking by oxygenases coupled to the action of peptide synthetases. Chembiochem 6(2):267–272
8. Haslinger K, Peschke M, Brieke C, Maximowitsch E, Cryle MJ (2015) X-domain of peptide synthetases recruits oxygenases crucial for glycopeptide biosynthesis. Nature 521(7550):105–109
9. Greule A, Stok JE, De Voss JJ, Cryle MJ (2018) Unrivalled diversity: the many roles and reactions of bacterial cytochromes P450 in secondary metabolism. Nat Prod Rep 35(8):757–791
10. Rittle J, Green MT (2010) Cytochrome P450 compound I: capture, characterization, and C-H bond activation kinetics. Science 330(6006):933–937
11. Denisov IG, Makris TM, Sligar SG, Schlichting I (2005) Structure and chemistry of cytochrome P450. Chem Rev 105(6):2253–2277
12. Cryle MJ, Brieke C, Haslinger K (2013) Oxidative transformations of amino acids and peptides catalysed by cytochromes P450. Amino Acids Pept Proteins 30:1–36
13. Butz D, Schmiederer T, Hadatsch B, Wohlleben W, Weber T, Sussmuth RD (2008) Module extension of a non-ribosomal peptide synthetase of the glycopeptide antibiotic balhimycin produced by Amycolatopsis balhimycina. Chembiochem 9(8):1195–1200

14. Schoppet M, Tailhades J, Kulkarni K, Cryle MJ (2018) Precursor manipulation in glycopeptide antibiotic biosynthesis: are beta-amino acids compatible with the oxidative cyclization cascade? J Org Chem 83(13):7206–7214
15. Izore T, Cryle MJ (2018) The many faces and important roles of protein-protein interactions during non-ribosomal peptide synthesis. Nat Prod Rep 35(11):1120–1139
16. Quadri LE, Weinreb PH, Lei M, Nakano MM, Zuber P, Walsh CT (1998) Characterization of Sfp, a Bacillus subtilis phosphopantetheinyl transferase for peptidyl carrier protein domains in peptide synthetases. Biochemistry 37(6): 1585–1595
17. Tailhades J, Zhao Y, Schoppet M, Greule A, Goode RJA, Schittenhelm RB et al (2019) Enzymatic cascade to evaluate the tricyclization of glycopeptide antibiotic precursor peptides as a prequel to biosynthetic redesign. Org Lett 21(21):8635–8640
18. Kaniusaite M, Tailhades J, Marschall EA, Goode RJA, Schittenhelm RB, Cryle MJ (2019) A proof-reading mechanism for non-proteinogenic amino acid incorporation into glycopeptide antibiotics. Chem Sci 10(41):9466–9482
19. Greule A, Izore T, Iftime D, Tailhades J, Schoppet M, Zhao Y et al (2019) Kistamicin biosynthesis reveals the biosynthetic requirements for production of highly crosslinked glycopeptide antibiotics. Nat Commun 10(1): 2613
20. Aldemir H, Shu S, Schaefers F, Hong H, Richarz R, Harteis S et al (2022) Carrier protein-free enzymatic biaryl coupling in arylomycin A2 assembly and structure of the cytochrome P450 AryC. Chemistry 28(2): e202103389
21. Culp EJ, Waglechner N, Wang W, Fiebig-Comyn AA, Hsu YP, Koteva K et al (2020) Evolution-guided discovery of antibiotics that inhibit peptidoglycan remodelling. Nature 578(7796):582–587
22. Tailhades J, Zhao Y, Ho YTC, Greule A, Ahmed I, Schoppet M et al (2020) A chemoenzymatic approach to the synthesis of glycopeptide antibiotic analogues. Angew Chem Int Ed Engl 59(27):10899–10903
23. Bell SG, Xu F, Johnson EO, Forward IM, Bartlam M, Rao Z et al (2010) Protein recognition in ferredoxin-P450 electron transfer in the class I CYP199A2 system from *Rhodopseudomonas palustris*. J Biol Inorg Chem 15(3): 315–328
24. Zhao Y, Goode RJA, Schittenhelm RB, Tailhades J, Cryle MJ (2020) Exploring the tetracyclization of teicoplanin precursor peptides through chemoenzymatic synthesis. J Org Chem 85(3):1537–1547
25. Sunbul M, Marshall NJ, Zou Y, Zhang K, Yin J (2009) Catalytic turnover-based phage selection for engineering the substrate specificity of Sfp phosphopantetheinyl transferase. J Mol Biol 387(4):883–898
26. Izore T, Candace Ho YT, Kaczmarski JA, Gavriilidou A, Chow KH, Steer DL et al (2021) Structures of a non-ribosomal peptide synthetase condensation domain suggest the basis of substrate selectivity. Nat Commun 12(1):2511
27. Ishizuka M, Abe F, Sano Y, Takahashi K, Inoue K, Nakajima M et al (2011) Novel development of 5-aminolevurinic acid (ALA) in cancer diagnoses and therapy. Int Immunopharmacol 11(3):358–365

Chapter 10

Cross-Linking of the Nonribosomal Peptide Synthetase Adenylation Domain with a Carrier Protein Using a Pantetheine-Type Probe

Akimasa Miyanaga, Fumitaka Kudo, and Tadashi Eguchi

Abstract

Adenylation domains (A-domains) are responsible for the selective incorporation of carboxylic acid substrates in the biosynthesis of nonribosomal peptides and related natural products. The A-domain transfers an acyl substrate onto its cognate carrier protein (CP). The proper interactions between an A-domain and the cognate CP are important for functional substrate transfer. To stabilize the transient interactions sufficiently for structural analysis of A-domain–CP complex, vinylsulfonamide adenosine inhibitors have been traditionally used as molecular probes. Recently, we have developed an alternative strategy using a synthetic pantetheine-type probe that enables site-specific cross-linking between an A-domain and a CP. In this chapter, we describe the laboratory protocols for this cross-linking reaction.

Key words Nonribosomal peptide synthetase, Adenylation domain, Carrier protein, Cross-linking, Pantetheine-type probe

1 Introduction

Nonribosomal peptides are a major family of natural products, which have a broad range of biological activities. They are biosynthesized by nonribosomal peptide synthetases (NRPSs) [1]. NRPSs are multifunctional enzymes with several catalytic domains that assemble carboxylic acid building blocks, such as amino acids, into peptides. The adenylation domain (A-domain) activates a carboxylic acid substrate using adenosine 5′--triphosphate (ATP) to form an acyl adenylate intermediate, and then transfers the acyl group onto the terminal thiol group of the 4′-phosphopantetheine arm of the carrier protein (CP) domain to form a thioester conjugate (Fig. 1). The resulting CP-bound acyl group is used in a condensation reaction, catalyzed by the condensation domain, to elongate the peptide chain. Type I NRPSs are comprised of modules that contain a set of catalytic domains for

Michael Burkart and Fumihiro Ishikawa (eds.), *Non-Ribosomal Peptide Biosynthesis and Engineering: Methods and Protocols*, Methods in Molecular Biology, vol. 2670, https://doi.org/10.1007/978-1-0716-3214-7_10,

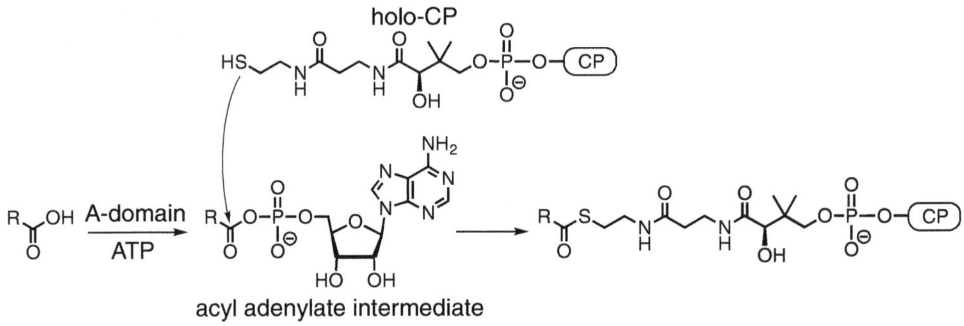

Fig. 1 Reaction of A-domain

incorporation of an acyl building block into a peptide product, while type II NRPSs consist of a complex of subunits, such as a stand-alone A-domain (adenylation enzyme) and a stand-alone CP [1, 2]. NRPSs are frequently fused with polyketide synthases to produce amino acid/peptide–polyketide hybrid molecules, providing structural diversity in natural products [3].

The A-domains are responsible for the selective incorporation of carboxylic acid building blocks into the NRPS machinery. The substrates for A-domain are a variety of carboxylic acids, including proteinogenic α-amino acids, nonproteinogenic amino acids, and aryl acids [1, 4]. The A-domain shows a strict substrate specificity with the substrate-conferring code, which corresponds to a set of amino acid residues located at the substrate-binding pocket [4–6]. Because the A-domains are key determinants for carboxylic acid building block specificity in the biosynthesis of nonribosomal peptides, the A-domains are attractive targets for NRPS engineering. However, the substitution of an A-domain by a homologous A-domain possessing different substrate specificity has resulted in reduced or abolished production of the desired compounds in many cases, which might be caused by a lack of the proper A-domain–CP interactions [7]. Structural analysis of the A-domain–CP complex is required to elucidate the molecular basis of the interactions between the A-domain and the cognate CP during functional substrate transfer. However, structural information on A-domain–CP interactions is limited because the interactions are weak and transient. For structural characterization of the transient interactions between A-domains and CPs, acyl vinylsulfonamide adenosine inhibitors, which mimic acyl adenylate intermediates, have been developed as molecular probes [8, 9]. The thiol group of the 4′-phosphopantetheine arm of the CP forms a covalent linkage with the vinylsulfonamide adenosine inhibitor that is non-covalently bound at the active site of the A-domain, which results in capture of the transient A-domain–CP complex. This strategy using vinylsulfonamide adenosine inhibitors has enabled the structural determination of several A-domain–CP complexes. However, this strategy requires the chemical synthesis of a specific vinylsulfonamide adenosine inhibitor that contains an appropriate acyl group corresponding to a specific substrate of the target A-domain in every case.

Pantetheine-type cross-linking probes have been used for the structural analysis of functional interactions between CPs and partner catalytic domains in fatty acid and polyketide biosynthesis [8, 10]. Recently, we showed the applicability of a pantetheine-type cross-linking probe for structural characterization of the protein–protein interaction between an A-domain and CP. We determined the structure of the complex of HitB and HitD, which are the stand-alone A-domain and the stand-alone CP, respectively, involved in hitachimycin biosynthesis, using a pantetheine-type cross-linking probe [11]. In this strategy, an acyl pantetheine analog that has an electrophilic bromoacetamide group connected through a nonhydrolyzable amide bond was used as a cross-linking probe [12]. The synthetic pantetheine analog was loaded onto an apo-CP to generate a crypto-CP in a one-pot enzymatic reaction. *Escherichia coli* coenzyme A (CoA) biosynthetic enzymes, pantothenate kinase (CoaA), 4′--phosphopantetheine adenyltransferase (CoaD), and dephospho-CoA kinase (CoaE), were used for the conversion of the pantetheine analog to the corresponding CoA derivative [13]. The promiscuous phosphopantetheinyl transferase Sfp from *Bacillus subtilis* was used for the preparation of crypto-CP by attaching the phosphopantetheine analog moiety of the CoA derivative to apo-CP (Fig. 2) [14]. The resulting crypto-CP can efficiently react with a partner A-domain that has a Cys mutation at an appropriate position to form a covalently cross-linked A-domain–CP complex, enabling the structural determination of the A-domain–CP complex. Because this cross-linking reaction requires only the bromoacetamide pantetheine probe and a Cys mutation in the A-domain, this strategy could facilitate the structural determination of various A-domain–CP complexes. In this chapter, we describe the laboratory protocols for the preparation of the bromoacetamide pantetheine probe and the cross-linking reaction between the A-domain and CP.

2 Materials

2.1 Synthesis of the Bromoacetamide Pantetheine Probe

All of the materials and reagents described below can be purchased from commercial suppliers and used without further purification:

1. CH_2Cl_2.
2. $CHCl_3$.

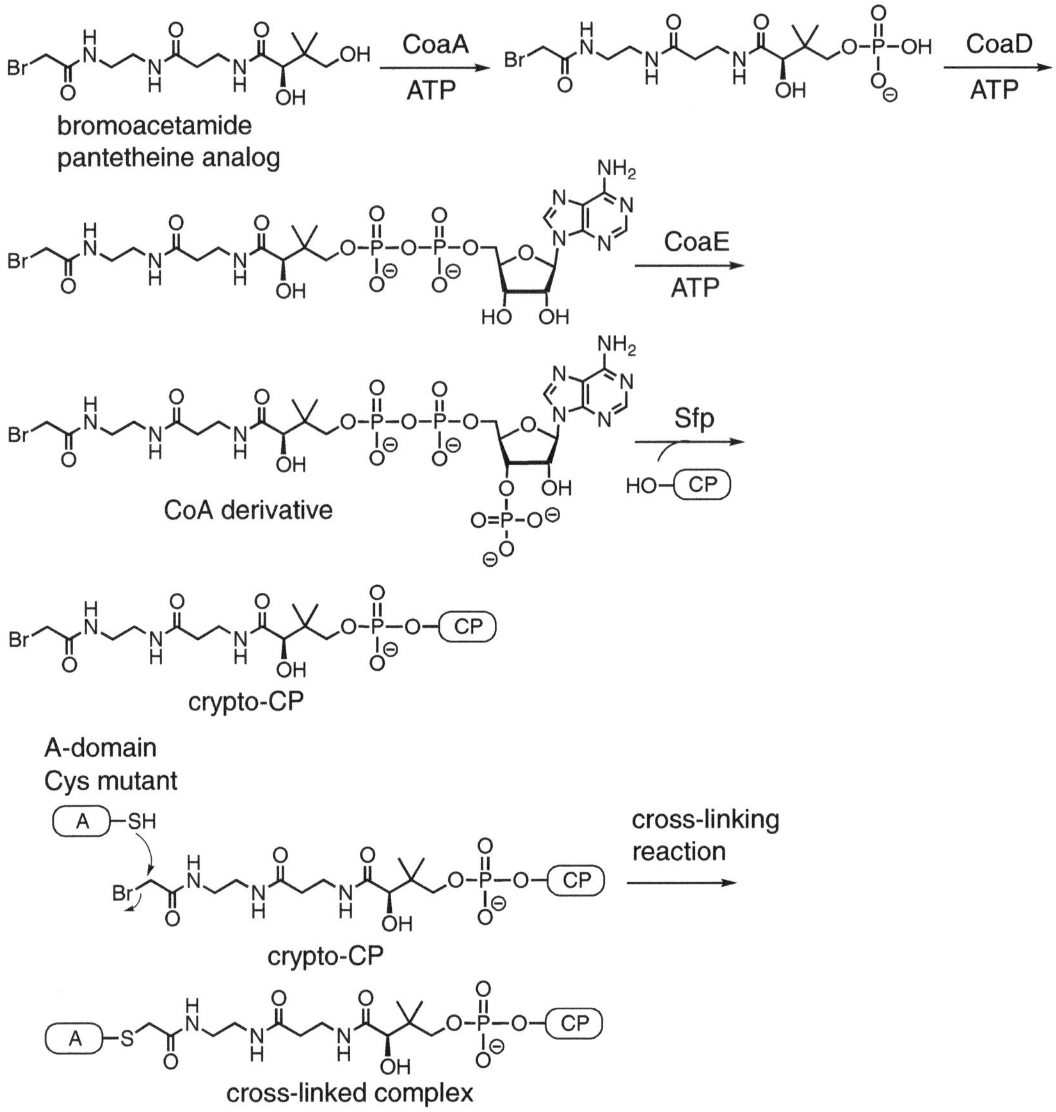

Fig. 2 Preparation of crypto-CP and the cross-linking reaction. Crypto-CP is prepared in a one-pot reaction using CoaA, CoaD, CoaE, and Sfp, and is then used for the cross-linking reaction with the A-domain Cys mutant

3. THF.
4. Methanol.
5. Ethyl acetate.
6. Diethyl ether.
7. Sodium D-pantothenate.
8. *p*-anisaldehyde dimethyl acetal.
9. Trimethylsilyl chloride.

10. 2-chloroethylamine hydrochloride.
11. Sodium azide.
12. *N*-methylmorpholine.
13. Isobutyl chloroformate.
14. Triphenylphosphine.
15. *N*-succinimidyl bromoacetate.
16. Acetic acid.

2.2 Recombinant Proteins

2.2.1 A-Domain Recombinant Protein

The A-domain genes are routinely cloned into a pCold I vector and transformed into *E. coli* BL21(DE3) for overexpression. Expressed His-tagged A-domain proteins are purified by His-tag affinity column chromatography, and then stored at −80 °C in a buffer containing 10% glycerol.

2.2.2 CP Recombinant Protein

The CP genes are routinely cloned into a pCold I vector or pET vector and transformed into *E. coli* BL21(DE3) for overexpression. Expressed His-tagged apo-CP proteins are purified by His-tag affinity column chromatography, and then stored at −80 °C in a buffer containing 10% glycerol.

2.2.3 CoaA, CoaD, and CoaE Recombinant Proteins

In our laboratory, plasmids encoding CoaA, CoaD, and CoaE provided by Professor M. D. Burkart [13] are used for preparation of CoaA, CoaD, and CoaE recombinant proteins, each of which is fused to a maltose-binding protein. CoaA, CoaD, and CoaE recombinant proteins are expressed in *E. coli* BL21(DE3), purified by amylose resin, and then stored at −80 °C in a buffer containing 10% glycerol.

2.2.4 Sfp Recombinant Protein

A pCold I plasmid encoding Sfp is transformed into *E. coli* Rosetta 2(DE3) for overexpression. Expressed His-tagged Sfp protein is purified by His-tag affinity column chromatography, and then stored at −80 °C in a buffer containing 10% glycerol.

2.3 Cross-Linking Reaction

1. Buffer A: 50 mM HEPES-NaOH (pH 8.0) and 10% (v/v) glycerol.
2. $MgCl_2$ stock solution: 500 mM $MgCl_2$ in water.
3. ATP stock solution: dissolve 11 mg of ATP disodium salt hydrate in 0.07 mL of water. Neutralize the ATP solution with NaOH aq. and confirm that the pH is 7–8 using pH paper. Adjust the final volume to 0.1 mL with water and store the solution at −30 °C. The final concentration is 200 mM.
4. Bromoacetamide pantetheine analog stock solution: dissolve 0.76 mg of bromoacetamide pantetheine analog in 0.1 mL of DMSO. Aliquot the pantetheine analog stock solution in 20 μL

portions and store at −30 °C. The final concentration is 20 mM.

5. SDS-PAGE gel loading buffer: 0.05 M Tris–HCl (pH 6.8), 1% sodium dodecyl sulfate (SDS), 1% (v/v) β-mercaptoethanol, 20% (v/v) glycerol, and 0.1% (w/v) bromophenol blue.

3 Methods

3.1 Synthetic Procedure for the Preparation of the Bromoacetamide Pantetheine Analog

The synthesis of the bromoacetamide pantetheine analog is shown in Fig. 3.

3.1.1 Synthesis of p-Methoxybenzylidene (PMB)-Protected Pantothenic Acid

1. Dissolve sodium D-pantothenate (10 g, 41 mmol) and *p*-anisaldehyde dimethyl acetal (8.4 mL, 50 mmol) in dry CH_2Cl_2 (200 mL) under an argon atmosphere.
2. Add trimethylsilyl chloride (6.5 mL) slowly in a drop-wise manner until the solution becomes acidic (pH 2–3).
3. Stir the resulting solution at room temperature for 3 h.
4. Filter the reaction mixture through a pad of celite.
5. Remove the organic solvent under reduced pressure to give a crude residue.
6. Store the crude residue at 4 °C overnight to obtain pure PMB-protected pantothenic acid as a white crystalline powder (7.6 g, 54%). Alternatively, the crude residue can be purified by silica gel chromatography ($CHCl_3$: methanol: NEt_3 = 30: 1:0.5).

3.1.2 Synthesis of 2-Azidoethylamine

1. Dissolve 2-chloroethylamine hydrochloride (4.0 g, 34 mmol) in water (42 mL).
2. Add sodium azide (5.2 g, 80 mmol) in a drop-wise manner.
3. Stir the resulting solution at 80 °C overnight.
4. Quench the reaction by adding 1–2 M KOH aq. solution until the solution becomes basic (pH 10–11).
5. Extract the solution with CH_2Cl_2, separate the organic layer, and dry over $MgSO_4$.
6. Remove the solvent under reduced pressure to give 2-azidoethylamine (2.3 g, 78%) as a clear oil (*see* **Note 1**).

3.1.3 Synthesis of PMB-Protected Pantetheine Azide

1. Dissolve PMB-protected pantothenic acid (5.0 g, 15 mmol) in dry THF (140 mL) under an argon atmosphere.

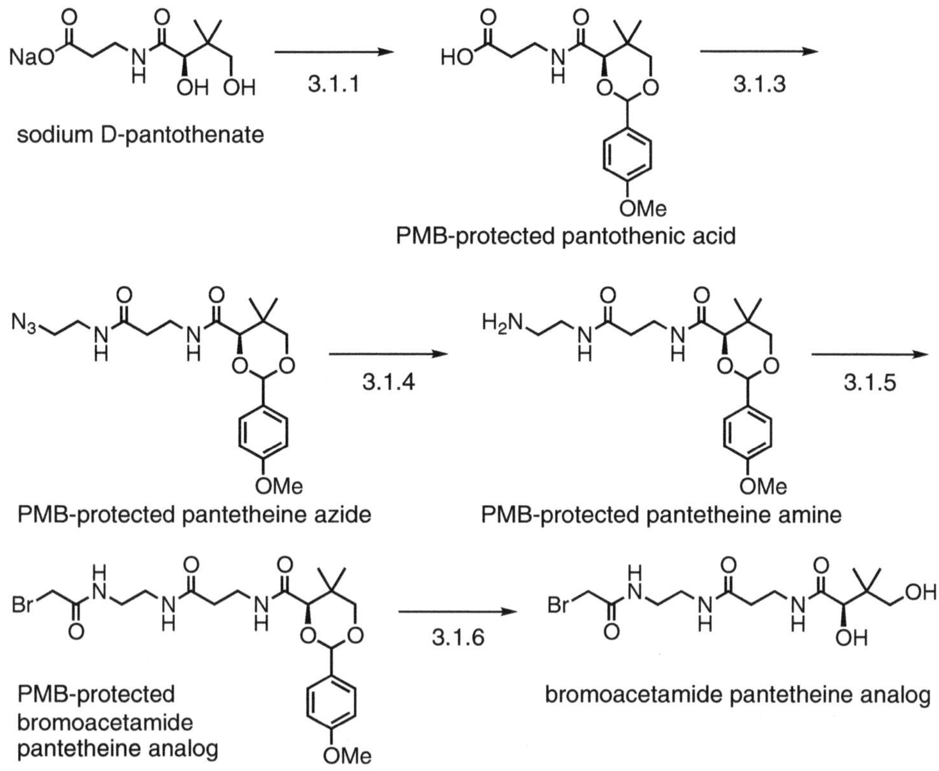

Fig. 3 Synthetic route to the bromoacetamide pantetheine analog. The numbering refers to the corresponding section in the text

2. Add *N*-methylmorpholine (3.2 mL, 29 mmol) slowly in a drop-wise manner.
3. Add isobutyl chloroformate (1.9 mL, 15 mmol) in a drop-wise manner.
4. Stir the resulting solution at room temperature for 20 min.
5. Add 2-azidoethylamine (1.5 g, 17 mmol) dissolved in CH_2Cl_2/THF solution (3–4 mL) slowly in a drop-wise manner and stir the resulting solution at room temperature for 3 h (*see* **Note 2**).
6. Quench the reaction by adding methanol (10 mL).
7. Remove the organic solvent under reduced pressure and redissolve the crude residue in ethyl acetate.
8. Wash the organic layer with water, saturated sodium bicarbonate, and brine, and dry over $MgSO_4$.
9. Remove the organic solvent under reduced pressure.

10. Purify the crude residue by silica gel chromatography ($CHCl_3$: methanol: NEt_3 = 20:1:0.1) to give PMB-protected pantetheine azide as a clear oil (3.7 g, 62%).

3.1.4 Synthesis of PMB-Protected Pantetheine Amine

1. Dissolve PMB-protected pantetheine azide (0.51 g, 1.3 mmol) in THF (10 mL).
2. Add triphenylphosphine (0.65 g, 2.5 mmol).
3. Add water (2.5 mL) slowly in a drop-wise manner.
4. Stir the resulting solution at room temperature for 14 h.
5. Remove the organic solvent under reduced pressure.
6. Purify the crude residue by silica gel chromatography ($CHCl_3$: methanol: NEt_3 = 85:15:1) to give pure PMB-protected pantetheine amine as a pale yellowish solid (340 mg, 72%).

3.1.5 Synthesis of PMB-Protected Bromoacetamide Pantetheine Analog

1. Dissolve PMB-protected pantetheine amine (83 mg, 0.22 mmol) and *N*-succinimidyl bromoacetate (61 mg, 0.26 mmol) in dry THF (2.0 mL) under an argon atmosphere (*see* **Note 3**).
2. Stir the solution at room temperature for 2 h.
3. Remove the organic solvent under reduced pressure.
4. Purify the crude residue by silica gel chromatography (ethyl acetate: methanol = 20:1) to yield the PMB-protected bromoacetamide pantetheine analog as a clear oil (80 mg, 73%).

3.1.6 Synthesis of Bromoacetamide Pantetheine Analog

1. Dissolve the PMB-protected bromoacetamide pantetheine analog (36 mg, 0.072 mmol) in 80% aqueous acetic acid (1.0 mL) at 0 °C (*see* **Note 3**).
2. Stir the solution at room temperature for 3 h.
3. Remove the solvent under reduced pressure (*see* **Note 4**).
4. Purify the crude residue by silica gel chromatography (ethyl acetate: methanol = 20:1).
5. Recrystallize the purified product in a mixture of methanol and diethyl ether. Add a few drops of methanol to the purified product and then add diethyl ether (5–10 mL) until a precipitate appears. After keeping the flask at 4 °C overnight, remove the diethyl ether layer by decanting and evaporation to yield the pure bromoacetamide pantetheine analog as a white powder (10 mg, 36%).

3.2 Procedure for the Cross-Linking Reaction

3.2.1 Preparation of Crypto-CP

1. In a 1.5 mL microcentrifuge tube, mix the components given in Table 1 in the order shown.
2. Incubate the reaction mixture at 28 °C for 3 h (*see* **Note 5**). The generated crypto-CP can be used for the cross-linking reaction (*see* Subheading 3.2.2) without purification.

Table 1
Solutions for preparing crypto-CP

Component	Component volume (μL)	Final concentration
Buffer A	35.325	
200 mM ATP	1.25	5 mM
500 mM $MgCl_2$	0.75	7.5 mM
50 μM CoaA	0.5	0.5 μM
50 μM CoaD	0.7	0.7 μM
50 μM CoaE	0.6	0.6 μM
250 μM Sfp	0.5	2.5 μM
500 μM apo-CP	10	100 μM
20 mM bromoacetamide pantetheine analog	0.375	150 μM
Reaction volume	50	

3.2.2 Cross-Linking Reaction

1. Initiate the cross-linking reaction by adding 50 μL of 100 μM A-domain solution (final concentration: 50 μM) (*see* **Note 6**) directly into 50 μL of 100 μM crypto-CP solution (final concentration: 50 μM). The total volume is 100 μL (*see* **Note 7**).
2. Incubate at 10 or 20 °C for 60 min to 18 h (*see* **Note 8**).
3. Take 4 μL samples of the reaction mixture at several time points (0 min to 18 h) and quench the reaction by the addition of 4 μL of SDS-PAGE gel loading buffer.
4. Run an aliquot of the reaction mixture on 15% SDS-PAGE gel and stain the gel with Coomassie Brilliant Blue to monitor the reaction (Fig. 4) (*see* **Note 9**).

4 Notes

1. Perform the evaporation carefully. Because the boiling temperature of 2-azidoethylamine is low, longer evaporation times might result in loss of 2-azidoethylamine.
2. Check the conversion rate by TLC. If necessary, use additional 2-azidoethylamine for the reaction.
3. Use a halogen-free solvent in the reaction and purification steps to avoid substitution of the bromoacetamide moiety.
4. Evaporation with toluene (3 mL), which allows the formation of an azeotrope with acetic acid, is repeated three times to remove residual acetic acid. Residual acetic acid might acetylate the hydroxy group of the bromoacetamide pantetheine analog.

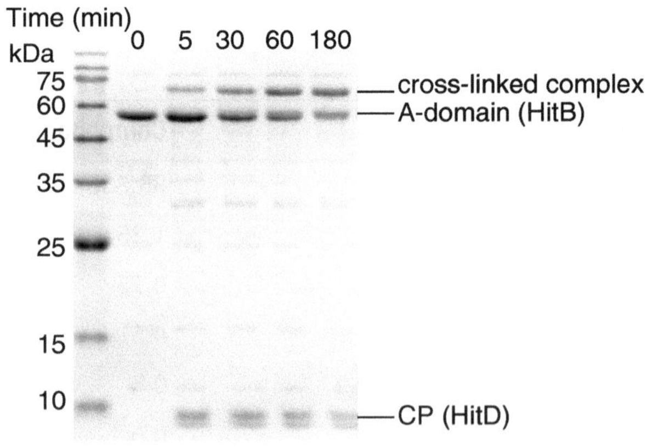

Fig. 4 Example of SDS-PAGE analysis of the cross-linking reaction of the A-domain with CP using a synthetic bromoacetamide pantetheine analog. Covalent cross-linking between the A-domain and CP results in shifts in the band for the A-domain. For example, in the cross-linking reaction of HitB and HitD, the 55 kDa (HitB) band was shifted to 70 kDa (HitB–HitD cross-linked complex)

5. Before the addition of the A-domain, the complete conversion of apo-CP to crypto-CP should be confirmed by HPLC or LC-ESI-MS analysis. If complete conversion is not observed, the concentration of Sfp or the reaction time can be increased.
6. The Cys residue is an appropriate nucleophile for the electrophilic bromoacetamide moiety in the cross-linking reaction. Generally, the A-domain does not have a Cys residue at the active site. Therefore, a Cys mutation is required. The aspartate residue that interacts with the amino group of the amino acid substrate can be used as a target for a Cys mutation.
7. For analysis, the reaction volume is typically 50–100 μL. In the large-scale reaction for crystallization of the cross-linked complex, the reaction volume is increased up to 2–3 mL. For purification of the cross-linked complex, anion-exchange column and/or size-exclusion column chromatography can be used to separate the product from CoaA, CoaD, CoaE, Sfp, unreacted CP, and unreacted A-domain proteins.
8. The reaction temperature and reaction time should be optimized for each A-domain–CP pair. For example, in the cross-linking reaction of HitB and HitD [11], the incubation of HitB with crypto-HitD at 10 °C for 15 h efficiently generated the HitB–HitD cross-linked complex.
9. The results of the cross-linking reaction using the Cys mutant should be compared with that using the wild-type enzyme. Sometimes crypto-CP forms a covalent complex not only with the Cys mutant but also with the wild-type. In this case, the bromoacetamide probe might be too reactive for the target

A-domain–CP pair, resulting in an undesired cross-linking reaction. To reduce the undesired reaction, the nucleophilic residue that might cause an undesired cross-linking reaction can be mutated to Ala or Gly.

Acknowledgments

This work was supported in part by Grants-in-Aid for Scientific Research (C) (17K07747 to AM) and Grants-in-Aid for Scientific Research (B) (20H02911 to AM) from Japan Society for the Promotion of Science, Grants-in-Aid for Scientific Research in Innovative Areas (16H06451 to TE) from the Ministry of Education, Culture, Sports, Science and Technology, and the Uehara Memorial Foundation (to FK).

References

1. Süssmuth RD, Mainz A (2017) Nonribosomal peptide synthesis-principles and prospects. Angew Chem Int Ed 56:3770–3821
2. Jaremko MJ, Davis TD, Corpuz JC et al (2020) Type II non-ribosomal peptide synthetase proteins: structure, mechanism, and protein-protein interactions. Nat Prod Rep 37:355–379
3. Miyanaga A, Kudo F, Eguchi T (2018) Protein–protein interactions in polyketide synthase–nonribosomal peptide synthetase hybrid assembly lines. Nat Prod Rep 35: 1185–1209
4. Kudo F, Miyanaga A, Eguchi T (2019) Structural basis of the nonribosomal codes for non-proteinogenic amino acid selective adenylation enzymes in the biosynthesis of natural products. J Ind Microbiol Biotechnol 46:515–536
5. Stachelhaus T, Mootz HD, Marahiel MA (1999) The specificity-conferring code of adenylation domains in nonribosomal peptide synthetases. Chem Biol 6:493–505
6. Challis GL, Ravel J, Townsend CA (2000) Predictive, structure-based model of amino acid recognition by nonribosomal peptide synthetase adenylation domains. Chem Biol 7:211–224
7. Brown AS, Calcott MJ, Owen JG et al (2018) Structural, functional and evolutionary perspectives on effective re-engineering of non-ribosomal peptide synthetase assembly lines. Nat Prod Rep 35:1210–1228
8. Gulick AM, Aldrich CC (2018) Trapping interactions between catalytic domains and carrier proteins of modular biosynthetic enzymes with chemical probes. Nat Prod Rep 35:1156–1184
9. Corpuz JC, Sanlley JO, Burkart MD (2022) Protein-protein interface analysis of the non-ribosomal peptide synthetase peptidyl carrier protein and enzymatic domains. Synth Syst Biotech 7:677–688
10. Finzel K, Lee DJ, Burkart MD (2015) Using modern tools to probe the structure-function relationship of fatty acid synthases. ChemBioChem 16:528–547
11. Miyanaga A, Kurihara S, Chisuga T et al (2020) Structural characterization of complex of adenylation domain and carrier protein by using pantetheine cross-linking probe. ACS Chem Biol 15:1808–1812
12. Miyanaga A, Ouchi R, Ishikawa F et al (2018) Structural basis of protein-protein interactions between a trans-acting acyltransferase and acyl carrier protein in polyketide disorazole biosynthesis. J Am Chem Soc 140:7970–7978
13. Haushalter RW, Worthington AS, Hur GH et al (2008) An orthogonal purification strategy for isolating crosslinked domains of modular synthases. Bioorg Med Chem Lett 18: 3039–3042
14. Beld J, Sonnenschein EC, Vickery CR et al (2014) The phosphopantetheinyl transferases: catalysis of a post-translational modification crucial for life. Nat Prod Rep 31:61–108

Chapter 11

A Practical Guideline to Engineering Nonribosomal Peptide Synthetases

Nadya Abbood, Leonard Präve, Kenan A. J. Bozhueyuek, and Helge B. Bode

Abstract

The bioengineering of nonribosomal peptide synthetases (NRPSs) is a rapidly developing field to access natural product derivatives and new-to-nature natural products like scaffolds with changed or improved properties. However, the rational (re-)design of these often gigantic assembly-line proteins is by no means trivial and needs in-depth insights into structural flexibility, inter-domain communication, and the role of proofreading by catalytic domains—so it is not surprising that most previous rational reprogramming efforts have been met with limited success. With this practical guide, the result of nearly one decade of NRPS engineering in the Bode lab, we provide valuable insights into the strategies we have developed during this time for the successful engineering and cloning of these fascinating molecular machines.

Key words DNA assembly, NRPS engineering, Synthetic leucine zippers, Synthetic biology, Natural products

1 Introduction

Microbial natural products (NPs) biosynthesized via large modular enzymatic assembly lines are of great value for the identification of novel lead compounds, for example, for the treatment of infectious diseases and cancer [1, 2]. To become therapeutically useful, many NP leads having pharmacological or biological activity require further modification [3–5]. Their chemical scaffolds serve as a starting point for synthetic modifications to improve potency, selectivity, drug likeness, or pharmacokinetic parameters [6]. Nevertheless, the complexity of NP structures makes it difficult to generate synthetic NP derivatives, for example, to explore structure–activity relationships and to develop hits into leads for preclinical development, and further into clinical candidates. To make matters worse,

Nadya Abbood and Leonard Präve contributed equally with all other contributors.

Michael Burkart and Fumihiro Ishikawa (eds.), *Non-Ribosomal Peptide Biosynthesis and Engineering: Methods and Protocols*, Methods in Molecular Biology, vol. 2670, https://doi.org/10.1007/978-1-0716-3214-7_11,

if the chemical synthesis of NP analogs is possible at all, they often cannot be produced on a large scale at economically justifiable costs [7]. Thus, the development of efficient synthetic biology methods to bioengineer microbial natural products is of great scientific, economic, and societal importance—to overcome the limitations of synthetic chemistry as well as the bottlenecks of traditional NP research, such as the supply issue at economical cost of goods. In this hands-on protocol, we will describe the methods and strategies we have developed in recent years to enable more efficient and reproducible engineering of **n**on**r**ibosomal **p**eptide **s**ynthetases (NRPSs) to obtain genetically engineered NP analogs as well as new-to-nature tailor-made NP-like peptide scaffolds.

NRPSs are modular, multifunctional mega-enzymes that act as molecular machines in an assembly-line manner to biosynthesize highly functionalized peptides independently of the ribosome. In an NRPS assembly line, multiple repeating modules are responsible for selection, activation, programmed functional group modifications, and coupling of an amino acid to the growing peptide chain. These modules, in turn, consist of individual domains that are connected to each other via linkers, like beads on a string. Although many optional *in cis* or *in trans* acting modification domains can be present, introducing structurally complex motives into the peptide chain, for instance, epimerization (E), methylation (MT), hydroxylation, and glycosylation patterns, a canonical NRPS elongation module depends on the activity of three core domains: an adenylation (A) domain, which selects and activates a substrate; a thiolation (T) domain, onto which the activated amino acid is covalently attached to; and a condensation (C) domain, which catalyzes peptide bond formation of the bound amino acid to the growing peptide chain. To terminate the NRPS catalytic cycle and to regenerate the assembly line, most NRPS termination modules have a thioesterase (TE) domain, which is usually responsible for the release of linear, cyclic, or branched cyclic peptides. For a comprehensive introduction to the state of the art in NRPS, we would like to refer to the review paper—Nonribosomal Peptide Synthesis—Principles and Prospect—by Süßmuth and Mainz [8].

The dream of bioengineering biosynthetic gene clusters (BGCs), genomic bacterial islands encoding NRPS, is almost as old as the discovery of NRPS itself in the 1960s—and was realized for the first time in 1995 [9]. Since then the NRPS community has envisaged using these enzymes like a molecular toolbox to produce customized NP analogs. For many decades, there have been three general strategies to bioengineer NRPSs, such as (1) the modification of the A domains active site residues, (2) the exchange of A and A-T didomains, and (3) the exchange of C-A didomains or whole modules (C-A-T tri-domains)—but most attempts to achieve assembly-line pathway engineering resulted in biosynthetic machineries that are either greatly impaired in their activity or nonfunctional [10, 11]. To overcome these problems, we shifted our focus to structure-based approaches, as opposed to earlier attempts, which were mostly heavily sequence-based.

As a result of our efforts, we initially developed the concept of eXchange **u**nits (**XU**) and further developed it into the eXchange **u**nit **c**ondensation domain (**XUC**) concept to overcome the observed problems, such as the C domains inferred acceptor-site substrate-specificity and the resulting inability to easily create peptide libraries [12, 13]. However, due to the very complementary workflows, we will focus here only on the XU concept, followed by instructions on how to create the recently introduced **type S NRPSs**, which for the first time have opened up avenues for large-scale, high-throughput, and truly biocombinatorial approaches to NPs research [14, 15]. For a detailed description of the XUC and how the exact splicing position within the linker of the pseudo-dimeric C domain can be identified, we would like to refer to the original work [13]. However, as we have seen examples in literature not correctly applying the XUC method, at this point we would like to put special emphasis on the major XUC design principle: *C domains of different phylogenetic clades,* such as *$^{L}C_{L}$, $^{D}C_{L}$, Cy (heterocyclization), dual C/E, C starter, and C termination domains must never be combined* [16]. In particular, we would like to highlight that $^{L}C_{L}$ domains are structurally not compatible with $^{D}C_{L}$ domains.

1.1 The eXchange Unit (XU)

Within the framework of the XU concept, efficient NRPS engineering was achieved by taking advantage of the modular structure of NRPSs—but as opposed to the historical definition of a module (C-A-T units), here A-T-C units along with a precisely defined fusion site within the C-A linker region is leveraged to achieve the de-novo (re-)design of high peptide amount producing NRPSs with high confidence, if certain prerequisites are met, such as an appropriate selection of XUs and the correct splicing/fusion site [12]. The latter follows an α-helix within the C-A linker region [17] and was originally described as a WNATE consensus motif based on *Photorhabdus* sp. and *Xenorhabdus* sp. NRPS sequence alignments. As in particular the WNATE motif caused a lot of confusion within the NRPS community, because it is not always present, the end of the α-helix easily can be determined by a conserved sequence of up to three leucines. Nevertheless, linker regions excised from our model system, the GameXPeptide A-D producing synthetase (GxpS) from *Photorhabdus luminescens* subsp. *laumondii* TT01 [18], can be utilized as a template to identify this fusion site in other NRPSs by performing a simple sequence alignment (Fig. 1). Besides recombining two different NRPSs at one position, the XU fusion site offers the possibility to combine several XUs in a sequential manner. Consequently, not only single

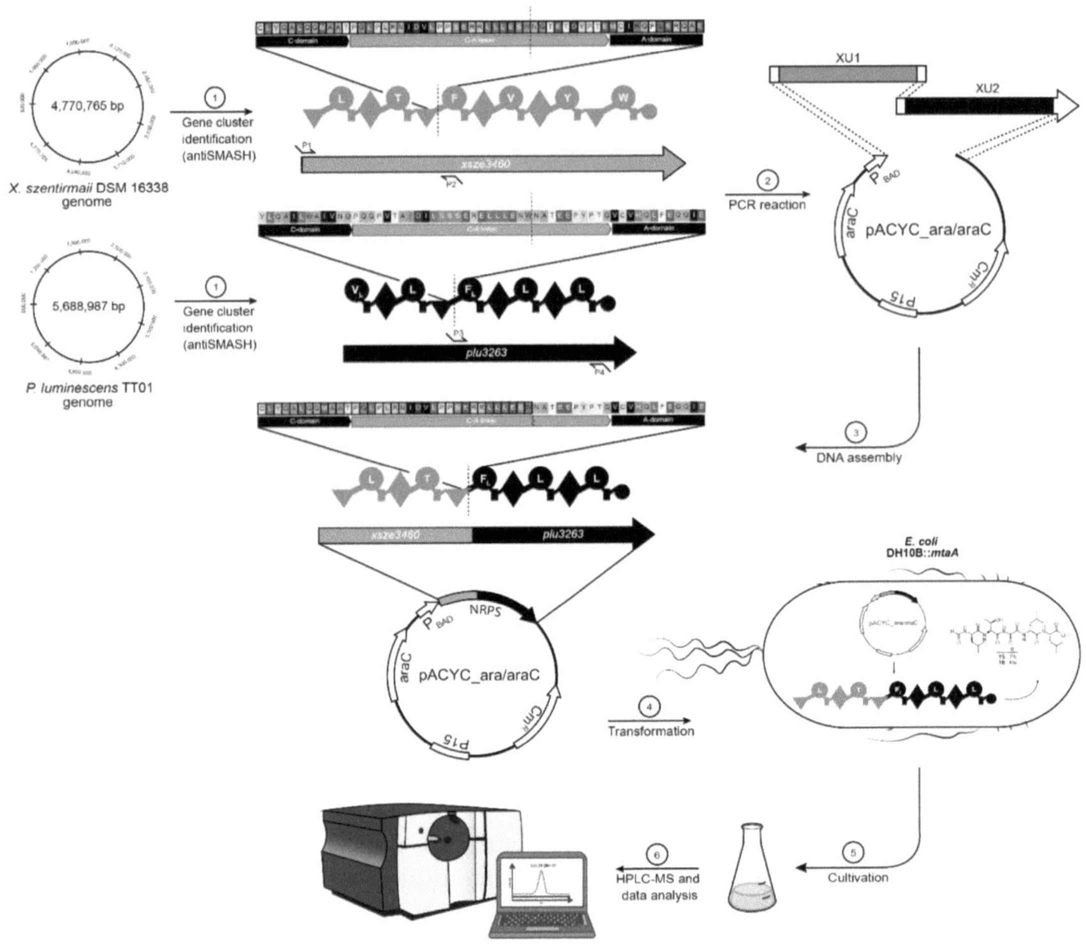

Fig. 1 *Workflow eXchange unit concept.* The procedure starts with the in silico identification of the target BGCs (1) followed by the identification of the XU fusion site by aligning the C_3-linker-A_3 sequence of GxpS from *P. luminescens* subsp. *laumondii* TT01 with the C-linker-A sequences extracted from the target BGCs. XU1 and XU2 are amplified with oligonucleotides P1:P2 and P3:P4 containing overlaps of 25–30 bps (2). DNA fragments are assembled into the linearized plasmid pACYC_ara/araC via HiFi or Hot Fusion (3) to obtain the plasmid encoded artificial NRPS assembly line. The plasmid is transformed into *E. coli* DH10B::*mtaA* (4) and the generated strain is cultivated in growth media supplemented with appropriate antibiotics and arabinose for induction (5). After cultivating 72 h at 22 °C, the culture is prepared for HPLC-MS and generated data are analyzed (6)

positions within NRPSs can be exchanged but also entirely new NRPSs can be assembled de novo [12].

The experience we have gained in recent years to increase our success rate in developing functional recombinant pathways via XU shuffling can be summarized in three rules:

1. *Only mix-and-match XUs within the same genera.* Although we were able to recombine XUs of gram-negative and gram-positive origin, we found that most attempts to recombine

NRPSs of distinct genera resulted in nonfunctional or impaired assembly lines with low peptide yields [12].

2. *Consider the C domain type.* We found that changing the stereochemistry of a particular substrate (i.e., changing A-T-C units against A-T-C/E units and vice versa), significantly effects the productivity of engineered NRPSs [14].
3. *Consider the acceptor-site specificity of the upstream C domain.* The original XU concept initially inferred a strong specificity of C domains at the acceptor site. However, a number of recent insightful studies have led to results that at least question the "proofreading" role of C domains during NRP synthesis [19]. Nevertheless, the XU rule ("the specificity of the upstream C domain must be respected") can still be used as a rule of thumb to directly guide engineering attempts. Sequence and structural analysis showed that it is more likely that C domains upstream of A domains with the same or similar specificity can form a more wild-type like and thus functional interface.

Of note, extensive phylogenetic and experimental analyses showed that the relevance of these rules decreases with descending numbering. However, we would like to emphasize that these rules only help improve the efficiency of engineering, but do not necessarily have to be considered to generate functional engineered NRPSs.

1.2 Type S NRPS

Due to the mere size of NRPSs, reprogramming is usually associated with many practical limitations. To simplify and accelerate the cloning of reprogrammed NRPSs, we established type S NRPSs with a lower complexity and smaller size compared to modular one-protein NRPSs [14, 15]. For this purpose, the NRPS encoding BGC is divided and PCR-amplified in two individual smaller subunits. Both fragments (subunit 1 and subunit 2) are subsequently assembled on two distinct plasmids carrying different origins of replication (ORI) and resistance cassettes. By splitting NRPSs into two fragments, PCR amplification and plasmid assembly can be simplified, making standard cloning strategies such as Hot Fusion, Gibson, and HiFi cloning sufficient. This reduces the cloning effort about half (from 6 to 3 days) compared to the previous applied **t**ransformation-**a**ssociated **r**ecombination (TAR)-based yeast cloning [12, 15]. In addition, the throughput of correctly generated plasmids is significantly increased, as the plasmids are constructed from a smaller number of inserts, thus reducing the likelihood of incorrectly assembled plasmids. To reconstitute the in vivo functionality of the original full-length NRPS system heterologously in *Escherichia coli*, we introduced complementary pairs of synthetic leucine zippers that interact with high affinity (10 nM)

[20] and mediate communication of the two independently expressed subunits 1 and 2. For our initial proof of concept study, we selected the shortest available pair (SZ17:18) from a pool of 27 physicochemical-characterized SYNZIPs [20]. Here we inserted subunits 1 and 2 upstream and downstream of SZ17 and SZ18, respectively. After transformation of plasmids encoding subunit 1_SZ17 and subunit 2_SZ18 into *E. coli*, protein expression is induced and NRPS subunits with attached SYNZIPs are synthesized. In the living cell, the type S NRPS is functionally reconstituted and consequently produces the expected peptides, which we usually detect by HPLC-MS [12, 14].

In addition to simplified cloning, SYNZIP-mediated engineering of NRPSs offers an increased bio-combinatorial potential as another advantage. Once assembled, the plasmids coding for type S NRPSs can be reused and recombined in myriad ways to reconstitute wild-type-like assembly lines or to create entirely new-to-nature NRPSs. The number of new NRPSs that can be constructed increases exponentially with each newly generated type S NRPS encoding plasmid. To further expand the scope of type S NRPSs, we have also demonstrated that SYNZIPs can be inserted into both the A-T and T-C linker regions (*see* Subheading "Fusion Sites for SYNZIP Insertion") and can be used to generate tri-partite type S NRPSs—in this case, the assembly line is split into three individually expressible subunits. In the latter, a second SYNZIP pair (SZ1: 2) is introduced that cannot interact with SZ17:18 and forms an orthogonal SYNZIP network [14].

2 Materials

2.1 Plasmids and Strains

1. For cloning, we use *E. coli* DH10B [21, 22] that shows high DNA transformation efficiency and allows ideal cloning and maintaining of large plasmids. For heterologous expression of NRPS genes, we inserted *mtaA* from *Stigmatella aurantiaca* DW4/3 [23] into *entD* of *E. coli* DH10B to obtain *E. coli* DH10B::*mtaA*. MtaA is a Sfp-like phosphopantetheinyl transferase, required to efficiently posttranslationally modify T domains into the active *holo* form.
2. We use plasmids pACYCDuet^{TM}-1 (chloramphenicol resistance gene, P15A replicon), pCOLADuet^{TM}-1 (kanamycin resistance gene, ColA replicon), and pCDFDuet^{TM} (streptomycin/spectinomycin resistance gene, pCloDF13 replicon). In all three plasmids, we replaced regulator/promotor genes LacI/P_{T7Lac} by AraC/P_{BAD} to obtain pACYC_ara/araC, pCOLA_ara/tacI, and pCDF_ara/tacI, respectively.

2.2 Strain Cultivation and Stock Solutions

1. Low-salt LB medium: 10 g tryptone, 5 g yeast extract, 5 g NaCl, pH 7.5 in 1 L double-distilled water (ddH_2O). Add 1% (w/v) agar if solid medium is required. Sterilize by autoclaving.
2. XPPM: 10 g glycerin, 20 mL M9 salt A, 20 mL M9 salt B, 2 g L-amino acid mix, 1 g Na-pyruvate in 1 L ddH_2O. Sterilize by autoclaving and add 2 mL vitamin solution and 1 mL trace element solution afterward.
3. M9 salt A: 350 g K_2HPO_4, 100 g KH_2PO_4 in 1 L ddH_2O. Sterilize by autoclaving.
4. M9 salt B: 29.4 g sodium citrate, 50 g ammonium sulfate, 5 g magnesium sulfate in 1 L ddH_2O. Sterilize by autoclaving.
5. L-amino acid mix: 2 g of all proteinogenic amino acids (L-alanine, L-asparagine, L-cysteine, L-glutamate, L-isoleucine, L-leucine, L-lysine, L-histidine, L-serine, L-tyrosine, L-tryptophan, L-methionine, L-arginine, L-aspartate, L-glutamine, L-glycine, L-proline, L-threonine, L-valine, and L-phenylalanine).
6. Vitamin solution: 10 mg folic acid, 6 mg biotin, 200 mg p-aminobenzoic acid, 1 g thiamine-HCl, 1.2 g pantothenic acid, 100 mg riboflavin, 2.3 g nicotinic acid, 12 g pyridoxine HCl, and 20 mg vitamin B12 in 1 L ddH_2O. Sterilize by 0.22 μm filter.
7. Trace element solution: 40 mg $ZnCl_2$, 200 mg $FeCl_3 \times 6\ H_2O$, 10 mg $CuCl_2 \times 2\ H_2O$, 10 mg $MnCl_2 \times 4\ H_2O$, 10 mg $Na_2B_4O_7\ Cl_2 \times 10\ H_2O$, and 10 mg $(NH_4)_6Mo_7O_{24} \times 4\ H_2O$ in 1 L ddH_2O. Sterilize by autoclaving.
8. 20% L-arabinose solution: weigh 10 g L-arabinose and fill up to 50 mL with ddH_2O. Sterilize by 0.22 μm filter.
9. 60% glucose solution: weigh 60 g D-glucose and fill up to 100 mL with ddH_2O. Sterilize by 0.22 μm filter.
10. 10% glycerin solution: measure 100 mL glycerin and fill up to 1 L with ddH_2O. Sterilize by autoclaving.

2.3 Kits and Equipment

1. For genomic DNA extraction the Puregene Yeast/Bact. Kit B (Qiagen) was used.
2. PCR fragments were purified using the Monarch® DNA gel extraction kit (New England Biolabs) or Monarch PCR & DNA Cleanup Kit (New England Biolabs).
3. Plasmid DNA was extracted using the PureYield™ Plasmid Miniprep System (Promega).

2.4 PCR Amplification

1. For PCR amplification, we use Q5®Hot Start High-Fidelity DNA Polymerases (New England Biolabs) or Phusion® Hot Start Flex DNA Polymerase (New England Biolabs).

2. 5 M betaine solution: dissolve 29.27 g betaine in 50 mL ddH_2O. Sterilize by 0.22 μm filter.

2.5 DNA Assembly

1. For plasmid DNA assembly, we either use NEBuilder® HiFi Assembly Master Mix or Hot Fusion [24] master mix.
2. 2× Hot-Phusion Mix: 200 μL 5× isothermal reaction buffer, 0.4 μL T5 exonuclease 10 U/μL (Epicentre), 12.5 μL Phusion DNA-Polymerase 2 U/μL (New England Biolabds), and 287.1 μL ddH_2O.
3. 5× isothermal reaction buffer: 1.5 g PEG-8000, 3 mL of 1 M Tris–HCl pH 7.5, 150 μL of 2 M $MgCl_2$, 300 μL of 1 M DTT, 60 μL of 100 mM dGTP, 60 μL of 100 mM dATP, 60 μL of 100 mM dTTP, 60 μL of 100 mM dCTP, and 300 μL of 100 mM NAD.

3 Methods

3.1 Plasmid Design

3.1.1 XU Concept Fusion Site

1. For identifying the XU fusion site, use the C_3-**linker**-A_3 sequence (YLQAILWAIVNQ**PQQPVTAIDILSSSERELL-LENWNATEEPYPTQ**VCVHQLFEQQIE) of GxpS from *P. luminescens* subsp. *laumondii* TT01 GxpS (locus_tag: PLUMV2_16690) and align with a target sequence of choice. Choose the splicing position right after the α-helical secondary structure (leucine sequence). Use the splicing position W][NATE in GxpS as guide to find the right fusion site (Fig. 1).
2. Extract NRPS encoding gene fragments and reassemble them according to the XU fusion site, for example, using cloning software such as Geneious Prime or Clone Manager. Design primers according to standard procedures. We commonly use a T_a of the binding site of >60 °C and ~25 bp overhangs with a T_a of at least >50 °C.

3.1.2 SYNZIPs

SYNZIP Plasmid Construction

1. SYNZIP protein and DNA sequences can be found on the Keating Lab homepage (retrieved on 28.04.22, https://www.keatinglab.mit.edu/synzips). SYNZIP encoding plasmids are available from addgene. We used SYNZIP17 (pENTR-SYNZIP17 was a gift from Amy Keating, Addgene plasmid # 80671; http://n2t.net/addgene:80671; RRID: Addgene_80671), SYNZIP18 (pENTR-SYNZIP18 was a gift from Amy Keating, Addgene plasmid # 80672; http://n2t.net/addgene:80672; RRID:Addgene_80672), SYNZIP1 (pQLinkHD-SYNZIP1 was a gift from Amy Keating, Addgene plasmid # 80647; http://n2t.net/addgene:80647; RRID: Addgene_80647), and SYNZIP2 (pQLinkHD-SYNZIP2 was a gift from Amy Keating, Addgene plasmid # 80658; http://n2t.net/addgene:80658; RRID:Addgene_80658).

Table 1
Primer pairs utilized to insert SYNZIPs into expression plasmids pACYC_ara/araC and pCOLA_ara/tacI

Target plasmid	Description	Name	Sequence 5′ → 3′
pACYC_ara/araC	Insertion of SZ17	KB-pACYC-fw	GAACAGTTAAAACAGAAGCGTGAACAA TTAAAGCAAAAGATCGCCAATC TGCGTAAGGAGATCGAAGCCTACAAGTGACAA TTAATCATCGGCTCG
		KB-pACYC-rv	TTCACGCTTCTGTTTTAACTGTTCGATGCGA TTACGCAATTCAGCCTTTTTC GATTTTAATTCCTCCTTCTCGTTCATGGAATTCC TCCTGTTAGC
pACYC_ara/araC_SZ17	Plasmid linearization	KB-pACYC-II-fw	AACGAGAAGGAGGAATTAAAATCG
		KB-pACYC-II-rv	CATGGAATTCCTCCTGTTAGC
pCOLA_ara/tacI	Insertion of SZ18	KB-pCOLA-fw	CATTGACAAAGAGCTGCG TGCCAACGAAAACGAACTTCGCGCCCTTG ATAACGAGCTGACTGCAGCTATCTCATGACAA TTAATCATCGGCTCG
		KB-pCOLA-rv	TTGGCACGCAGCTCTTTGTCAATGGCATTTAAC TCGCGGTCCAAGGC TTTCAGTTCACGCTCTTCAGCATAGAACA TGGAATTCCTCCTGTTAGC
pCOLA_ara/tacI_SZ18	Plasmid linearization	KB-pCOLA-II-fw	TGACAATTAATCATCGGCTCG
		KB-pCOLA-II-rv	TGAGATAGCTGCAGTCAGCTCG

2. We inserted SYNZIP17 into pACYC_ara/araC with oligonucleotides KB-pACYC-FW and KB-pACYC-RV. After PCR amplification and plasmid assembly we linearized obtained pACYC_ara/araC_SZ17 with oligonucleotides KB-pACYC-II-FW and KB-pACYC-II-RV (Table 1). NRPS subunit 1 of choice is assembled into the linearized plasmid backbone (Fig. 2, 1.1 and 1.2).
3. For the insertion of SYNZIP18 into pCOLA_ara/tacI, we used oligonucleotides KB-pCOLA-FW and KB-pCOLA-RV. To linearize obtained pCOLA_ara/tacI_SZ18, we used KB-pCOLA-II-FW and KB-pCOLA-II-RV. NRPS subunit 2 of choice is assembled into the linearized plasmid backbone.

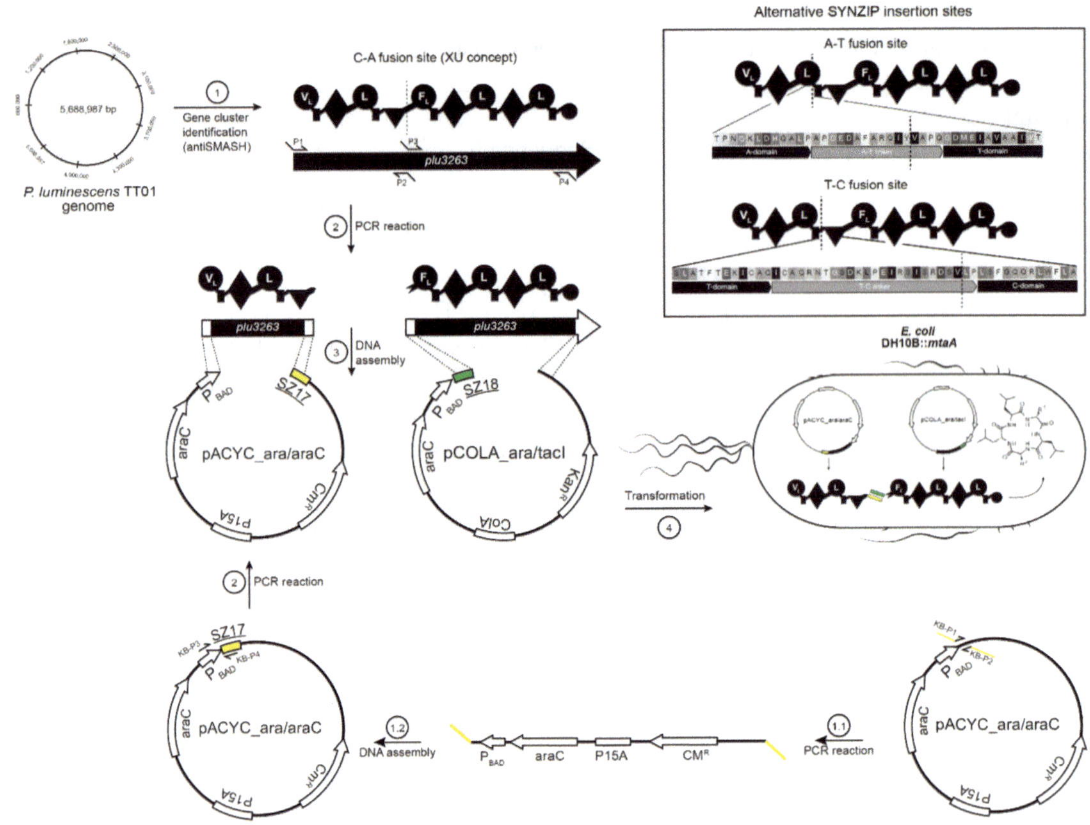

Fig. 2 Workflow type S NRPS. The procedure starts with the in silico identification of the target BGC (1). SZ17 and SZ18 are introduced into pACYC_ara/araC and pCOLA_ara/tacl with oligonucleotides KB_pACYC_fw/KB_pACYC_rv and KB_pCOLA_fw/KB_pCOLA_rv, respectively (1.1 and 1.2). The NRPS encoding BGC is divided and PCR amplified in two individual smaller subunits with oligonucleotides P1:P2 and P3:P4. Plasmids pACYC_ara/araC_SZ17 and pCOLA_ara/tacl_SZ18 are linearized with oligonucleotides KB-pACYC-II-fw/KB-pACYC-II-rv and KB-pCOLA-II-fw/KB-pCOLA-II-rv, respectively (2). PCR amplified subunits 1 and 2 are cloned into the linearized plasmids (3). Verified plasmids are co-transformed into *E. coli* DH10B::*mtaA* for protein expression and nonribosomal peptide (NRP) biosynthesis (4). Produced NRPs are analyzed via HPLC-MS

Fusion Sites for SYNZIP Insertion

1. For identifying the XU, A-T, or T-C insertion site, use the C_3-**linker**-A_3 (YLQAILWAIVNQ**PQQPVTAIDILSSSER-ELLLENWNATEEPYPTQ**VCVHQLFEQQIE), A_2-**linker**-T_2 (TPNGKLDHQALP**APGEDAFARQIYVAPQ**GD-MEIAVAAIWC), or T_2-**linker**-C_3 (SLATFTEKICAQ**I-CAQRNTGSDKLPEIRSISRDSVLP**LSFGQQRLWFLA) sequence of GxpS from *P. luminescens* subsp. *laumondii* TT01 GxpS (locus_tag: PLUMV2_16690) and align with a target sequence of choice. Use the splicing position W][NATE (XU), Y][VAPO (A-T) or V][LPLS (T-C) in GxpS as guide to find the right insertion site (Fig. 2).

Table 2
PCR protocol for fragment amplification of DNA fragments from gDNA or plasmids

1-step PCR				2-step PCR			
Initializing	98 °C	30 s		Initializing	98 °C	30 s	
Denaturation	98 °C	10 s		Denaturation	98 °C	10 s	
Annealing	$T_{m\ (bind\ region)}$ − 1 °C	15 s	34×	Annealing	$T_{m\ (bind\ region)}$ − 1 °C	15 s	4×
Elongation	72 °C	30 s per 1 kb		Elongation	72 °C	30 s per 1 kb	
Final elongation	72 °C	5 min		Denaturation	98 °C	10 s	
				Annealing	$T_{m\ (full\ primer)}$ − 1 °C	15 s	30×
				Elongation	72 °C	30 s per 1 kb	
				Final elongation	72 °C	5 min	

3.2 PCR Amplification

1. To efficiently assemble different NRPS gene fragments into the plasmid backbone, we preferably perform DNA assembly via HiFi or Hot Fusion. NRPS gene fragments are amplified from gDNA or verified plasmid DNA by a 2-step PCR using primers with 5′ homologs overhangs of 20–30 bp length to the plasmid backbone and/or another NRPS fragment (Fig. 1). The T_a in the second annealing step of the 2-step PCR reaction is set to max. 72 °C. Plasmid backbones are amplified via a 1-step PCR reaction. Standard protocols for 1- and 2-step PCR reactions are depicted in Tables 2 and 3. The PCR efficiency can be improved by the addition of 1 mM $MgCl_2$, 0.8–3.2% DMSO, or 1 M betaine. Adjust the quantity of water accordingly.

3.3 DpnI and Gel Extraction

1. To remove plasmid template DNA, we digest PCR products with DpnI (New England Biolabs) according to the manufacturer's instructions.
2. We purify PCR products either with Monarch PCR & DNA Cleanup Kit or, in the case of by-products, with the Monarch® DNA gel extraction kit from a 1% agarose gel.

3.4 DNA Assembly

1. For DNA assembly, we use Hot Fusion Master Mix or NEBuilder® HiFi DNA Assembly Master Mix (New England Biolabs). NEBuilder® is highly efficient for multi-fragment assembly of up to 7 fragments of ≤8 kb size.

Table 3
Standard PCR setup

Buffer (5×)	5 μL
Primer fw (2 μM)	5 μL
Primer rev (2 μM)	5 μL
dNTPs (10 mM)	0.5 μL
DMSO (100%)	0.4 μL
$MgCl_2$ (50 mM)	0.5 μL
Template	25 pg/μL
Polymerase	0.2 μL
DNase free water to total volume of 25 μL	

2. We typically apply 50 ng linearized plasmid backbone and depending on the size and ratio 2–300 ng PCR insert. For optimized cloning, we apply threefold molar excess of insert in a 1-fragment assembly, twofold molar excess in a 2-fragment assembly, and 1:1 ratio (insert:vector) if more than 4 fragments are applied. However, it is worth mentioning that in most cases optimization is not necessary, but can help in cases where the assembly fails or assembly efficiency is too low.
3. If the total volume of DNA exceeds 1.5 μL, we reduce the volume to 1.5 μL using a vacuum concentrator. Afterward, we add the similar volume HotFusion Master Mix or NEBuilder® to the DNA, which we incubate for 1 h at 50 °C. We add 1 μL of the assembled DNA mix to 50 μL of competent *E. coli* DH10B::*mtaA* cells and perform transformation via electroporation (*see* Subheadings 3.5 and 3.6). After incubation for 1 h at 37 °C, all cells are plated on LB with appropriate antibiotics.

3.5 Competent Cells

1. For competent *E. coli* cells, inoculate LB-medium 1:50 with an overnight culture of the respective *E. coli* strain.
2. Grow cells to an OD_{600} 0.6–0.8 and then incubate on ice for 15 min.
3. Centrifuge for 12 min at 4 °C and 4000 *g* and discard supernatant afterward.
4. Wash cells two times with 1/5 and 1/50 culture volume ice-cold 10% glycerin solution.
5. Then, incubate cells for 30 min on ice.
6. Pooled with 1/200 culture volume 10% glycerin solution and aliquot into 50 μL volumes.

3.6 Transformation via Electroporation

1. Add 1 μL assembly mix or 1 μL plasmid (5 ng/μL) to 50 μL electro competent *E. coli* DH10B or *E. coli* DH10B::*mtaA*.
2. Transfer cells to 1 mm diameter electroporation cuvette.
3. Perform electroporation with 1 mm gap width cuvettes at 25 μF, 1250 V, 200 Ω, and 1 pulse.
4. Directly recover cells with 800 μL low-salt LB medium and incubate for 1 h at 37 °C gently shaking.
5. Plate all cells on low-salt LB-agar plates with antibiotics for selection.
6. Incubate for 12 h at 37 °C.

3.7 Plasmid Verification and Sequencing

1. For plasmid verification, inoculate five clones per construct into LB supplemented with appropriate antibiotics. Additionally, transfer each clone to an LB agar plate, which can be used after verification (master plate).
2. After overnight incubation, isolate plasmids from 2 mL overnight culture. We use the PureYield™ Plasmid Miniprep System (Promega).
3. Perform a restriction enzyme test digestion with all isolated plasmids using type II restriction endonucleases according to the manufacturer's instruction in a 10 μL approach to verify the correct assembly of the plasmid. Plasmids with correct DNA restriction patterns on an analytical 1% agarose gel are then verified via sequencing.
4. XU concept: inoculate three correctly verified clones from the previously prepared master plate in 3 mL LB with antibiotics to generate biological triplicates for further analyses (c.f., Subheading 3.8).
5. SYNZIP concept: we recommend to use only one of the verified plasmids for the subsequent transformation into *E. coli* DH10B::*mtaA* together with all complementary plasmids. Otherwise, the throughput of recombined type S NRPS increases too rapidly.

3.8 Production Cultures—Heterologous Protein Expression in E. coli

1. Cultivate recombinant *E. coli* DH10B::*mtaA* strains in 50 mL Erlenmeyer flasks for peptide production.
2. Supplement 10 mL LB medium (or XPPM) with the respective antibiotics for selection. Furthermore, add 0.02% L(+)-arabinose to the medium for induction.
3. Inoculate the medium with 1:100 of the overnight culture from Subheading 3.7 in biological triplicates. Cells containing an empty vector are used as a negative control while a verified nonribosomal peptide (NRP)-producing recombinant *E. coli* strain should be included as positive control.

4. Cultivate cells for 72 h at 22 °C and 180 rpm.
5. After cultivation, mix 0.5 mL culture with HPLC-pure methanol and incubate for 0.5–1 h at RT, shaking.
6. Afterward, prepare the sample for HPLC-MS analysis.

To analyze whether the NRP is accumulated within the cell or secreted to the supernatant, supernatant and cells can be separated by centrifugation at 17,000 *g* for 3 min. The cells are resuspended in the same culture volume of methanol and no further solvents need to be added to the supernatant. Both samples are prepared for HPLC-MS to analyze the NRP localization. When the produced NRP is secreted to the supernatant and is rather hydrophobic, 2% XAD™ 16N Polymeric Absorbent (Amberlite™) are added to the production culture which is used to absorb hydrophobic molecules from polar solvents. For extraction, XAD™ 16N Polymeric Absorbent is collected by decanting or sieving and one culture volume methanol is added. Extraction is performed for 1 h. Subsequently, samples are prepared for HPLC-MS analysis. This kind of extraction can be used to enrich the yields of the NRP for analysis or further purification.

4 Notes

1. Preferentially, Q5® High-Fidelity DNA Polymerase (New England Biolabs) is used for amplification due to its low error rate, which is crucial when amplifying larger DNA fragments.
2. When amplifying more than one module in a single PCR, we strongly recommend to perform a gel extraction (*see* Subheading 3.3) instead of a less time-consuming PCR purification to ensure efficient DNA assembly in the subsequent steps—fusion sites are located in highly conserved regions of the NRPS, unwanted smaller fragments are often generated, which are preferentially incorporated during DNA assembly.
3. Since fusion sites are often located within conserved regions [12, 15, 25], primers can bind unspecifically at several positions within the cluster, resulting in low or no yields of the desired amplified fragment. In this case, we recommend a nested PCR. Here, a first set of very specific primers binding at a unique region in the close vicinity of the target region are used to amplify a larger region around the target region. The resulting PCR product is used as a template in a second round of PCR to enhance the yields of the desired DNA fragment.
4. In case the alignment is insufficient, when identifying the XU (A-T or T-C) fusion site [12], we recommend to include more extracted and annotated C-linker-A sequences from other BGCs, in which the XU fusion site can be identified.

5. It should be noted that the efficiency of DNA assembly decreases with the length and number of different fragments. Nevertheless, plasmid sizes up to 30 kb can be efficiently generated [12–15, 25, 26].
6. Introductions of SYNZIPs [20] via oligonucleotides can lower the PCR reaction efficiency due to self- and cross-primer dimerization. In case PCR does not work, *see* Subheading 3.2. Alternatively, the SYNZIP encoding sequences of interest can be chemically synthesized or purchased from addgene.
7. We recommend the use of short SYNZIPs, preferably SZ17:18, as short pairs have been shown to have the least effect on the activity of the enzyme [14, 15, 26].
8. For type S NRPSs [14, 15], multiple plasmids are transformed into *E. coli* DH10B::*mtaA*. Successful transformation of two plasmids is achieved by transforming at least 50 ng of each plasmid and plating all cells on LB agar plates to obtain sufficient colonies. If more than two plasmids are required (tri-partite NRPSs), prepare competent cells after transformation of the first two plasmids, then perform transformation of the third plasmid.

Acknowledgments

The authors are grateful to all current and past members from the Bode lab being involved in optimizing the NRPS engineering rules and protocols. Work in the Bode lab was supported by an ERC Advanced Grant (835108) and the LOEWE TBG research center.

References

1. Davies J (2013) Specialized microbial metabolites: functions and origins. J Antibiot 66:361–364
2. Newman DJ, Cragg GM (2020) Natural products as sources of new drugs over the nearly four decades from 01/1981 to 09/2019. J Nat Prod 83:770–803
3. Bérdy J (2012) Thoughts and facts about antibiotics: where we are now and where we are heading. J Antibiot 65:385–395
4. Lawson ADG, MacCoss M, Heer JP (2018) Importance of rigidity in designing small molecule drugs to tackle protein-protein interactions (PPIs) through stabilization of desired conformers. J Med Chem 61:4283–4289
5. Payne DJ, Gwynn MN, Holmes DJ et al (2007) Drugs for bad bugs: confronting the challenges of antibacterial discovery. Nat Rev Drug Discov 6:29–40
6. Hughes JP, Rees S, Kalindjian SB et al (2011) Principles of early drug discovery. Br J Pharmacol 162:1239–1249
7. Wright PM, Seiple IB, Myers AG (2014) The evolving role of chemical synthesis in antibacterial drug discovery. Angew Chem Int Ed 53: 8840–8869
8. Süssmuth RD, Mainz A (2017) Nonribosomal peptide synthesis-principles and prospects. Angew Chem Int Ed 56:3770–3821
9. Stachelhaus T, Schneider A, Marahiel MA (1995) Rational design of peptide antibiotics by targeted replacement of bacterial and fungal domains. Science 269:69–72
10. Brown AS, Calcott MJ, Owen JG et al (2018) Structural, functional and evolutionary perspectives on effective re-engineering of non-ribosomal peptide synthetase assembly lines. Nat Prod Rep 35:1210–1228

11. Calcott MJ, Ackerley DF (2014) Genetic manipulation of non-ribosomal peptide synthetases to generate novel bioactive peptide products. Biotechnol Lett 36:2407–2416
12. Bozhüyük KAJ, Fleischhacker F, Linck A et al (2018) De novo design and engineering of non-ribosomal peptide synthetases. Nat Chem 10:275–281
13. Bozhüyük KAJ, Linck A, Tietze A et al (2019) Modification and de novo design of non-ribosomal peptide synthetases using specific assembly points within condensation domains. Nat Chem 11:653–661
14. Abbood N, Duy Vo T, Watzel J et al (2022) Type S non-ribosomal peptide synthetases for the rapid generation of tailormade peptide libraries. Chemistry 28:e202103963
15. Bozhüyük KAJ, Watzel J, Abbood N et al (2021) Synthetic zippers as an enabling tool for engineering of non-ribosomal peptide synthetases. Angew Chem Int Ed 60:17531–17538
16. Rausch C, Hoof I, Weber T et al (2007) Phylogenetic analysis of condensation domains in NRPS sheds light on their functional evolution. BMC Evol Biol 7:78
17. Tanovic A, Samel SA, Essen LO et al (2008) Crystal structure of the termination module of a nonribosomal peptide synthetase. Science 321:659–663
18. Nollmann FI, Dauth C, Mulley G et al (2015) Insect-specific production of new GameXPeptides in photorhabdus luminescens TTO1, widespread natural products in Entomopathogenic bacteria. Chembiochem 16:205–208
19. Calcott MJ, Owen JG, Ackerley DF (2020) Efficient rational modification of non-ribosomal peptides by adenylation domain substitution. Nat Commun 11:4554
20. Thompson KE, Bashor CJ, Lim WA et al (2012) SYNZIP protein interaction toolbox: in vitro and in vivo specifications of heterospecific coiled-coil interaction domains. ACS Synth Biol 1:118–129
21. Durfee T, Nelson R, Baldwin S et al (2008) The complete genome sequence of *Escherichia coli* DH10B: insights into the biology of a laboratory workhorse. J Bacteriol 190:2597–2606
22. Terpe K (2006) Overview of bacterial expression systems for heterologous protein production: from molecular and biochemical fundamentals to commercial systems. Appl Microbiol Biotechnol 72:211–222
23. Gaitatzis N, Hans A, Müller R et al (2001) The MtaA gene of the myxothiazol biosynthetic gene cluster from *Stigmatella aurantiaca* DW4/3-1 encodes a phosphopantetheinyl transferase that activates polyketide synthases and polypeptide synthetases. J Biochem 129: 119–124
24. Fu C, Donovan WP, Shikapwashya-Hasser O et al (2014) Hot fusion: an efficient method to clone multiple DNA fragments as well as inverted repeats without ligase. PLoS One 9: 115318
25. Bozhüyük KAJ, Präve L, Kegler C, Kaiser S, Shi Y-N, Kuttenlochner W, Schenk L, Mohiuddin TM, Groll M, Hochberg GKA, Bode HB (2022) Evolution Inspired Engineering of Megasynthetases. https://doi.org/10.1101/2022.12.02.518901
26. Abbood N, Effert J, Bozhueyuek KAJ, Bode HB (2023) Guidelines for Optimizing Type S Non-Ribosomal Peptide Synthetases. https://doi.org/10.1101/2023.03.21.533600

Chapter 12

Probing Substrate-Loaded Carrier Proteins by Nuclear Magnetic Resonance

Neeru Arya, Kenneth A. Marincin, and Dominique P. Frueh

Abstract

Carrier proteins (CPs) are central actors in nonribosomal peptide synthetases (NRPSs) as they interact with all catalytic domains, and because they covalently hold the substrates and intermediates leading to the final product. Thus, how CPs and their partner domains recognize and engage with each other as a function of CP cargos is paramount to understanding and engineering NRPSs. However, rapid hydrolysis of the labile thioester bonds holding substrates challenges molecular and biophysical studies to determine the molecular mechanisms of domain recognition. In this chapter, we describe a protocol to counteract hydrolysis and study loaded carrier proteins at the atomic level with nuclear magnetic resonance (NMR) spectroscopy. The method relies on loading CPs in situ, with adenylation domains in the NMR tube, to reach substrate-loaded CPs at steady state. We describe controls and experimental readouts necessary to assess the integrity of the sample and maintain loading on CPs. Our approach provides a basis to conduct subsequent NMR experiments and obtain kinetic, thermodynamic, dynamic, and structural parameters of substrate-loaded CPs alone or in the presence of other domains.

Key words Nonribosomal peptide synthetases, Carrier proteins, Nuclear magnetic resonance, In situ substrate loading, Reaction monitoring

1 Introduction

NRPSs are organized in modules comprised of conserved domains that function synergistically to activate, attach, and condense substrates thereby controlling the composition of the final product [1]. Carrier proteins (CPs) are involved in all steps of synthesis, such that NRPS synthesis may largely be described as a succession of covalent modifications of carrier proteins. Apo CPs are first activated by a phosphopantetheinyl transferase (PPTase) through covalent attachment of a 4′-phosphopantetheine arm (PP) onto a conserved serine to become holo-CPs. Next, adenylation

Neeru Arya and Kenneth A. Marincin contributed equally with all other contributors.

Michael Burkart and Fumihiro Ishikawa (eds.), *Non-Ribosomal Peptide Biosynthesis and Engineering: Methods and Protocols*, Methods in Molecular Biology, vol. 2670, https://doi.org/10.1007/978-1-0716-3214-7_12,

(A) domains activate substrates as adenylates and attach them to the PP arm of holo-CPs through a labile thioester bond. A domains display specificity for substrates and CPs, with relaxation in some systems, and provide the first line of fidelity [2–4]. Condensation (C) domains then catalyze the peptide bond formation between the substrates of an upstream donor CP and a downstream acceptor CP, restoring the upstream CP to its holo form and extending the peptide intermediate onto the downstream CP. The latter now acts as a donor for a downstream C domain, and the process is iterated until the final peptide is obtained and released in a termination module, most often through thioesterase (TE) domains. Thus, understanding NRPS synthesis and engineering NRPSs require understanding how CPs interact both with their partner domains and with their attached prosthetic groups.

CPs are not mere substrate carriers but instead participate actively in every stage of peptide synthesis through sequential interactions with partner domains. Initially, CPs were thought to serve as anchors for swinging phosphopantetheine arms sampling the active sites of catalytic domains [5–7]. However, crystallographic studies have since captured a variety of domain interactions revealing that CPs navigate between domains [8–13] and that the PP arm is used to reach buried active sites [8, 9, 12, 13]. It is currently unknown whether these snapshots reflect a sequence of stable domain–domain interactions or a selection of domain arrangements captured from a permanently dynamic domain architecture. The latter appears more likely as negative stain electron microscopy and cryo-electron microscopy (cryo-EM) display a variety of domain conformations [13–16] and NMR reveals transient, competitive domain interactions [17]. However, the NRPS dynamic landscape appears to be remodeled during synthesis, and it is hence not random, since domain interactions are promoted, at least in part, by the state of the carrier proteins: PPTases interact preferentially with apo-CPs [18], A domains with holo-CPs [19] and C domains, or the related cyclization domains, engage with loaded donor CPs only [20, 21]. The emerging picture is that CPs and their partners probe each other and only fully engage for productive interactions, when the CP harbors the right cargo, a description reminiscent of encounter complexes in protein-binding studies. Indeed, it is increasingly recognized that free and bound states often provide a limited molecular description of binding and critical residues are only identified when accounting for fleeting encounters preceding fully engaged complexes [22]. This condition is likely true for CPs as their PP arms must reach into buried active sites for all catalytic partners [23].

CPs not only interact with partner domains but also with their substrates and the phosphopantetheine arm. We and the Burkart lab observed that CPs interact with their substrates and PP arms in a transient manner and exist in an equilibrium between an undocked state, where the PP arm is disordered, and a docked state, where the arm and substrate are interacting with the core of the protein [24, 25]. Thus, either catalytic domains probe CP cargos through the undocked form or recognize CPs and their cargo simultaneously through the docked form. In closing, to engineer efficient NRPS systems, one must understand how catalytic domains and their partner CP domains recognize each other through fleeting interactions before they engage in the stable productive states observed in crystal and cryo-EM structures. Notably, we must determine the molecular features of CPs harboring their cargos and how their partners recognize these features to reject or promote communication [26]. However, this objective is challenged by labile thioester bonds tethering substrates or intermediates, as the cargos rapidly fall off during structural and binding studies. This chapter presents a strategy that offers relief to this drawback for studies of CPs holding single substrates.

Nuclear magnetic resonance (NMR) can characterize fleeting molecular responses of CPs holding their substrates through a native thioester bond. NMR can provide kinetics, thermodynamics, dynamics, and structural information of proteins, all at atomic resolution [27–31]. This versatility has been key to advancing our understanding of encounter complexes [22, 32], protein dynamics [33], or allosteric (remote) responses [34] and results from the abundance of experimental readouts provided by a plethora of pulse sequences. In its most simple application, NMR readily informs on moieties involved in molecular communication as the frequency of NMR signals are direct reporters of changes in molecular environments. For proteins, ^{1}H-^{15}N correlation maps, such as ^{1}H-^{15}N-HSQC [35] or ^{1}H-^{15}N-TROSY [36], provide a readout at the residue level since every residue, save prolines, is represented by a correlation through its amide bond. For example, the comparison of 2D NMR spectra of apo, holo, and substrate-loaded CPs in the presence and the absence of other catalytic domains highlights residues involved in binding, both at binding sites and in regions changing conformations, for example, due to conformational changes of the PP arm. To overcome the lability of the thioester bond, we exploit the noninvasive nature of NMR and incubate holo-CPs with their cognate adenylation domains, their substrates, and adenosine triphosphate (ATP), until we reach a steady state of substrate-loaded CPs [37] (Fig. 1). Under these conditions, holo-CP domains are continuously loaded with substrates such that adenylation domains compensate for hydrolysis until depletion of ATP. A similar strategy was more recently implemented for an acyl carrier protein (ACP) where the authors used an acyl acyl-carrier protein synthetase (AasS) to counteract hydrolysis [38]. Once obtained, the substrate-loaded CP is available for subsequent NMR studies such as probing interactions with their phosphopantetheine arms and substrates [24], determining how partner

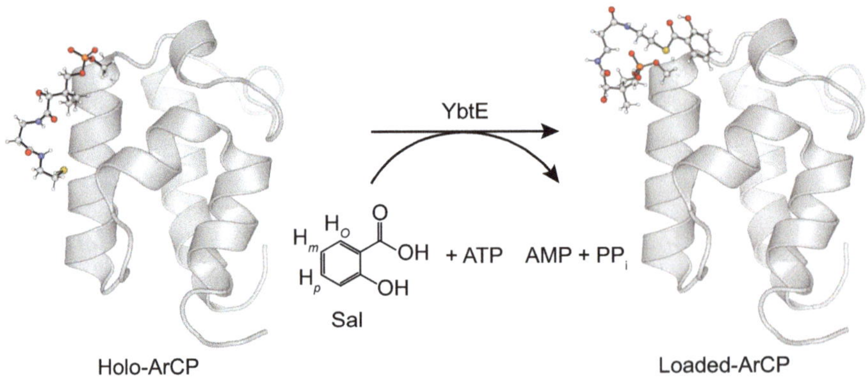

Fig. 1 In situ loading of carrier proteins with substrates. To overcome hydrolysis of substrates covalently tethered to the phosphopantetheine arm (ball-and-stick), holo-ArCP (left) is incubated with its Sal substrate, ATP, and the adenylation domain YbtE to generate Sal-loaded-ArCP (right). YbtE then outcompetes substrate hydrolysis. The positions of the ortho (H_o), meta (H_m), and para (H_p) protons on Sal are labeled. ArCP solution structures used PDB IDs 2N6Y and 2N6Z [24]

domains differentiate holo and loaded forms [20] or determining their dynamics [24]. Isotope editing is a mainstay in NMR strategies and allows for focusing on one protein in a complex mixture. Thus, the protocol provided here can be adapted through various isotopic labeling approaches so that catalytic partners may be studied as they are presented to loaded-CPs. Notably, using an adaptation of the protocol presented here, we recently discovered an allosteric response in a cyclization domain upon engagement with a loaded carrier protein [20]. Overall, the strategy we present here enables detailed molecular studies of loaded-CPs and their communication with partner domains through an arsenal of existing NMR experiments.

Here, we describe in detail the protocol used in our lab to reproducibly obtain loaded-CPs, and we discuss the controls necessary to maximize loading or monitor the integrity of the sample during measurements. We illustrate the method with the aryl carrier protein (ArCP) of HMWP2 [39], an NRPS used in yersiniabactin synthesis. Here, the stand-alone A domain YbtE catalyzes the attachment of a salicylate (Sal) substrate onto holo-ArCP through a native thioester bond. As an application, we show how chemical shift perturbations between spectra before and after loading readily provide molecular-level insights into substrate loading.

2 Materials

2.1 Protein Expression and Purification

It is critical to obtain homogeneous starting protein samples. Notably, holo-CPs can be prepared through various routes. First, purified apo-CPs can be converted to holo-CPs by incubation with

Coenzyme A (CoA) and a PPTase, such as Sfp [40]. Here, the first challenge is to obtain homogeneous samples of apo-CPs as *Escherichia coli* (*E. coli*) natively contains a PPTase, EntD, thus leading to background levels of holo-CPs, possibly in acylated forms due to endogenous CoA derivatives in the cell. We use the ΔEntD *E. coli* variant obtained from Drs. Challut and Guilhot (CNRS, Toulouse, France) to overcome this challenge [37]. Alternatively, homogeneous holo-ArCP is obtained by co-expression with a PPTase such as Sfp, for example, by using a pET-Duet-1 vector. Here, we overexpressed GB1-TEV-ArCP14-93 together with Sfp within a pET-Duet-1 vector as described in [37]. Purification is critical, and a simple affinity purification is in general insufficient as contaminants such as disordered proteins or small peptides may be detected even at low concentrations due to favorable NMR relaxation properties when compared to the 9–12 kDa carrier proteins. For example, our sample is purified through a nickel-nitrilotriacetic acid (Ni-NTA) resin, ion exchange, a reverse Ni-NTA elution following digestion with a tobacco etch virus (TEV) protease, and finally size-exclusion chromatography, as described in [37].

The sample of the stock solution of the A domain is also prepared to high purity. Thus, the YbtE A domain is purified through Ni-NTA and size exclusion, as described in [37].

Optionally, a solution of a type II thioesterase (TEII) may be prepared to perform controls in key experiments. We refer the reader to [20, 37] to see examples where the versatile SrfA-D TEII was used to restore the CP to its holo form.

Holo-CPs readily dimerize through disulfide bond formation and reducing agents are needed before data acquisition. Phosphines such as tris(2-carboxyethyl)phosphine (TCEP) are preferred over thiols, such as dithiothreitol (DTT) or 2-mercaptoethanol (BME), as the latter two will lead to parasitic transthioesterifications releasing loaded substrates [41].

2.2 Stock Solutions

All stock solutions are prepared in in situ loading buffer (in our case, 50 mM 2-[(carbamoylmethyl)amino]ethane-1-sulfonic acid (ACES), 150 mM NaCl, 1 mM $MgCl_2$, 1 mM TCEP, 0.05% NaN_3, pH = 6.8), and the pH of the final solutions are readjusted to 6.8 at 22 °C (*see* **Note 1**).

1. 100 mM Sal.
2. 100 mM ATP.
3. 26.3 μM YbtE.

To increase the lifetime of TCEP, the in situ loading buffer is freshly degassed before the addition of TCEP, and TCEP is added on the same day that NMR acquisitions begin. For longer acquisitions, to further delay oxidation, the NMR sample and stock solutions are bubbled with argon and the tube is sealed with parafilm after the final addition of YbtE.

2.3 NMR Samples

1. Neat methanol sample for temperature calibration.
2. Controls: (i) 445 μL of in situ loading buffer with 10% deuterium oxide (D_2O) and 200 μM sodium trimethylsilylpropanesulfonate (DSS) for reference, (ii) 2 mM Sal in in situ loading buffer (10 μL of Sal 100 mM + 445 μL of (i)) with 9.8% D_2O and 196.07 μM DSS, and (iii) 2 mM Sal and 2 mM ATP in in situ loading buffer (10 μL of Sal 100 mM + 10.5 μL of ATP 100 mM + 445 μL of (i)) with 9.6% D_2O and 192.12 μM DSS (*see* **Note 2**).
3. NMR sample for in situ loading of CPs: (i) 445 μL of isotopically labeled holo-CP sample with 10% D_2O and 200 μM DSS in in situ loading buffer. Here, we used 324 μM ^{15}N holo-ArCP, (ii) ^{15}N-labeled holo-ArCP +2 mM Sal (445 μL of (i) + 10 μL of Sal 100 mM) with 9.8% D_2O and 196.07 μM DSS at 317.6 μM, and (iii) ^{15}N-labeled holo-ArCP +2 mM Sal +2 mM ATP (445 μL of (i) + 10 μL of Sal 100 mM + 10.5 μL of ATP 100 mM) with 9.6% D_2O and 192.12 μM DSS at 311.2 μM; (iv) ^{15}N-labeled holo-ArCP +2 mM Sal +2 mM ATP + 100 nM YbtE (445 μL of (i) + 10 μL of Sal 100 mM + 10.5 μL of ATP 100 mM + 2 μL of 26.3 μM YbtE) with 9.5% D_2O and 191.38 μM DSS at 310 μM (*see* **Note 3**).

2.4 NMR Spectrometer

Any spectrometer may be used, but the user should account for reduced resolution and sensitivity when using lower fields or for reduced sensitivity when using room-temperature probes. All data presented in this chapter were recorded on a Bruker Avance III 600 MHz spectrometer equipped with a QCI triple resonance cryoprobe.

3 Methods

Here, we describe the protocol we employ to load Sal onto ^{15}N-labeled holo-ArCP and to monitor the loaded form during data acquisition. The protocol includes controls carried out before starting the in situ loading reaction. In addition to verifying the integrity of chemicals and reagents, these controls provide a reference to monitor the reaction. Figure 1 shows a schematic of the reaction. Notably, ATP is converted into adenosine monophosphate (AMP) and the molecules have distinct and characteristic NMR signals (vide infra). Similarly, ATP may degrade to adenosine diphosphate (ADP) at room temperature over longer periods, and ADP displays characteristic signals as well (vide infra). More generally, NMR spectra will reflect any changes in conditions, such as changes in pH, and we exploit this advantage at all stages of the measurements. Key Bruker TopSpin commands are reported in italic. We will assume that all NMR experiments have been installed by an expert NMR spectroscopist such that all power levels are safe and parameters are adequate, and we only underline aspects specifically relevant to monitoring the reaction. We will also assume that prosol was set up on the spectrometer. If not, simply update all relevant parameters manually.

3.1 Temperature Calibration

NMR chemical shifts change with temperature and deviations between the temperature targeted for measurements, as set by the user, and the true temperature of the sample will jeopardize data comparisons. When possible, we recommend measuring all data to be compared on the same spectrometer and in the same acquisition run. Regardless, the true temperature of the sample should be calibrated or at least determined before starting any experiment. Many standards are available for which differences between signal chemical shifts have been calibrated to temperatures [42–45]. Here, we describe the procedure using a standard neat methanol sample, where the chemical shift difference between the hydroxyl and methyl proton NMR signals are used to determine the temperature.

1. Insert the neat methanol sample into the spectrometer and turn off the lock and the frequency sweep using the BSMS Control Suite of the Bruker TopSpin software (*bsmsdisp*).
2. Record a 1D spectrum and, if needed, shim using the lineshape of the methanol signals (use *gs* and the shim control unit of the BSMS display).
3. Using the temperature control display (*edte*), set the desired sample temperature and allow at least 15 min for the sample to reach this temperature.
4. Record a 1D proton NMR spectrum.
5. Measure the chemical shift difference between the two detected methanol signals. To translate this difference into a temperature, we use the online NMR thermometer found at http://chem.ch.huji.ac.il/nmr/software/thermometer.html. If the calculated temperature value deviates from the set value, calibrate the temperature through at least five measurements at different temperatures encompassing the target sample temperature. Using your favorite software, establish a temperature calibration plot by reporting the calculated true temperatures as a function of their corresponding set temperatures and perform a linear regression. In TopSpin, define and apply a linear correction using the slope and offset values from your fit in the Correction folder of the Temperature Control Suite. Refer to the Bruker VTU user manual for a detailed procedure [46].

3.2 Control Experiments

We will first acquire individual 1D proton spectra for each component of the loading reaction, including the sample buffer itself. These spectra serve a dual purpose. First, they report on any impurities or degradation for each component of the buffer and stock solutions. Second, they serve as a reference to follow the reaction through reagents and substrates (*see* below).

1. Insert the NMR sample containing only the in situ loading buffer in the magnet.
2. Lock the deuterium signal using the appropriate solvent (here, $H_2O + D_2O$), tune and match the probe, and shim.
3. Calibrate the ^{1}H carrier frequency, O1, with the Bruker pulse sequence *zgpr*. Array the value of O1 with *popt* around 4.7 ppm, or a value previously determined at the buffer's pH and temperature, and find the optimal O1 through minimization of the residual water signal. We prefer to process the spectrum in magnitude mode, as it is then easy to identify the smallest residual water signal. Update the value of O1 for all experiments used. We used O1 = 2819.9 Hz.
4. Calibrate the ^{1}H pulse width using a pulse sequence of your choice or through the macro *pulsecal*. We used a ^{1}H pulse width of 10.71 μs.
5. In a different experiment, set the pulse sequence to *zgesgp* or your favorite 1D proton experiment with water suppression. Update O1 and the ^{1}H pulse widths and adapt the parameters TD, DS, NS, and SW so they reflect your sample and spectrometer. Here, we used TD = 64 k, DS = 16, NS = 16, and SW = 16 ppm, with a recycling delay d1 = 2 s. Optimize solvent suppression as needed for your experiment. Most 1D proton pulse sequences incorporate a water-selective shaped pulse, so it is critical to optimize the amplitude, length, and phase of these pulses. The length of the pulse will define the region covered by the pulse and is about 1 ms at 600 MHz for a sinc-shaped pulse. For *zgesgp*, or *zggpwg*-based experiments, we launch the interactive acquisition *gs* and optimize first the amplitude through the power level, spdb1, followed by the phase through the phase corrections phcor2 and phcor4 for *zgesgp* and phcor2 for *zggpwg*, and we iterate to minimize the residual water signal. The water signal should be sufficiently well suppressed to prevent ADC overflow with a gain of 128 (this value relates to an AVANCE III spectrometer). Record the NMR spectrum and verify the integrity of your buffer. In our case, we verify that the signals of ACES match those we are familiar with and inspect for the oxidation state of TCEP (Fig. 2a, *see* also Fig. 6b).
6. Copy this experiment set to a new one (*wra*) and go to this new experiment set (*re*).

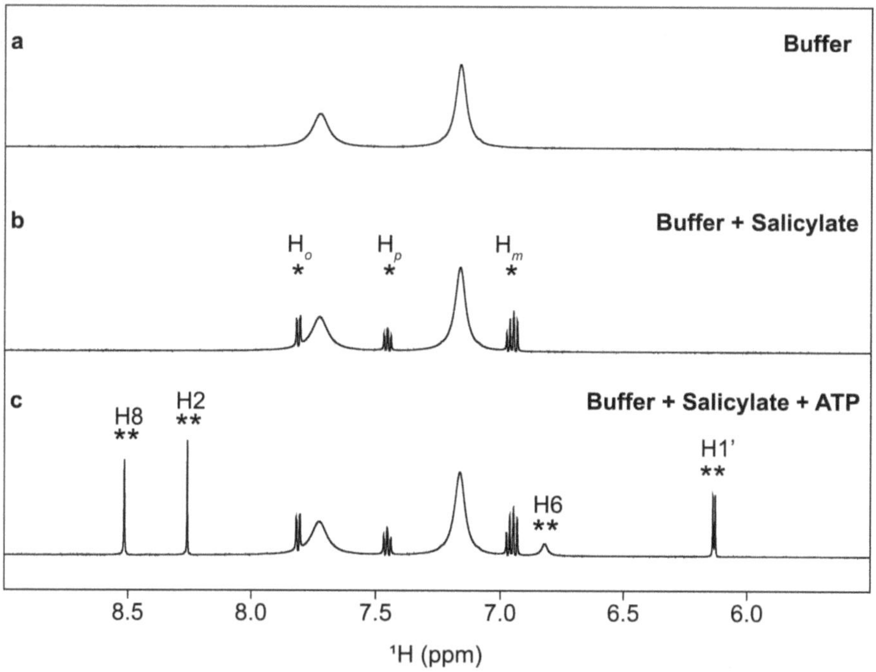

Fig. 2 NMR controls of buffer and reagents. 1D ^1H NMR provides a means to both assess the quality of buffer components prior to loading and to identify signals of each chemical. A comparison of in situ loading buffer (**a**) (50 mM ACES, 150 mM NaCl, 1 mM $MgCl_2$, 1 mM TCEP, 0.05% NaN_3, pH 6.8 at 22 °C) with spectra containing 2 mM Sal (**b**) and 2 mM ATP (**c**) clearly identify peaks of substrates used in the loading reaction. Asterisks denote signals of salicylate and double asterisks denote signals of ATP. In (**b**), the ortho (H_o), meta (H_m), and para (H_p) protons of salicylate are labeled. In (**c**), the protons belonging to the adenosine ring (H8, H6, and H2) as well as the ribose proton (H1') in ATP are labeled [47]. Labeling follows standard nomenclature

7. Eject the sample, add Sal to a final concentration of 2 mM and insert the sample. Acquire another 1D proton spectrum (Fig. 2b) (*see* **Note 4**). Verify that all new signals are accounted for. We assume that those signals have already been assigned. In our case, Sal only provides aromatic resonances [47]. Copy this experiment set to a new one and go there.
8. Eject the sample, add ATP to 2 mM and insert the sample back. Acquire another 1D proton spectrum (Fig. 2c). Verify the integrity of your ATP. ATP's shelf life is famously long at −80 °C or −20 °C but less so at room temperature, and repeated cycles of freeze/thaw of the stock sample may have compromised the stock solution. Fortunately, NMR spectra will reveal signals of ADP (vide infra).
9. Compare your 1D spectra. If the pH was adjusted in all stock solutions, and all contain the same salt concentration, the signals of the buffer should only change marginally. If you see large changes, check the pH of your stock solutions. If the pH is set to the expected value, prepare new stock solutions as the salt concentration may be wrong in one of them.

3.3 In Situ Loading of Holo-ArCP with Salicylate

To overcome the hydrolysis of the thioester bond holding the substrate onto the PP arm of holo-CP, we add the substrate, ATP, and an adenylation domain to the NMR tube along with holo-CP. A series of controls are recorded to verify that spectroscopic changes reflect substrate loading. Thus, we first add Sal to the ^{15}N holo-ArCP sample and compare ^{1}H-^{15}N-HSQCs before and after addition. We then add ATP and make a new comparison. Any changes in ^{1}H-^{15}N-HSQCs signals would either point at non-covalent binding or gross mistakes in the preparation of stock solutions. The latter would have likely been identified through the control experiments above. Clearly, binding of the substrate, or even more surprisingly ATP, would be of interest. Finally, we add catalytic amounts of the stand-alone A-domain YbtE. YbtE uses ATP to activate Sal into a Sal-AMP adenylate and load it onto the PP arm of holo-ArCP through thioesterification. Substrates are consumed differently during two distinct phases. First, holo-ArCP is converted to loaded-ArCP and stoichiometric amounts of Sal and ATP are consumed. Next, YbtE continually regenerates the loaded form within the NMR tube to outcompete hydrolysis. ATP is consumed at a much slower rate during this steady-state phase, and Sal is not consumed. Thus, during subsequent NMR data acquisitions, we monitor for the depletion of ATP through the disappearance of ATP signals. Over longer acquisitions, we must also monitor for ATP hydrolysis into ADP, which provides distinct characteristic signals. We recommend always verifying the integrity of the sample through sodium dodecyl-sulfate polyacrylamide gel electrophoresis (SDS-PAGE) and matrix-assisted laser desorption/ionization time-of-flight (MALDI-tof) mass spectrometry, particularly if they have been frozen or lyophilized.

1. Place the ^{15}N-labeled holo-ArCP sample in the magnet. Lock the deuterium signal with the proper solvent setting, $H_2O + D_2O$, tune and match the probe, and shim.
2. Calibrate the O1 value and ^{1}H and ^{15}N pulse widths. As the sample is enriched in ^{15}N, we calibrate the ^{1}H and ^{15}N pulse widths with a modified 1D HSQC pulse sequence where a single proton pulse and a single nitrogen pulse can be arrayed. Update the calibrated ^{1}H and ^{15}N pulse widths in the *prosol* table using the command *edprosol* and choosing the solvent $D_2O + H_2O$. We used a ^{1}H pulse width of 11.45 μs and a ^{15}N pulse width of 38.5 μs.
3. Acquire a 1D proton spectrum with the pulse sequence *zgesgp* or similar after optimizing solvent suppression as described under Subheading 3.2. We used the following parameters: O1 = 2819.9 Hz, ^{1}H pulse width = 11.45 μs, TD = 64 k, DS = 16, NS = 16, SW = 20 ppm, and recycling delay d1 = 1 s.

Compare this spectrum with that run for the control in Subheading 3.2, step 5. Note that we do not inspect protein signals, which will be impacted by isotope labeling. If you expect to compare protein signals in these 1D spectra, use pulse programs that include decoupling and decrease your TD to 4 k to preserve your probe.

4. In a different experiment set, use the pulse sequence *hsqcfpf3gpphwg* to record a 2D ^{1}H-^{15}N-HSQC spectrum. First use rpar *HSQCFPF3GPPHWG*, then run *getprosol* to update the ^{1}H and ^{15}N pulse widths. Type *ased* and update O1 and the spectral widths. Optimize water suppression as described under Subheading 3.2. It is necessary to optimize water suppression for each pulse sequence as the combination of radiation damping and pulse manipulations lead to different water magnetization trajectories. Acquire a spectrum and adjust the spectral width in the nitrogen dimension if needed. If this is the first time you record a ^{1}H-^{15}N-HSQC for this carrier protein, either record the ^{1}H-^{15}N-HSQC starting from half-dwell or record two HSQCs with two different spectral widths in ^{15}N to identify folded signals. We acquired the 2D ^{1}H-^{15}N-HSQC with the following parameters: O1 = 2819.8 Hz, ^{15}N carrier frequency = 117 ppm, ^{15}N decoupling pulse width = 240 μs, and 2048 × 384 complex points in proton and nitrogen dimensions, with spectral widths of 16 × 36 ppm, and with 8 scans and a recycling delay d1 = 1 s. This spectrum of holo-ArCP provides a reference for subsequent 2D spectra.

5. Eject the ^{15}N-labeled holo-ArCP sample and add Sal to 2 mM final concentration from the 100 mM stock solution. Insert the sample back into the magnet and acquire a 1D proton spectrum and a 2D ^{1}H-^{15}N-HSQC spectrum while keeping the same parameters as in points 3 and 4 (*see* **Note 4**). Inspect the 1D proton spectrum as described in Subheading 3.2. Figure 3a and b shows overlays of 2D ^{1}H-^{15}N-HSQCs of holo-ArCP and holo-ArCP + Sal. In Fig. 3a, the spectrum without Sal is in the background, whereas in Fig. 3b it is in the foreground. Always compare spectra in that manner as new signals may be overlooked if they appear in the spectrum in the foreground. Similarly, always compare spectra at high and low contour levels. At low levels, look for small, new signals. At high levels, look for shifts in resonances. Here, we did not observe any changes, indicating that Sal does not significantly interact with holo-ArCP under these conditions.

6. Eject the sample, add ATP to 2 mM, and insert the sample back into the spectrometer (*see* **Note 4**). Acquire a 1D proton and a 2D ^{1}H-^{15}N-HSQC spectrum. Inspect the 1D proton spectrum

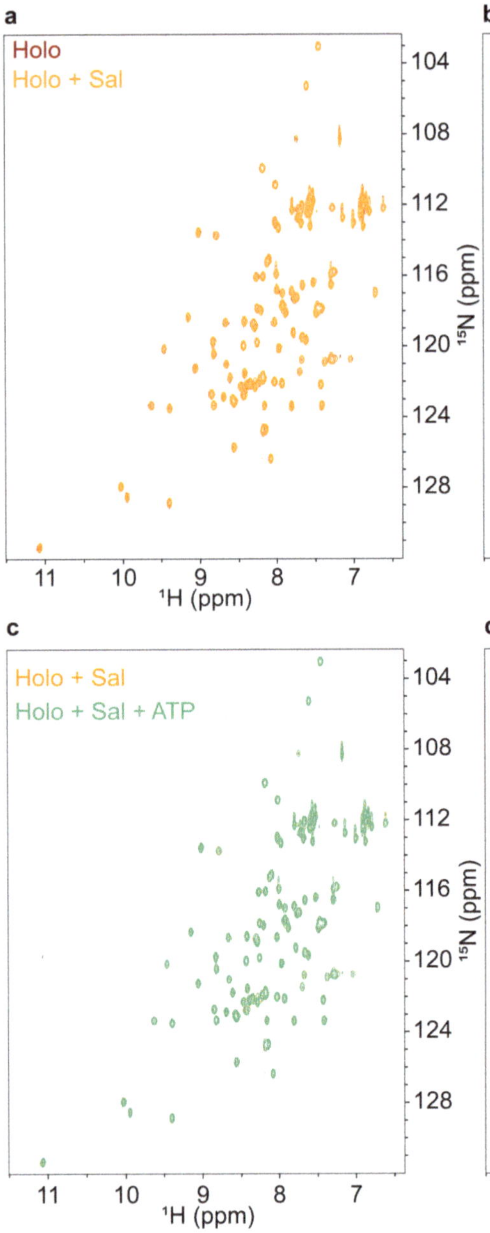

Fig. 3 Controls of the absence of interactions between carrier proteins and reagents. (**a**)–(**b**) Overlay of 2D ^{1}H-^{15}N-HSQCs of holo-ArCP (red) and holo-ArCP in the presence of 2 mM Sal (gold). No perturbations in holo-ArCP are observed. No perturbations are observed upon further addition of 2 mM ATP either ((**c**)–(**d**), green), consistent with no interactions between holo-ArCP and the reagents. Spectra in the background in (**a**) and (**c**) are in the foreground in (**b**) and (**d**), respectively

as described in Subheading 3.2. Here, no changes were observed in the ^{1}H-^{15}N-HSQC after the addition of ATP indicating that ATP does not interact with holo-ArCP under these conditions (Fig. 3c, d).

7. Copy and create a series of your 1D and 2D experiments. The loading reaction may take between 2 and 6 h depending on your concentration of CP. Make enough copies of experiments to monitor the entire reaction. Clearly, if subsequent measurements must be done, you would have set up these experiments ahead of time.

8. Eject the sample containing holo-ArCP, 2 mM Sal, and 2 mM ATP. Add YbtE to 100 nM and insert the sample back (*see* **Note 4**). Queue up the acquisition of your series of 1D proton experiments and 2D ^{1}H-^{15}N-HSQCs (*see* **Note 5**). We monitored the loading of Sal onto holo-ArCP by following changes in signal intensities for residue A77 as its signal appears in an isolated region of the 2D ^{1}H-^{15}N-HSQC spectrum and is easy to monitor (Fig. 4a). We also monitor the growth of AMP signals in 1D spectra as it provides another indicator of substrate loading (Fig. 4b). NMR is quantitative, and signals free of overlaps may be used to calculate the populations of holo and loaded carrier proteins (*see* **Note 6**).

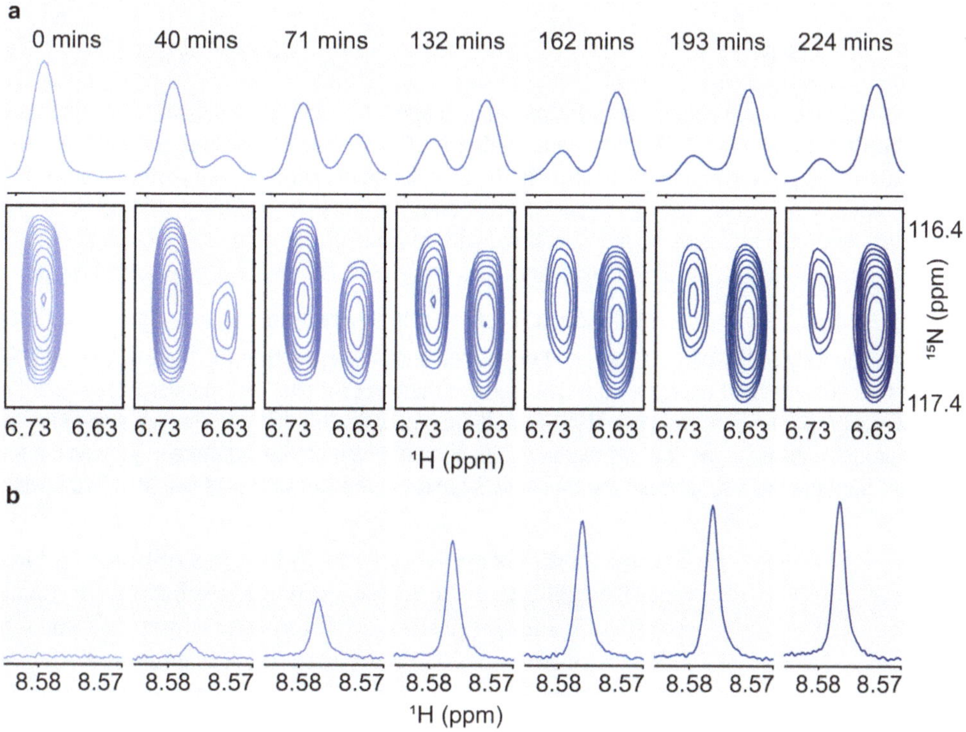

Fig. 4 Monitoring substrate-loading through 2D ^{1}H-^{15}N-HSQCs. (**a**) Changes in chemical shifts between holo-ArCP and Sal-loaded-ArCP allow for monitoring substrate loading (here, through the signal of A77). The elapsed time following the addition of YbtE is shown above each panel. The 1D slices on top highlight the buildup of the loaded peak and the decay of the holo signals during conversion. (**b**) AMP is formed as a by-product of substrate loading and can be monitored too, here through the signal of H8 (**b**). Each 1D ^{1}H spectrum in (**b**) is collected just before each 2D spectrum in (**a**)

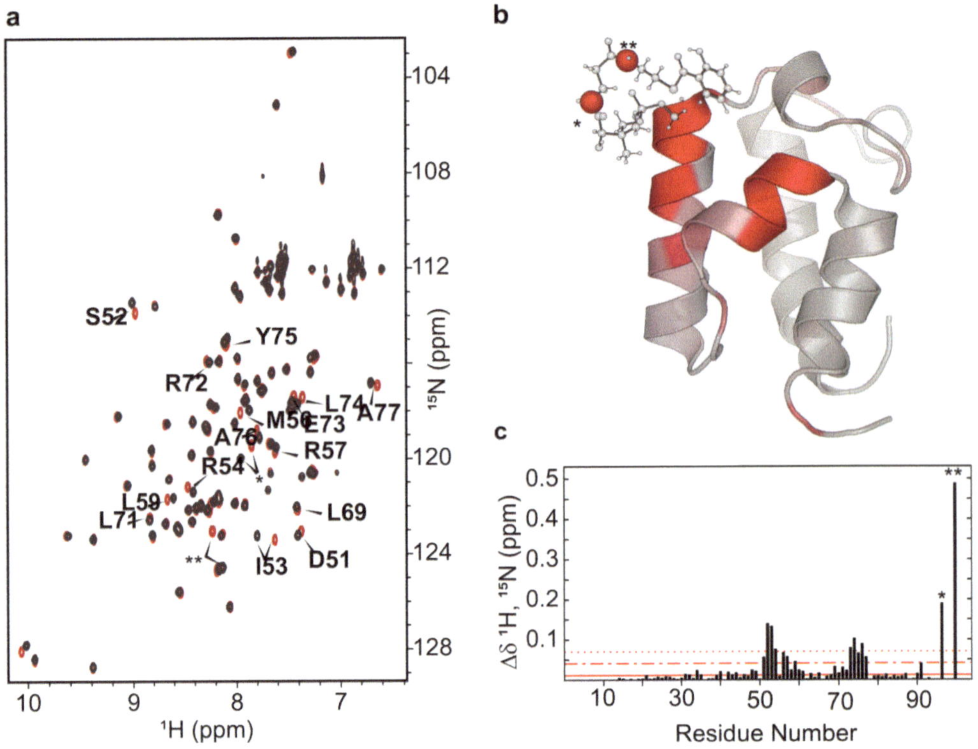

Fig. 5 Loaded-ArCP shows significant perturbations upon loading of a salicylate substrate. (**a**) Comparison of 2D ^{1}H-^{15}N-HSQCs between holo-ArCP (black) and loaded-ArCP (red) reveals residues sensitive to the loading of Sal at the end of the phosphopantetheine arm. (**b**)–(**c**) Calculation of chemical shift perturbations (CSPs, $\Delta\delta$) of backbone amides between holo- and loaded-ArCP emphasizes that loading of Sal to ArCP primarily alters chemical environments due to the conformational change of the arm upon loading. CSPs shown on the structure of loaded-ArCP in (**b**) and the bar plot in (**c**) were calculated as $\Delta\delta^{1}$H, where$^{15}N = \sqrt{\frac{1}{2}[\Delta\delta H^2 + \frac{1}{5}\Delta\delta N^2]}$, where ΔH and ΔN are the amide chemical shift differences (in ppm) in ^{1}H and ^{15}N, respectively [49]. Solid, dashed, and dotted red lines correspond to the median CSP (0.0124), the median CSP plus one standard deviation (0.0416), and the median CSP plus two standard deviations (0.0709) calculated over all residues from the core of ArCP. Bars marked with * and ** correspond to the proximal and distal amide groups (labeled in (**a**)–(**c**)), respectively. The color gradient on the structure in (**b**) was made with a linear gradient from gray to red between the median CSP and the median CSP plus two standard deviations

9. Figure 5a shows an overlay of 2D HSQCs of holo- and loaded-ArCP highlighting residues showing significant chemical shift perturbations as a result of covalent attachment of Sal onto the PP arm of holo-ArCP. As an immediate application, the chemical shift perturbations (CSPs, Fig. 5b-c) can be reported on the structural model of the carrier protein to highlight regions where attachment of the substrate leads to changes in chemical environments on the protein core and the PP arm. These hot spots may reflect new contacts between the CP and its substrate, changes in the PP or CP conformations, or changes in structural fluctuations [24]. Figure 5b shows the CSPs mapped

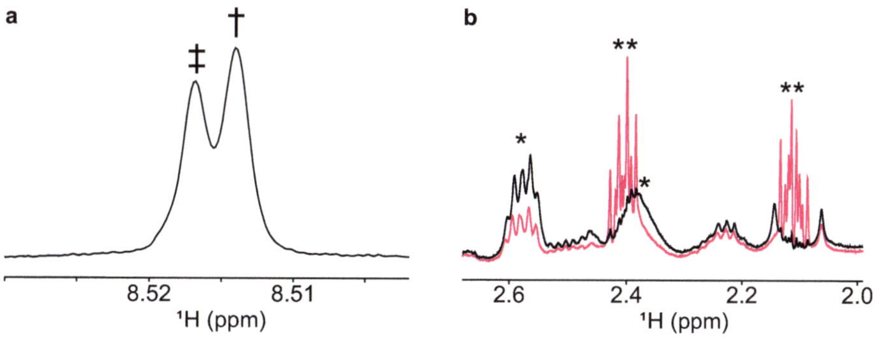

Fig. 6 1D NMR detects hydrolytic and oxidative products of buffer and reaction components. (**a**) 1D spectrum of an in situ loaded-ArCP sample six days after loading. ATP hydrolyses into ADP over time and the zoomed spectral region shows the H8 signals of both ATP (†) and ADP (‡). On this day, the percentage of loaded-ArCP was 57.2%. (**b**) Overlay of 1D spectra of a fresh in situ NMR sample (black) and seven days after loading (pink). Single asterisks identify fresh TCEP signals and double asterisks identify oxidized TCEP signals

on the solution structure of loaded-ArCP. Here, CSPs principally reflect the curling of the PP arm upon Sal attachment from a previously extended conformation.

10. When a steady state is reached, begin your targeted NMR data acquisition, for example, NOESY experiments for structure determination, T_1, T_2, heteronuclear NOE to probe dynamics at fast timescales (nanosecond to picosecond) [24], or begin titrations with other catalytic domains to inspect domain–domain interactions [20].
11. During subsequent experiments, monitor the integrity of the reaction. Insert ^{1}H-^{15}N-HSQCs and 1D spectra in your acquisition queues. The 1D spectra are particularly important as they help you anticipate actions that must be taken. Thus, hydrolysis of ATP into ADP can be seen through their signals (Fig. 6a) (*see* **Note 7**) and oxidation of TCEP leads to characteristic spectral changes (Fig. 6b). The ^{1}H-^{15}N-HSQCs will report on the protein stability. Unfolded proteins will display new signals with nearly degenerate ^{1}H frequencies. Aggregates will lead to severely broadened signals. Degradation will lead to characteristic signals in a region around 6 ppm to 8 ppm in ^{1}H and 128 ppm to 132 ppm in ^{15}N, a region typically otherwise free of signals. MALDI-tof spectra are collected to further assess sample stability.

4 Notes

1. Concentrated stock solutions are preferable to minimize dilution of the protein concentration upon addition to the NMR sample. At these concentrations, the pH must be adjusted after the addition of the solutes (Sal, ATP). Although phosphate

buffers are popular in NMR studies, they are inadequate for in situ loading of CPs as magnesium is present at 1 mM and magnesium phosphate will then precipitate. DTT or BME cannot be used as they will lead to transthioesterification with loaded substrates [41].

2. Only a single NMR tube is needed for all controls. After verifying the integrity of the buffer, Sal is first added to a final concentration of 2 mM, spectra are recorded, and ATP is then added to 2 mM to record the final control spectra. Dilution of D_2O is acceptable as long as the percentage vol/vol of D_2O remains sufficient for deuterium lock of the NMR signal (between 5% and 10%, depending on your spectrometer's generation).

3. 2 mM Sal, 2 mM ATP, and 100 nM YbtE are added in a stepwise manner to the NMR sample containing holo-ArCP. The percentage vol/vol of D_2O as well as the concentrations of DSS and holo-ArCP that we report here account for dilution. The labeling scheme will be adapted to the NMR experiments to be conducted. The minimal concentration of CP depends on the spectrometer's features and performances and on the NMR measurements to be performed. With a cryogenically cooled probe, simple experiments (e.g., ^{1}H-^{15}N-HSQC) can be performed with concentrations as low as 10 μM, albeit at the cost of acquisition times. For an *N*-fold dilution, the number of scans has to be increased N^2 fold to maintain signal to noise. A 10-min ^{1}H-^{15}N-HSQC with a 100 μM sample will take 1000 min or about 16 h for comparable signal to noise with a 10 μM sample. As of 2022, for more demanding experiments, such as NOESY, samples of CP concentrations above 300 μM are recommended. For long experiments, once you have reached a steady state in loaded-CP, degas the sample, cover with argon, and seal with parafilm before acquiring your next data to delay oxidation of TCEP.

4. Check if the shims need to be re-optimized after inserting the sample back into the spectrometer. Some spectrometers are more stable than others. If needed, also re-optimize solvent suppression. Other parameters need not be changed.

5. Loading of Sal onto holo-ArCP starts as soon as YbtE is added, so set up the series of 1D proton and 2D ^{1}H-^{15}N-HSQC experiments in advance, otherwise you might miss the initial time points of the loading reaction.

6. We have observed conversions varying between 65% and 95%. Reasons we have identified for incomplete yields include the following: TCEP has oxidized and holo carrier proteins dimerize. ATP had already degraded into ADP in the stock solution. The carrier protein unfolds over time or has a large population

in an unfolded state. The protein is subject to hydrolysis and truncated forms do not interact with the A domain. We have also hypothesized that a secondary hydrolase activity in cyclization domains limited the conversion of ArCP [20]. Although we did not investigate this proposition in detail, the Cryle lab has reported hydrolase activity in condensation domains [48]. Finally, mis-acylated holo-CPs would also limit the conversion to substrate-loaded forms, although we have not encountered this issue. To ensure maximum loading of CPs, always prepare the NMR sample in fresh buffer, add fresh TCEP just before preparing the sample, and use freshly prepared stocks of Sal and ATP.

7. Add more ATP to the NMR sample when it has been depleted, whether through conversion to AMP (Fig. 4b) or degradation into ADP (Fig. 6a). Monitor 1D spectra interleaved with other spectra.

Acknowledgments

Research in the Frueh lab is supported by NIH grant R01GM104257. The Frueh lab thanks Drs. Challut and Guilhot for the ΔEntD strain and the laboratory of Dr. C.T. Walsh for the Srf-TEII employed in controls and for Sfp. The Frueh lab uses NMRBox.

References

1. Reimer JM, Haque AS, Tarry MJ, Schmeing TM (2018) Piecing together nonribosomal peptide synthesis. Curr Opin Struct Biol 49: 104–113
2. Challis GL, Ravel J, Townsend CA (2000) Predictive, structure-based model of amino acid recognition by nonribosomal peptide synthetase adenylation domains. Chem Biol 7:211–224
3. Zhu M, Wang L, He J (2019) Chemical diversification based on substrate promiscuity of a standalone adenylation domain in a reconstituted NRPS system. ACS Chem Biol 14:256–265
4. Gulick AM (2009) Conformational dynamics in the Acyl-CoA synthetases, adenylation domains of non-ribosomal peptide synthetases, and firefly luciferase. ACS Chem Biol 4:811–827
5. Lipmann F (1973) Nonribosomal polypeptide synthesis on polyenzyme templates. Acc Chem Res 6:361–367
6. Koglin A, Mofid MR, Löhr F et al (2006) Conformational switches modulate protein interactions in peptide antibiotic synthetases. Science 312(1979):273–276
7. de Crécy-Lagard V, Marlière P, Saurin W (1995) Multienzymatic non ribosomal peptide biosynthesis: identification of the functional domains catalysing peptide elongation and epimerisation. C R Acad Sci III 318:927–936
8. Sundlov JA, Shi C, Wilson DJ et al (2012) Structural and functional investigation of the intermolecular interaction between NRPS adenylation and carrier protein domains. Chem Biol 19:188–198
9. Mitchell CA, Shi C, Aldrich CC, Gulick AM (2012) Structure of PA1221, a nonribosomal peptide synthetase containing adenylation and peptidyl carrier protein domains. Biochemistry 51:3252–3263
10. Samel SA, Schoenafinger G, Knappe TA et al (2007) Structural and functional insights into a peptide bond-forming bidomain from a

nonribosomal peptide synthetase. Structure 15:781–792

11. Tanovic A, Samel SA, Essen L-O, Marahiel MA (2008) Crystal structure of the termination module of a nonribosomal peptide synthetase. Science 321(1979):659–663
12. Reimer JM, Eivaskhani M, Harb I et al (2019) Structures of a dimodular nonribosomal peptide synthetase reveal conformational flexibility. Science 366(1979):eaaw4388
13. Drake EJ, Miller BR, Shi C et al (2016) Structures of two distinct conformations of holo-non-ribosomal peptide synthetases. Nature 529:235–238
14. Tarry MJ, Haque AS, Bui KH, Schmeing TM (2017) X-ray crystallography and electron microscopy of cross-and multi-module nonribosomal peptide synthetase proteins reveal a flexible architecture. Structure 25:783–793
15. Katsuyama Y, Sone K, Harada A et al (2021) Structural and functional analyses of the tridomain-nonribosomal peptide synthetase FmoA3 for 4-methyloxazoline ring formation. Angew Chem Int Ed 60:14554–14562
16. Wang J, Li D, Chen L et al (2022) Catalytic trajectory of a dimeric nonribosomal peptide synthetase subunit with an inserted epimerase domain. Nat Commun 13:1–12
17. Frueh DP, Arthanari H, Koglin A et al (2008) Dynamic thiolation–thioesterase structure of a non-ribosomal peptide synthetase. Nature 454:903–906
18. Tufar P, Rahighi S, Kraas FI et al (2014) Crystal structure of a PCP/Sfp complex reveals the structural basis for carrier protein posttranslational modification. Chem Biol 21:552–562
19. Goodrich AC, Meyers DJ, Frueh DP (2017) Molecular impact of covalent modifications on nonribosomal peptide synthetase carrier protein communication. J Biol Chem 292: 10002–10013
20. Mishra SH, Kancherla AK, Marincin KA, et al (2021) Global dynamics as communication sensors in peptide synthetase cyclization domains. bioRxiv
21. Shi C, Miller BR, Alexander EM et al (2020) Design, synthesis, and biophysical evaluation of mechanism-based probes for condensation domains of nonribosomal peptide synthetases. ACS Chem Biol 15:1813–1819
22. Ubbink M (2009) The courtship of proteins: understanding the encounter complex. FEBS Lett 583:1060–1066
23. Corpuz JC, Sanlley JO, Burkart MD (2022) Protein-protein interface analysis of the non-ribosomal peptide synthetase peptidyl carrier protein and enzymatic domains. Synth Syst Biotechnol 7:677–688
24. Goodrich AC, Harden BJ, Frueh DP (2015) Solution structure of a nonribosomal peptide synthetase carrier protein loaded with its substrate reveals transient, well-defined contacts. J Am Chem Soc 137:12100–12109
25. Jaremko MJ, Lee DJ, Opella SJ, Burkart MD (2015) Structure and substrate sequestration in the pyoluteorin type II peptidyl carrier protein PltL. J Am Chem Soc 137:11546–11549
26. Kittilä T, Mollo A, Charkoudian LK, Cryle MJ (2016) New structural data reveal the motion of carrier proteins in nonribosomal peptide synthesis. Angew Chem Int Ed 55:9834–9840
27. Kleckner IR, Foster MP (2011) An introduction to NMR-based approaches for measuring protein dynamics. Biochim Biophys Acta Proteins Proteomic (1814):942–968
28. Henzler-Wildman K, Kern D (2007) Dynamic personalities of proteins. Nature 450:964–972
29. Ishima R, Bagby S (2017) Protein dynamics revealed by CPMG dispersion. In: Webb G (ed) Modern magnetic resonance, 2nd edn. Springer, pp 435–452
30. Kanelis V, Forman-Kay JD, Kay LE (2001) Multidimensional NMR methods for protein structure determination. IUBMB Life 52: 291–302
31. Cavalli A, Salvatella X, Dobson CM, Vendruscolo M (2007) Protein structure determination from NMR chemical shifts. Proc Natl Acad Sci 104:9615–9620
32. Iwahara J, Clore GM (2006) Detecting transient intermediates in macromolecular binding by paramagnetic NMR. Nature 440:1227–1230
33. Palmer AG III (2014) Chemical exchange in biomacromolecules: past, present, and future. J Magn Reson 241:3–17
34. Lisi GP, Loria JP (2017) Allostery in enzyme catalysis. Curr Opin Struct Biol 47:123–130
35. Bodenhausen G, Ruben DJ (1980) Heteronuclear 2D correlation spectra with double in-phase transfer steps. Chem Phys Lett 69: 185–189
36. Pervushin K, Riek R, Wider G, Wüthrich K (1997) Attenuated T 2 relaxation by mutual cancellation of dipole–dipole coupling and chemical shift anisotropy indicates an avenue to NMR structures of very large biological macromolecules in solution. Proc Natl Acad Sci 94:12366–12371
37. Goodrich AC, Frueh DP (2015) A nuclear magnetic resonance method for probing molecular influences of substrate loading in

nonribosomal peptide synthetase carrier proteins. Biochemistry 54:1154–1156

38. Sztain T, Bartholow TG, McCammon JA, Burkart MD (2019) Shifting the hydrolysis equilibrium of substrate loaded acyl carrier proteins. Biochemistry 58:3557–3560

39. Gehring AM, Mori I, Perry RD, Walsh CT (1998) The nonribosomal peptide synthetase HMWP2 forms a thiazoline ring during biogenesis of yersiniabactin, an iron-chelating virulence factor of Yersinia pestis. Biochemistry 37:11637–11650

40. Lambalot RH, Gehring AM, Flugel RS et al (1996) A new enzyme superfamily—the phosphopantetheinyl transferases. Chem Biol 3: 923–936

41. Sou Z, Walsh CT, Miller DA (1999) Tandem heterocyclization activity of the multidomain 230 kDa HMWP2 subunit of Yersinia pestis yersiniabactin synthetase: interaction of the 1-1382 and 1383-2035 fragments. Biochemistry 38:14023–14035

42. Raiford DS, Fisk CL, Becker ED (1979) Calibration of methanol and ethylene glycol nuclear magnetic resonance thermometers. Anal Chem 51:2050–2051

43. Karschin N, Krenek S, Heyer D, Griesinger C (2022) Extension and improvement of the methanol-d4 NMR thermometer calibration. Magn Reson Chem 60:203–209

44. Hoffman RE (2006) Standardization of chemical shifts of TMS and solvent signals in NMR solvents. Magn Reson Chem 44:606–616

45. Ammann C, Meier P, Merbach A (1982) A simple multinuclear NMR thermometer. J Magn Reson 46:319–321

46. Tyburn J-M (1998) VTU User Manual, Version 001. Bruker SA, Wissenbourg

47. Wishart DS, Knox C, Guo AC et al (2009) HMDB: a knowledgebase for the human metabolome. Nucleic Acids Res 37: D603–D610

48. Schoppet M, Peschke M, Kirchberg A et al (2019) The biosynthetic implications of late-stage condensation domain selectivity during glycopeptide antibiotic biosynthesis. Chem Sci 10:118–133

49. Williamson MP (2013) Using chemical shift perturbation to characterise ligand binding. Prog Nucl Magn Reson Spectrosc 73:1–16

Chapter 13

Ribosomal Synthesis of Peptides Bearing Noncanonical Backbone Structures via Chemical Posttranslational Modifications

Yuki Goto and Hiroaki Suga

Abstract

Noncanonical peptide backbone structures, such as heterocycles and non-α-amino acids, are characteristic building blocks present in peptidic natural products. To achieve ribosomal synthesis of designer peptides bearing such noncanonical backbone structures, we have devised translation-compatible precursor residues and their chemical posttranslational modification processes. In this chapter, we describe the detailed procedures for the in vitro translation of peptides containing the precursor residues by means of genetic code reprogramming technology and posttranslational generation of objective noncanonical backbone structures.

Key words Engineered translation, Peptide modification, Chemical posttranslational modification, Noncanonical peptide backbone, Genetic code reprogramming, FIT system, Flexizyme

1 Introduction

Various noncanonical peptide backbone structures, such as heterocycles and non-α-amino acids, are found in peptidic natural products [1–3]. Such exotic backbone structures endow the peptides with unique properties (e.g., improved proteolytic resistance and cell membrane permeability, rigidified local and global conformation, and less immunogenicity), thus contributing to their potent bioactivities. In the biosynthesis of such natural products, these backbone structures are exclusively generated by nonribosomal peptide synthetases (NRPSs) [1] or dedicated peptide-modifying enzymes involved in ribosomally synthesized and posttranslationally modified peptide (RiPP) pathways [2, 3]. Because of its capability of template-dependent production of various peptide sequences, the translation machinery has been in vitro engineered to expand the repertoire of usable nonproteinogenic building blocks [4–7]. However, despite recent technical advances in

Michael Burkart and Fumihiro Ishikawa (eds.), *Non-Ribosomal Peptide Biosynthesis and Engineering: Methods and Protocols*, Methods in Molecular Biology, vol. 2670, https://doi.org/10.1007/978-1-0716-3214-7_13,

engineered translation systems, it remains challenging to directly incorporate noncanonical backbone structures via ribosomes. To achieve the synthesis of designer peptides bearing exotic backbones in an mRNA template-dependent manner, we recently devised selective chemical modifications of translation-compatible precursor residues, allowing for posttranslational formation of various noncanonical backbone structures [8–13] (Fig. 1a). In this strategy, appropriately designed artificial residues are charged onto a suppressor tRNA by an artificial tRNA-acylating ribozyme, flexizyme [14] (*see* **step 3.1**). The resulting artificial acyl-tRNA is supplemented to a custom-made cell-free translation, so-called flexible in vitro translation (FIT) system [15], to express peptides bearing the precursor residues (*see* **step 3.2**). The expressed peptides are then subjected to posttranslational chemical modification conditions, in which the precursor residues are selectively converted to the objective noncanonical backbone structures. In this chapter, we describe protocols for the ribosomal synthesis of peptides containing two representative noncanonical backbone structures, azoles [12] (*see* **step 3.3**) and hydroxyhydrocarbon (Hhc) units[13] (*see* **step 3.4**).

Azoles, such as oxazoles and thiazoles, are five-membered heterocycles that enhance the structural rigidity and lipophilicity of peptides, contributing to potent target binding as well as improved membrane permeability [16–18]. The azole moieties in peptides generally originate from cysteine (Cys), serine (Ser), and threonine (Thr) residues in the biosynthetic pathways. While Cys/Ser/Thr residues are converted into azolines by cyclase domains [19] and subsequently aromatized by oxidation domains[20] in NRPS pathways, azoles are analogically produced from Cys/Ser/Thr by YcaO cyclodehydratases [21, 22] and dehydrogenases [23] in RiPP biosynthesis. Although it has been reported that a dedicated mutant ribosome can incorporate azole-containing dipeptide units into the nascent peptide chain, the expression efficiency was low [24, 25], indicating that direct incorporation of the azole moiety by ribosome-mediated peptidyl-transfer reaction is still intractable. To express azole-containing peptides in vitro, we designed a 4-bromovinylglycine derivative (BrvG) and developed its chemoselective conversion to the corresponding azole structure [12] (Fig. 1b). In this method, FIT-expressed peptides containing BrvG are subjected to tetrabutylammonium fluoride (TBAF)-induced dehydrobromination conditions, in which the BrvG residue can be specifically converted to the corresponding β,γ-alkynylglycine derivative (AkyG). The β,γ-alkynyl group spontaneously reacted with an adjacent amide carbonyl group to yield a backbone oxazole moiety. Moreover, the combination of FIT-mediated thioamide formation [26] at the -1 position of BrvG enables the posttranslational formation of thiazoles.

Hhc units exhibit unique biochemical and folding propensities and play vital roles in the potent bioactivities of natural products

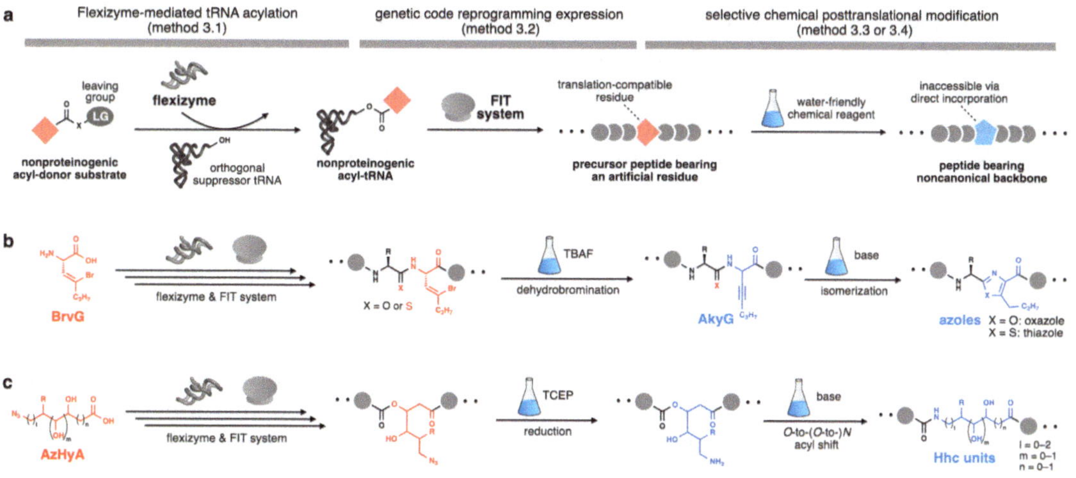

Fig. 1 Chemical posttranslational modification of artificial residues to yield noncanonical backbones. (**a**) Schematic depiction of the chemical posttranslational modification strategy constructing noncanonical backbones. (**b**) Ribosomal synthesis of peptides bearing backbone azoles via posttranslational dehydrobromination of BrvG. (**c**) Ribosomal synthesis of peptides bearing Hhc units via posttranslational reduction of AzHyAs followed by backbone acyl shift. The noncanonical backbone structures not amenable to direct ribosomal incorporation are shown in blue. The artificial residues that can be incorporated via genetic code reprogramming and used as the chemical precursor of the objective noncanonical backbone structure are shown in red

and peptidomimetics [1]. For instance, γ-amino-β-hydroxy acids, so-called statine derivatives, can act as mimetics of the oxyanion in the tetrahedral transition state during peptide bond hydrolysis [27, 28] and are thus often present in protease inhibitors [29–31]. In nature, Hhc-containing peptides are produced via polyketide (PKS)–NRPS hybrid pathways, in which Hhc units are biosynthesized by PKS and directly used for peptide elongation catalyzed by NRPS [1]. Ribosomal incorporation of Hhc units is an arduous task due to rapid self-deacylation of Hhc-tRNAs [10, 32] and their poor compatibility with the ribosome active site [32–34]. To circumvent these issues, we designed a series of azide/hydroxy-acids (AzHyA), which can be posttranslationally reduced and undergo peptide backbone rearrangement via an *O*-to-(*O*-to-)*N* acyl shift reaction to generate the target Hhc units [13] (Fig. 1b).

These methods allow for one-pot ribosomal synthesis of designer peptides bearing backbone azoles as well as various Hhc units. They have broad potential applications in in vitro synthetic biology and the discovery of de novo peptidic agents.

2 Materials

Prepare all materials in an RNase-free manner. Use RNase-free tubes, pipettes, pipette tips, and water. Wear gloves at all times.

2.1 Flexizyme-Mediated Acylation to Prepare Nonproteinogenic acyl-tRNAs

1. 1.7 mL microcentrifuge tubes.
2. 500 mM HEPES-KOH buffer (pH 7.5): This is our standard buffer for the flexizyme reaction. But, for some acyl-donor substrates that give better acylation yields at higher pH, different buffers [500 mM HEPES-KOH buffer (pH 8.0) or 500 mM bicine-KOH buffer (pH 8.5)] are recommended. (*see* **Note 1** and Fig. 2 for general recommendation).
3. Flexizyme: Prepare an artificial tRNA acylating ribozyme, eFx or dFx, according to the previously described method [14, 15] (*see* **Note 2** and Fig. 2 for general recommendation).
4. Suppressor tRNAs: Prepare an appropriate suppressor tRNA according to the previously described method [14, 15] (*see* **Note 3** and Fig. 2 for general recommendation).
5. A heating block.
6. 3 M $MgCl_2$ *aq*.
7. 25 mM acyl-donor substrate in dimethyl sulfoxide (DMSO): Prepare an amino acid derivative of choice, whose carboxylic group is activated with an appropriate leaving group (*see* **Note 2** and Fig. 2 for general recommendation). Dissolve the acyl-donor substrate into DMSO to make a 25 mM solution.
8. 0.3 M sodium acetate (NaOAc) (acidic): Dilute 1 mL of 3 M NaOAc-acetic acid (AcOH) buffer (pH 5.2) with 9 mL of water.
9. Ethanol (pure reagent grade).
10. A centrifuge.
11. 70% ethanol containing 0.1 M acidic NaOAc: Mix 333 μL of 3 M NaOAc-AcOH buffer (pH 5.2) and 2.67 mL of water with 7 mL of ethanol.
12. 70% ethanol: Mix 3 mL of water with 7 mL of ethanol.

2.2 Expression of Peptides Containing Noncanonical Amino Acids

1. 0.6 mL low-binding microcentrifuge tubes.
2. 10× TS (tRNAs and small molecules) solution: 500 mM HEPES-KOH (pH 7.6), 1 M potassium acetate, 120 mM magnesium acetate, 20 mM ATP, 20 mM GTP, 10 mM CTP, 10 mM UTP, 200 mM creatine phosphate, 1 mM 10-formyl-5,6,7,8-tetrahydrofolic acid, 20 mM spermidine, 10 mM DTT, 15 mg/mL *E. coli* total tRNA.
3. 10× RP (ribosome and protein factors) solution: 3 mM magnesium acetate, 12 μM *E. coli* ribosome, 6 μM MTF, 27 μM IF1, 4 μM IF2, 15 μM IF3, 2.6 μM EF-G, 100 μM EF-Tu, 6.6 μM EF-Ts, 2.5 μM RF2, 1.7 μM RF3, 5 μM RRF, 40 μg/mL creatine kinase, 30 μg/mL myokinase, 1 μM inorganic pyrophosphatase, 1 μM nucleotide diphosphate kinase, and 1 μM T7 RNA polymerase. Some proteins (e.g., creatine kinase

nonproteinogenic residue		leaving group	flexizyme	suppressor tRNA	flexizyme reaction conditions	concentration of acyl-tRNA used in the FIT system	noncanonical backbone structures after chemical posttranslational modification
BrvG	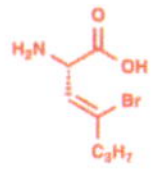	DBE	dFx	$tRNA^{AsnE2}$	pH 7.5, 4 h	50 μM	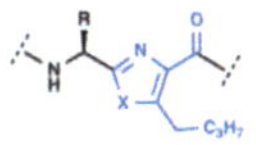
Ala^{S}	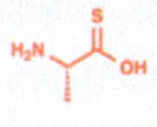	DBE	dFx	$tRNA^{GluE2}$	pH 7.5, 0.5 h	50 μM	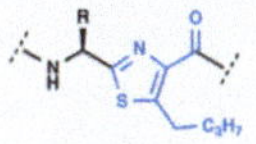
$^{\gamma}N_3{}^{\beta}OH$	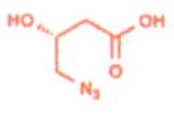	CBT	eFx	$tRNA^{Pro1E2\#3}$	pH 8.0, 40 h	25 μM	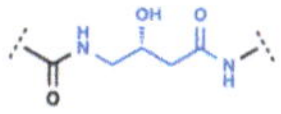
$^{\beta}N_3{}^{\alpha}OH$	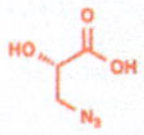	DBE	dFx	$tRNA^{Pro1E2\#3}$	pH 8.5, 48 h	25 μM	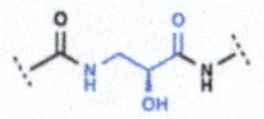
$^{\gamma}N_3{}^{\alpha}OH$	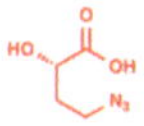	DBE	dFx	$tRNA^{Pro1E2\#3}$	pH 7.5, 40 h	25 μM	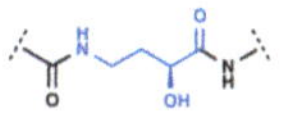
$^{\delta}N_3{}^{\alpha}OH$	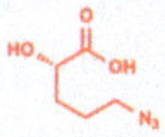	DBE	dFx	$tRNA^{Pro1E2\#3}$	pH 8.5, 24 h	25 μM	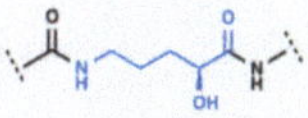
$^{\gamma}N_3{}^{\alpha}OH^{\beta}OH$	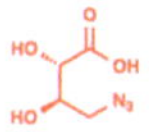	CBT	eFx	$tRNA^{Pro1E2\#3}$	pH 8.5, 16 h	25 μM	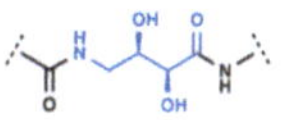
$^{\delta}N_3{}^{\alpha}OH^{\beta}OH$	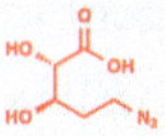	DBE	dFx	$tRNA^{Pro1E2\#3}$	pH 7.5, 40 h	25 μM	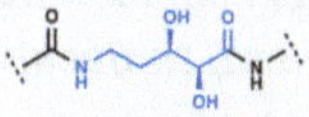
Sta-$^{\gamma}N_3$	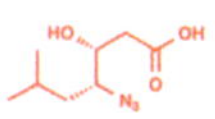	CME	eFx	$tRNA^{Pro1E2\#3}$	pH 8.0, 40 h	25 μM	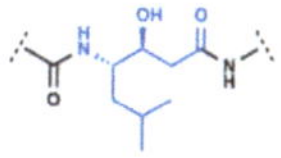
PhSta-$^{\gamma}N_3$	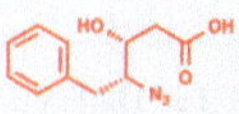	CME	eFx	$tRNA^{Pro1E2\#3}$	pH 8.0, 48 h	25 μM	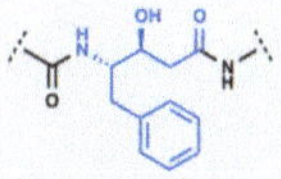
^{ClAc}Tyr	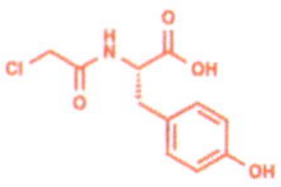	CME	eFx	$tRNA^{fMetE}$	pH 7.5, 2 h	25 μM	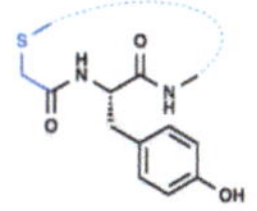

Fig. 2 List of selected artificial residues used in this chapter. Recommended acylation/translation conditions for respective substrates are also shown

and myokinase) can be obtained from commercial sources. Other individual protein factors are generally overexpressed in *E. coli* with appropriate tags such as polyhistidine, and purified by affinity chromatography according to the standard methodology [35–37].

4. 10× ARS (aminoacyl-tRNA synthetases) solution: 7.3 μM AlaRS, 0.3 μM ArgRS, 3.8 μM AsnRS, 1.3 μM AspRS, 0.2 μM CysRS, 0.6 μM GlnRS, 2.3 μM GluRS, 0.9 μM GlyRS, 0.2 μM HisRS, 4.0 μM IleRS, 0.4 μM LeuRS, 1.1 μM LysRS, 0.3 μM MetRS, 6.8 μM PheRS, 1.6 μM ProRS, 0.4 μM SerRS, 0.9 μM ThrRS, 0.3 μM TrpRS, 0.2 μM TyrRS, and 0.2 μM ValRS. The proteins are generally overexpressed in *E. coli* with appropriate tags such as polyhistidine, and purified by affinity chromatography according to the standard methodology [35–37].
5. 5 mM each amino acids solution: Mix proteinogenic amino acids required for the expression of the objective peptide sequence to make 5 mM each solution. Do not add amino acids that are canonically encoded in the codons to be reprogrammed with the artificial residues.
6. 0.8 μM DNA template: Prepare DNA duplex coding a desirable peptide sequence according to the previously described method [14, 15] (*see* **Note 4**).
7. 1 mM NaOAc (acidic): Dilute 2 μL of 3 M NaOAc-AcOH buffer (pH 5.2) with 5998 μL of water.
8. An air incubator: Use an air incubator for the reactions with small reaction volumes (~100 μL).

2.3 Posttranslational Chemical Modification of BrvG to Backbone Azole Moieties

1. Acetone (pure reagent grade).
2. *N*,*N*-dimethylformamide (DMF) (pure reagent grade).
3. 1.1 M tetrabutylammonium fluoride (TBAF) in DMF: Dissolve 34 mg of TBAF trihydrate solid into 100 μL of DMF. Prepare it freshly.
4. 5% trifluoroacetic acid (TFA) *aq.*: Mix 50 μL of TFA with 950 μL of water.
5. An LC-MS instrument: Use an appropriate detection/characterization method for the products. Any alternative methods/instruments can be used (*see* **Note 5**).

2.4 Posttranslational Chemical Modification of AzHyA to Hhc Units

1. Tris(2-carboxyethyl)phosphine hydrochloride (TCEP) reaction buffer: Prepare 200 mM TCEP-HCl solution, whose pH is adjusted to 10 by the addition of NaOH. Prepare 500 mM Na_2HPO_4-NaOH buffer (pH 10). Mix and dilute these solutions to make a solution containing 50 mM TCEP (pH 10) and 350 mM Na_2HPO_4-NaOH buffer (pH 10). Prepare it freshly.

3 Methods

Carry out all procedures in an RNase-free manner on ice unless otherwise specified.

3.1 Flexizyme-Mediated Acylation to Prepare Nonproteinogenic acyl-tRNAs

1. Mix 2 μL of 500 mM HEPES-KOH buffer (pH 7.5, *see* **Note 1**), 2 μL of 250 μM flexizyme of choice (*see* **Note 2**), and 2 μL of 250 μM suppressor tRNA of choice (*see* **Note 3**) with 6 μL of water.
2. Heat the sample at 95 °C for 2 min, then slowly cool it at room temperature over 5 min.
3. Add 4 μL of 3 M $MgCl_2$ into the sample, then incubate it at room temperature for 5 min followed by on ice for 3 min.
4. Add 4 μL of 25 mM acyl-donor substrate of choice in DMSO (*see* **Note 2**) into the sample and mix well.
5. Incubate the acylation reaction mixtures on ice for appropriate reaction time (0.5–48 h, *see* **Note 1** and Fig. 2 for general recommendation).
6. Add 80 μL of 0.3 M NaOAc (acidic) and 200 μL of ethanol into the reaction mixtures to quench the acylation reaction (*see* **Note 6**).
7. Centrifuge the samples at 15,000 × *g* for 15 min at 25 °C. Then, remove the supernatant completely.
8. Add 100 μL of 70% ethanol containing 0.1 M acidic NaOAc to the tube and vortex the tube well to break the RNA pellet into pieces (*see* **Note 7**).
9. Centrifuge the sample at 15,000 × *g* for 5 min at 25 °C. Then, remove the supernatant completely.
10. Repeat **steps 8** and **9** one more time.
11. Add 80 μL of 70% ethanol to the tube.
12. Centrifuge the sample at 15,000 × *g* for 3 min at 25 °C. Then, remove the supernatant completely.
13. Open the tube lid and cover it with tissues, then dry the RNA at room temperature for 5 min. The obtained RNA pellet corresponds to 500 pmol of acyl-tRNA (*see* **Note 8**).

3.2 Expression of Peptides Containing Noncanonical Amino Acids

1. Mix 0.5 μL of 10× TS solution, 0.5 μL of 10× RP solution, 0.5 μL of 10× ARS solution, 0.5 μL of 5 mM each amino acids solution, and 0.25 μL of 0.8 μM DNA template with 1.75 μL of water.
2. Resuspend an appropriate amount (*see* **Note 9** and Fig. 2 for general recommendation) of the acyl-tRNA pellet (*see* **Note 10**) in 1 μL of 1 mM NaOAc (acidic) (*see* **Note 11**).

3. Add the acyl-tRNA solution (1 μL) to the sample.
4. Incubate the mixture at 37 °C for appropriate reaction time (0.5–3 h, *see* **Note 9** and Fig. 2 for general recommendation).
5. Use the resulting translation mixture directly for the downstream assay or reactions (*see* **Note 5**).

3.3 Posttranslational Chemical Modification of BrvG to Backbone Azole Moieties

1. Mix 2.5 μL of the translation mixture with 7.5 μL of acetone.
2. Incubate the solution at −20 °C for 30 min.
3. Centrifuge the sample at 15,000 × *g* for 5 min at 4 °C. Then, remove the supernatant completely.
4. Open the tube lid and cover it with tissues, then dry the peptide pellet at room temperature for 5 min.
5. Dissolve the pellet in 1 μL of water.
6. Add 1.5 μL of DMF and 22.5 μL of 1.1 M TBAF in DMF to the sample.
7. Incubate the sample at 60 °C for 2 or 16 h
8. Add 50 μL of 5% TFA *aq*. and 750 μL of acetone to quench the reaction.
9. Incubate the sample at −20 °C for 30 min.
10. Centrifuge the sample at 15,000 × *g* for 5 min at 4 °C. Then, remove the supernatant completely.
11. Open the tube lid and cover it with tissues, then dry the peptide pellet at room temperature for 5 min.
12. Dissolve the resulting product pellet into an appropriate solution and use it for downstream assays (*see* **Note 5**).
13. For example, dissolve the pellet into 30 μL of 5% TFA *aq*. for LC-MS analysis.
14. Inject 10 μL of the sample (*see* **Note 12**) into an appropriate UPLC-ESI-qTOF instrument equipped with a C18 reverse-phase UPLC column.

3.4 Posttranslational Chemical Modification of AzHyA to Hhc Units

1. Incubate 2 μL of the translation mixture on ice for 1 min.
2. Add 18 μL of cooled TCEP reaction buffer to the solution.
3. Incubate the sample on ice for 2 h, followed by at 42 °C for 2 h.
4. Aliquot 5 μL of the reaction mixture and mix it with 20 μL of 5% TFA *aq*. to quench the reaction.
5. Use the solution for downstream assays (*see* **Note 5**).
6. For example, to perform LC-MS analysis, centrifuge the sample at 15,000 × *g* for 10 min at 4 °C.

7. Inject 3 μL of the supernatant (*see* **Note 12**) into an appropriate UPLC-ESI-qTOF instrument equipped with a C18 reverse-phase UPLC column.

4 Notes

1. The optimal conditions (pH and reaction time) for the flexizyme reactions vary depending on the acyl-donor substrates. The empirically determined optimal conditions for representative substrates used in this chapter are summarized in Fig. 2. For other acyl-donor substrates, it is recommended to optimize the conditions by gel-shift assays [15, 38, 39].
2. Depending on the structure of the nonproteinogenic acid to be charged, an appropriate leaving group should be chosen to make the corresponding acyl-donor substrate. Additionally, an appropriate flexizyme variant compatible with the leaving group should be used. For the general guidance to select the leaving group and the flexizyme variant, see the previously described article [15]. The recommended leaving groups and flexizymes for representative nonproteinogenic acids used in this chapter are shown in Fig. 2.
3. Engineered tRNAs that are inert against endogenous ARSs, so-called orthogonal tRNAs, should be used for the suppressor tRNAs. The anticodon triplet sequence should be appropriately designed to suppress the corresponding reprogrammed codon. In addition, tRNA body sequences often affect the incorporation efficiencies of nonproteinogenic residues. In general, the use of either $tRNA^{AsnE2}$ [40], $tRNA^{GluE2}$ [41], or $tRNA^{Pro1E2}$ [42] is recommended for the suppression of elongation codons. To reprogram the initiation codon, $tRNA^{fMetE}_{CAU}$ [8, 43] is recommended. The recommended suppressor tRNAs for representative nonproteinogenic acids used in this chapter are shown in Fig. 2.
4. The DNA templates should be composed of the following elements in the 5′ to 3′ order: T7 promoter, GGG triplet (enhancer of transcription efficiency), epsilon sequence (AU-rich bacterial translation enhancer element that was originally found in nontranslated region of T7 phage gene) [44], Shine–Dalgano sequence (purine-rich sequence that functions as a ribosomal binding site), start codon (ATG), peptide coding sequence, stop codon (TAA or TGA), and additional 6–10 bp sequence at the downstream of stop codon. Alternatively, plasmids or mRNAs composed of these sequence elements can be used.

5. Peptides expressed using the FIT system or synthesized by the posttranslational chemical modification can be analyzed by various methods. In this chapter, characterization of the peptides by means of LC-MS is described as a representative analysis technique. For quantitative analysis, the peptides can be expressed in the presence of an appropriate radioisotope-labeled amino acid and analyzed by tricine-SDS PAGE followed by autoradiography. For qualitative identification of the products, the peptide samples can be desalted using an SPE column and analyzed by MALDI-TOF mass spectrometry.
6. Ethanol precipitation and the following washing steps (*see* Subheading 3.1, **steps 7** through **12**) should be carried out at room temperature, not at a lower temperature, to prevent undesirable salt precipitations.
7. The washing steps (*see* Subheading 3.1, **steps 8** through **12**) are critical for the following translation reaction. If the metal ions remain in the pellet, it decreases the efficiency of the downstream translation reaction.
8. Acyl-tRNAs synthesized by flexizyme are recovered just by the ethanol precipitation procedure described and no further purification is generally necessary.
9. The optimal conditions (concentrations of acyl-tRNAs and reaction time) for translation can vary depending on the non-proteinogenic residues to be incorporated. The empirically determined optimal conditions for representative residues are summarized in Fig. 2.
10. Two or more different acyl-tRNAs can be mixed and used for translation to incorporate multiple nonproteinogenic residues into the nascent peptide chain. For instance, the combinatorial use of *N*-chloroacetylated amino acids [8, 45] with BrvG/AzHyAs allows for posttranslational macrocyclization, yielding thioether-closed macrocyclic peptides bearing noncanonical backbones. Alternatively, if amino carbothioic acids [26] (e.g., thionated alanine (Ala^S)) and BrvG are consecutively incorporated, a backbone thiazole moiety can be generated.
11. Resuspend the acyl-tRNA pellet just before adding them to the translation reaction mixture because they are unstable and easily hydrolyzed in solution.
12. The injection volume may be changed according to the sensitivity of the instrument used.

Acknowledgments

We thank H. Tsutsumi and T. Kuroda for their major contributions to the development of the methods presented in this chapter. We also appreciate the financial supports from the Japan Society for the Promotion of Science, KAKENHI (JP16H06444 to H.S. and Y.G.; JP20H05618 to H.S.; JP17H04762, JP18H04382, JP19K22243, JP20H02866 to Y.G.).

References

1. Walsh CT, O'Brien RV, Khosla C (2013) Nonproteinogenic amino acid building blocks for nonribosomal peptide and hybrid polyketide scaffolds. Angew Chem Int Ed Engl 52: 7098–7124
2. Arnison PG, Bibb MJ, Bierbaum G et al (2013) Ribosomally synthesized and post-translationally modified peptide natural products: overview and recommendations for a universal nomenclature. Nat Prod Rep 30: 108–160
3. Montalban-Lopez M, Scott TA, Ramesh S et al (2020) New developments in RiPP discovery, enzymology and engineering. Nat Prod Rep 38:130–239
4. Goto Y, Suga H (2021) The RaPID platform for the discovery of pseudo-natural macrocyclic peptides. Acc Chem Res 54:3604–3617
5. Kofman C, Lee J, Jewett MC (2021) Engineering molecular translation systems. Cell Syst 12: 593–607
6. Hartman MCT (2022) Non-canonical amino acid substrates of *E. coli* aminoacyl-tRNA synthetases. ChemBioChem 23:e202100299
7. Hecht SM (2022) Expansion of the genetic code through the use of modified bacterial ribosomes. J Mol Biol 434:167211
8. Goto Y, Ohta A, Sako Y et al (2008) Reprogramming the translation initiation for the synthesis of physiologically stable cyclic peptides. ACS Chem Biol 3:120–129
9. Goto Y, Iwasaki K, Torikai K et al (2009) Ribosomal synthesis of dehydrobutyrine- and methyllanthionine-containing peptides. Chem Commun (Camb) 23:3419–3421
10. Nakajima E, Goto Y, Sako Y et al (2009) Ribosomal synthesis of peptides with C-terminal lactams, thiolactones, and alkylamides. ChemBioChem 10:1186–1192
11. Kato Y, Kuroda T, Huang Y et al (2020) Chemoenzymatic posttranslational modification reactions for the synthesis of Psi[CH2 NH]-containing peptides. Angew Chem Int Ed Engl 59:684–688
12. Tsutsumi H, Kuroda T, Kimura H et al (2021) Posttranslational chemical installation of azoles into translated peptides. Nat Commun 12:696
13. Kuroda T, Huang Y, Nishio S et al (2022) Posttranslational backbone-acyl shift yields natural product-like peptides bearing hydroxyhydrocarbon units. Nat Chem:in press
14. Murakami H, Ohta A, Ashigai H et al (2006) A highly flexible tRNA acylation method for non-natural polypeptide synthesis. Nat Methods 3:357–359
15. Goto Y, Katoh T, Suga H (2011) Flexizymes for genetic code reprogramming. Nat Protoc 6: 779
16. Wipf P, Fritch PC, Geib SJ et al (1998) Conformational studies and structure-activity analysis of lissoclinamide 7 and related cyclopeptide alkaloids. J Am Chem Soc 120:4105–4112
17. Siodlak D, Stas M, Broda MA et al (2014) Conformational properties of oxazole-amino acids: effect of the intramolecular N-H...N hydrogen bond. J Phys Chem B 118:2340–2350
18. Ahlbach CL, Lexa KW, Bockus AT et al (2015) Beyond cyclosporine A: conformation-dependent passive membrane permeabilities of cyclic peptide natural products. Future Med Chem 7:2121–2130
19. Marshall CG, Burkart MD, Keating TA et al (2001) Heterocycle formation in vibriobactin biosynthesis: alternative substrate utilization and identification of a condensed intermediate. Biochemistry 40:10655–10663
20. Schneider TL, Shen B, Walsh CT (2003) Oxidase domains in epothilone and bleomycin biosynthesis: thiazoline to thiazole oxidation during chain elongation. Biochemistry 42: 9722–9730
21. Dunbar KL, Melby JO, Mitchell DA (2012) YcaO domains use ATP to activate amide backbones during peptide cyclodehydrations. Nat Chem Biol 8:569–575
22. Koehnke J, Bent AF, Zollman D et al (2013) The cyanobactin heterocyclase enzyme: a processive adenylase that operates with a defined order of reaction. Angew Chem Int Ed Engl 52:13991–13996
23. Melby JO, Li X, Mitchell DA (2014) Orchestration of enzymatic processing by thiazole/oxazole-modified microcin dehydrogenases. Biochemistry 53:413–422

24. Chowdhury SR, Maini R, Dedkova LM et al (2015) Synthesis of fluorescent dipeptidomimetics and their ribosomal incorporation into green fluorescent protein. Bioorg Med Chem Lett 25:4715–4718
25. Chowdhury SR, Chauhan PS, Dedkova LM et al (2016) Synthesis and evaluation of a library of fluorescent dipeptidomimetic analogues as substrates for modified bacterial ribosomes. Biochemistry 55:2427–2440
26. Maini R, Kimura H, Takatsuji R et al (2019) Ribosomal formation of thioamide bonds in polypeptide synthesis. J Am Chem Soc 141: 20004–20008
27. Bott R, Subramanian E, Davies DR (1982) Three-dimensional structure of the complex of the Rhizopus chinensis carboxyl proteinase and pepstatin at 2.5-A resolution. Biochemistry 21:6956–6962
28. Rich DH, Bernatowicz MS, Agarwal NS et al (1985) Inhibition of aspartic proteases by pepstatin and 3-methylstatine derivatives of pepstatin. Evidence for collected-substrate enzyme inhibition. Biochemistry 24:3165–3173
29. Umezawa H, Aoyagi T, Morishima H et al (1970) Pepstatin, a new pepsin inhibitor produced by actinomycetes. J Antibiot (Tokyo) 23:259–262
30. Tumminello FM, Bernacki RJ, Gebbia N et al (1993) Pepstatins: aspartic proteinase inhibitors having potential therapeutic applications. Med Res Rev 13:199–208
31. Kuranaga T, Matsuda K, Takaoka M et al (2020) Total synthesis and structural revision of kasumigamide, and identification of a new analogue. ChemBioChem 21:3329–3332
32. Lee J, Schwarz KJ, Kim DS et al (2020) Ribosome-mediated polymerization of long chain carbon and cyclic amino acids into peptides in vitro. Nat Commun 11:4304
33. Trobro S, Aqvist J (2005) Mechanism of peptide bond synthesis on the ribosome. Proc Natl Acad Sci U S A 102:12395–12400
34. Voorhees RM, Weixlbaumer A, Loakes D et al (2009) Insights into substrate stabilization from snapshots of the peptidyl transferase center of the intact 70S ribosome. Nat Struct Mol Biol 16:528–533
35. Shimizu Y, Inoue A, Tomari Y et al (2001) Cell-free translation reconstituted with purified components. Nat Biotechnol 19:751–755
36. Charlton A, Zachariou M (2008) Immobilized metal ion affinity chromatography of histidine-tagged fusion proteins. Methods Mol Biol 421: 137–149
37. Block H, Maertens B, Spriestersbach A et al (2009) Immobilized-metal affinity chromatography (IMAC): a review. Methods Enzymol 463:439–473
38. Martinis SA, Schimmel P (1992) Enzymatic aminoacylation of sequence-specific RNA minihelices and hybrid duplexes with methionine. Proc Natl Acad Sci U S A 89:65–69
39. Putz J, Wientges J, Sissler M et al (1997) Rapid selection of aminoacyl-tRNAs based on biotinylation of alpha-NH2 group of charged amino acids. Nucleic Acids Res 25:1862–1863
40. Ohta A, Murakami H, Higashimura E et al (2007) Synthesis of polyester by means of genetic code reprogramming. Chem Biol 14: 1315–1322
41. Terasaka N, Hayashi G, Katoh T et al (2014) An orthogonal ribosome-tRNA pair via engineering of the peptidyl transferase center. Nat Chem Biol 10:555–557
42. Katoh T, Iwane Y, Suga H (2017) Logical engineering of D-arm and T-stem of tRNA that enhances d-amino acid incorporation. Nucleic Acids Res 45:12601–12610
43. Goto Y, Iseki M, Hitomi A et al (2013) Non-standard peptide expression under the genetic code consisting of reprogrammed dual sense codons. ACS Chem Biol 8:2630–2634
44. Olins PO, Devine CS, Rangwala SH et al (1988) The T7 phage gene 10 leader RNA, a ribosome-binding site that dramatically enhances the expression of foreign genes in *Escherichia coli*. Gene 73:227–235
45. Sako Y, Goto Y, Murakami H et al (2008) Ribosomal synthesis of peptidase-resistant peptides closed by a nonreducible inter-side-chain bond. ACS Chem Biol 3:241–249

Chapter 14

Thioester Capture Strategy for the Identification of Nonribosomal Peptide and Polyketide Intermediates

Yueying Li, Lauren A. Washburn, and Coran M. H. Watanabe

Abstract

Nonribosomal peptide synthetases (NRPSs) and polyketide synthases (PKSs) are multi-domainal megasynthases. While they are capable of generating a structurally diverse array of metabolites of therapeutic relevance, their mere size and complex nature of their assembly (intermediates are tethered and enzyme bound) make them inherently difficult to characterize. In order to facilitate structural characterization of these metabolites, a thioester capture strategy that enables direct trapping and characterization of the thioester-bound enzyme intermediates was developed. Specifically, a synthetic Biotin-Cys agent was designed and utilized, enabling direct analysis by LCMS/MS and NMR spectroscopy. In the long term, the approach might facilitate the discovery of novel scaffolds from cryptic biosynthetic pathways, paving the way for the development of drug leads and therapeutic initiatives.

Key words Nonribosomal peptides, Polyketides, Thioester capture strategy, Biotin-Cys, AziB, ClbN, AziA3

1 Introduction

Nonribosomal peptide synthetases (NRPS) and polyketide synthases (PKS) have contributed a wide array of secondary metabolites with potent bioactivities [1–3]. However, their architectural diversity and complex assembly, in turn, convolute their analysis, slowing progress toward their biosynthetic understanding before they can be made amenable to industrial applications [4–6].

Here, we developed a thioester capture strategy, utilizing a Biotin-Cys probe, to covalently link and trap the thioester-bound intermediates tethered to the NRPS/PKS enzymes in order to identify the structures by LCMS/MS and NMR spectroscopy [7]. The theory behind it is based upon the principles of protein ligation and intein chemistry, where the cysteine thiol group of the Biotin-Cys agent carries out a nucleophilic attack on the carboxy group of the enzyme-linked metabolite intermediate, forming

Michael Burkart and Fumihiro Ishikawa (eds.), *Non-Ribosomal Peptide Biosynthesis and Engineering: Methods and Protocols*, Methods in Molecular Biology, vol. 2670, https://doi.org/10.1007/978-1-0716-3214-7_14,

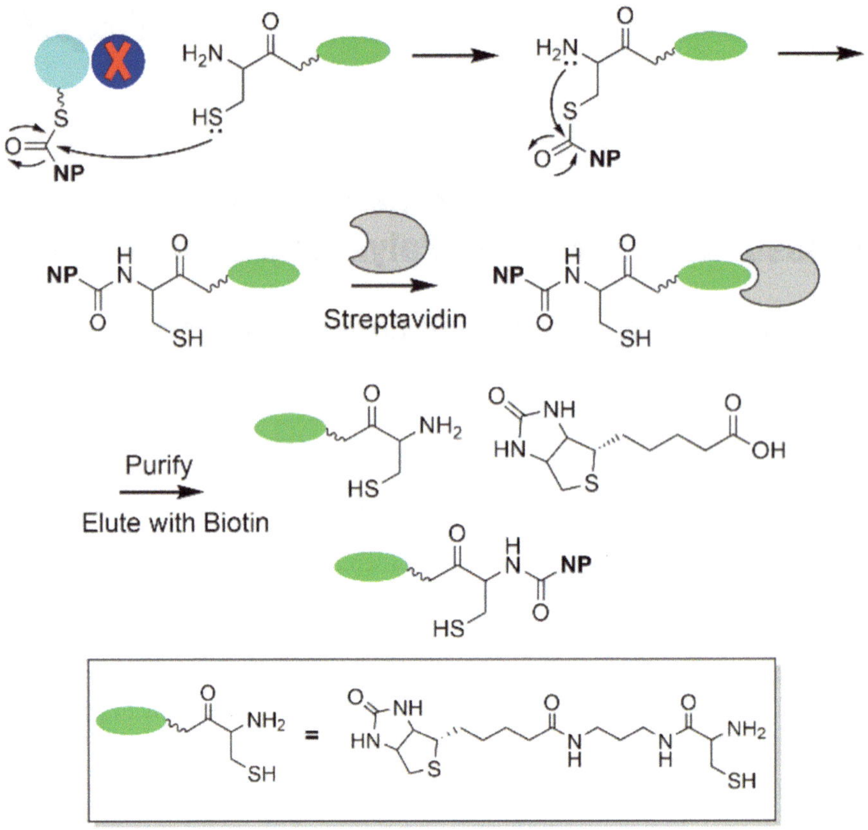

Fig. 1 Biotin-Cys capture strategy

initially a thioester, which subsequently rearranges to give an amide bond (Fig. 1) [8]. The biotin moiety within the capture agent allows for selective purification of target compounds as well as provides direct derivatization for structural characterization/spectroscopic analysis [9].

In this chapter, we introduce the synthesis of the Biotin-Cys capture agent and demonstrate its application in the evaluation of NRPS and PKS biosynthetic pathways. We selected AziB from the azinomycin biosynthetic pathway and ClbN from the colibactin biosynthetic pathway to serve as PKS examples and demonstrate the generality of the approach (Fig. 2). Additionally, AziA3 from the azinomycin biosynthetic pathway was selected to serve as an NRPS example and demonstrate the use of the strategy in the identification of an unknown metabolite (Fig. 2) [1, 2].

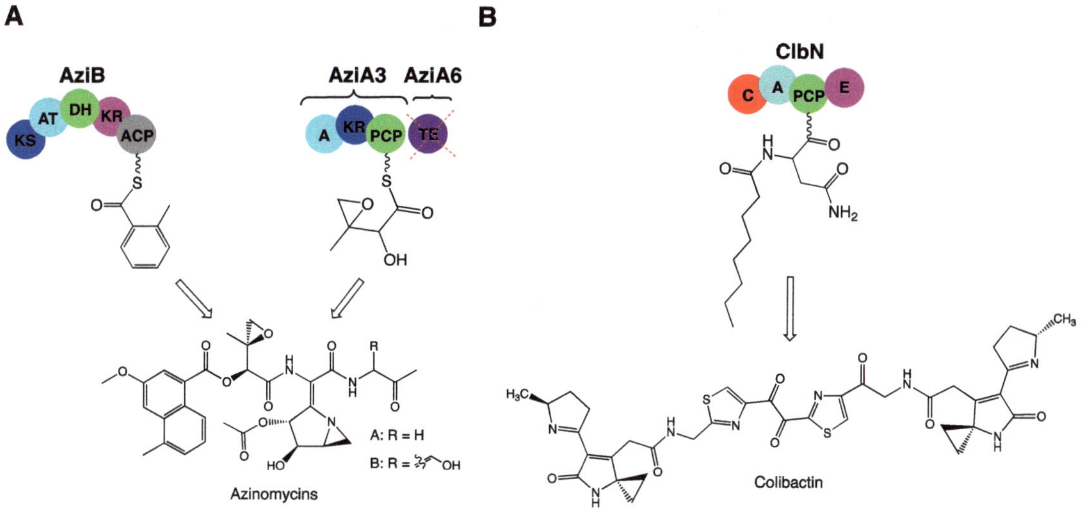

Fig. 2 (**a**) AziB (PKS) and AziA3 (NRPS) from the azinomycin biosynthetic pathway, (**b**) ClbN (NRPS) from the colibactin biosynthesis pathway

2 Materials

^{1}H and ^{13}C-NMR spectra were recorded on either a Bruker Avance 500 (equipped with a cryoprobe) or a Bruker Ascend 400. Mass spectra were obtained at the Laboratory for Biological Mass Spectrometry at the Department of Chemistry, Texas A&M University. All solutions were generated with deionized water (generated at 16 MΩ-cm at 25 °C) and stored at 4 °C unless indicated otherwise. Sterilization was performed on an autoclave at 121 °C for 20 min.

2.1 Synthesis of the Thioester Capture Agent Biotin-Cys

1. Materials used in synthesis scheme: N-(3-aminopropyl) biotinamide trifluoroacetate, N-(tert-Butoxycarbonyl)-S-trityl-L-cysteine, 4-dimethylaminopyridine, N-(3-(dimethylamino)propyl)-N-ethylcarbodiimide, trifluoroacetic acid, triethylsilane, dimethylformamide, and dichloromethane.
2. Solvent A: 0.1% formic acid solution in water. Pipet 1 mL of formic acid and make up to 1 L with water.
3. Solvent B: 75% methanol, 24.9% isopropanol, and 0.1% formic acid in water. Pipet 1 mL of formic acid to 750 mL of methanol and 249 mL of isopropanol to reach a volume of 1 L.

2.2 Capture Strategy Validation with PKS, AziB (from the Azinomycin Biosynthetic Pathway)

2.2.1 In Vitro Expression, Posttranslational Modification, and Purification of AziB

1. pET24_aziB: The plasmid was generated following reference [1].
2. LB liquid medium: 2.5% solution in water. Add 2.5 g of LB broth and dissolve in 100 mL of water in a 250 mL flask or add 25 g LB broth and dissolve in 1 L of water in a 2 L flask. Autoclave and cool to room temperature (*see* **Note 1**).
3. LB agar plates: 2.5% LB and 1.0% bacto agar in water. Add 2.5 g LB broth and 1.0 g bacto agar to 100 mL of water in a 250 mL flask. Autoclave and cool until it is warm to touch, then add antibiotic solution prior to pouring into plates, if applicable. Prepare four plates per 100 mL of solution (*see* **Note 1**).
4. 100 mg/mL kanamycin solution: Add 5 g kanamycin sulfate to a 50 mL Falcon tube and add 50 mL of water.
5. 100 mg/mL ampicillin solution: Add 5 g ampicillin sodium salt to a 50 mL Falcon tube and add 50 mL water.
6. 1 M isopropyl β-d-1-thiogalactopyranoside (IPTG) solution: Add 11.9 g IPTG to a 50 mL Falcon tube and add 50 mL water.
7. Cell harvest solution: 20 mM potassium phosphate pH 7.4, 500 mM NaCl, 1 mM dithiothreitol (DTT), 5 mM imidazole, and 20% glycerol. Add 2.424 g potassium phosphate dibasic and 828.1 mg potassium phosphate monobasic, 29.22 g NaCl, 154.3 mg DTT, 340.4 mg imidazole, and 200 mL glycerol to a 1 L glass bottle and bring the volume to 1 L with water.
8. 1 M beta-mercaptoethanol (ß-ME) solution: Add 3.5 mL ß-ME to a 50 mL Falcon tube, and bring the volume to 50 mL with water.
9. Media A: 1 mM DTT, 500 mM NaCl, 75 mM Tris–HCl, pH 8.0, and 20% glycerol. Add 154.3 mg DTT, 29.2 g NaCl, 9.1 g Tris base, and 200 mL glycerol to a 1 L glass bottle. Add 700 mL water and adjust the pH to 8.0. Adjust the volume to 1 L with water.
10. Media B: 500 mM imidazole, 1 mM DTT, 500 mM NaCl, 75 mM Tris–HCl, pH 8.0, and 20% glycerol. Add 34.0 g imidazole, 154.3 mg DTT, 29.2 g NaCl, 9.1 g Tris base, and 200 mL glycerol to a 1 L glass bottle. Add 700 mL water and adjust the pH to 8.0. Adjust the volume to 1 L with water.

2.2.2 Protein Validation with 15% Sodium Dodecyl Sulfate Polyacrylamide Gel Electrophoresis

1. Monomer solution: 30% acrylamide and 0.8% *N,N*′-methylenebisacrylamide in water. Add 150 g acrylamide and 4.0 g *N,N* ′-methylenebisacrylamide to a 500 mL glass bottle and add 500 mL water.

2. 4× running buffer: 1.5 M Tris base, pH 8.8. Add 181.7 tris base to a 1 L glass bottle and add 900 mL water. Adjust pH to 8.8 and adjust the volume to 1 L with water.
3. 4× stacking buffer: 0.5 M Tris base, pH 6.8. Add 60.6 g tris base to a 1 L glass bottle and add 1 L water. Adjust pH to 6.8 and adjust the volume to 1 L with water.
4. 10% ammonium persulfate solution: Add 0.1 g ammonium persulfate to a 1.5 mL Eppendorf tube and add 1 mL water (*see* **Note 3**).
5. 10% sodium dodecyl sulfate (SDS) solution: Add 0.1 g sodium dodecyl sulfate to a 1.5 mL Eppendorf tube and add 1 mL water (*see* **Note 3**).
6. 2× treatment buffer: 0.02% bromophenol blue, 3.1% dithiothreitol (DTT), 25% 4× stacking buffer, 40% SDS solution, and 20% glycerol. Add 10 mg bromophenol blue, 1.55 g DTT, 12.5 mL of 4× stacking buffer, 20 mL SDS solution, and 10 mL glycerol to a 50 mL Falcon tube and adjust the volume to 50 mL with water. Store at −20 °C.
7. Tank buffer: 25 mM Tris–HCl, 192 mM glycine, 0.1% SDS. Add 12.1 g Tris, 57.6 g glycine, and 4.0 g SDS, and add 4 L water. Adjust pH to 8.3. Store at room temperature.
8. Coomassie blue staining solution: 0.025% Coomassie brilliant blue R, 40% MeOH, and 7% HOAc. Add 0.25 g Coomassie brilliant blue R to 400 mL MeOH. Add 70 mL of HOAc after fully dissolved. Adjust the volume to 1 L with water. Store at room temperature in a cabinet hood.
9. Destain solution I: 40% methanol, 7% HOAc. Add 1.6 L methanol, 280 mL HOAc, and 2.12 L water. Store at room temperature in a cabinet hood.

2.2.3 Generation of AziB Intermediate In Vitro

1. Exchange buffer: 50 mM potassium phosphate, pH 7.5, and 20% glycerol. Add 6.406 g potassium phosphate dibasic and 1.799 g potassium phosphate monobasic and 200 mL glycerol to a 1 L flask bottle, and adjust the volume to 1 L with water.
2. 25 mg/mL coenzyme A solution: Add 25 mg coenzyme A to a 1.5 mL Eppendorf tube and add 1 mL water. Store at −20 °C.
3. 25 mg/mL acetyl coenzyme A solution: Add 25 mg acetyl coenzyme A to a 1.5 mL Eppendorf tube and add 1 mL water. Store at −20 °C.
4. 25 mg/mL malonyl coenzyme A solution: Add 25 mg malonyl coenzyme A to a 1.5 mL Eppendorf tube and add 1 mL water. Store at −20 °C.

5. 10 mM dihydronicotinamide-adenine dinucleotide phosphate (NADPH): Add 8.3 mg to a 1.5 mL Eppendorf tube and add 1 mL water. Store at −20 °C.
6. 1 M Dithiothreitol (DTT): Add 154 mg DTT to a 1.5 mL Eppendorf tube and add 1 mL water. Store at −20 °C.

2.2.4 Reaction Between the AziB Intermediates and the Thioester Capture Agent Biotin-Cys

1. 1 M Tris(2-carboxyethyl)phosphine hydrochloride (TCEP) solution: Add 286.7 mg TCEP to a 1.5 mL Eppendorf tube and add 1 mL water. Store at −20 °C.

2.2.5 Purification of Biotin-Cys-Captured AziB Intermediate

1. Solution 1.1 buffer: 1 M KH_2PO_4 solution in water. Add 2.04 g KH_2PO_4 and add 15 mL water.
2. Solution 1.2 buffer: 1 M K_2HPO_4 solution. Add 22.8 g $K_2HPO_4 \cdot 3H_2O$ and add 100 mL water.
3. Solution 2 washing buffer: 100 mM potassium phosphate, 150 mM NaCl, and pH = 7.2. Add 2.19 g NaCl in 100 mL water. Add 5.75 mL of solution 1.1 and 19.25 mL of solution 1.2. Adjust the volume to 250 mL with water.
4. Solution 3 equilibration buffer: 100 mM potassium phosphate, 150 mM NaCl, 400 mM ammonium sulfate, and pH = 7.2. Add 2.19 g NaCl and 13.2 g ammonium sulfate in 100 mL water. Add 2.13 mL of solution 1.1 and 22.88 mL of solution 1.2. Adjust the volume to 250 mL with water.
5. Solution 4 equilibration buffer (3x): 300 mM phosphate, 450 mM NaCl, 1.2 M ammonium sulfate, and pH 7.2. Add 0.66 g NaCl and 3.96 g ammonium sulfate in 15 mL water. Add 0.40 mL of solution 1.1 and 7.10 mL of solution 1.2. Adjust the volume to 25 mL with water.
6. Solution 5 elution buffer: 100 mM potassium phosphate, 150 mM NaCl, 2 mM D-biotin, and pH 7.2. Add 0.44 g NaCl and 24.4 mg D-biotin in 20 mL water. Add 1.16 mL of solution 1.1 and 3.86 mL of solution 1.2. Adjust the volume to 50 mL with water.

2.3 Capture Strategy Validation with NRPS (ClbN, Colibactin Biosynthesis Pathway)

2.3.1 In Vitro Expression, Posttranslational Modification, and Purification of ClbN

1. pET28_ClbN: The plasmid was purchased from Addgene (RRID: Addgene_51497).
2. 1 M phenylmethylsulfonyl fluoride (PMSF) solution: Add 8.71 g PMSF and add 50 mL dimethyl sulfoxide. The solution is stored at −20 °C.
3. LB liquid medium, LB agar plate, kanamycin solution, IPTG solution, cell harvest solution, ß-ME solution, media A, and media B were prepared as detailed in Subheading 2.2.1.
4. Coenzyme A solution was prepared as described in Subheading 2.2.3.

5. Sfp solution: Sfp protein was generated from the pET24_sfp plasmid. Overexpression and purification followed the same procedure as outlined for the overexpression of ClbN in *Escherichia coli* (*see* Subheading 3.3.1 **step 1–16**).
6. Protein validation solutions were prepared as detailed in Subheading 2.2.2.
7. Exchange buffer: 40 mM HEPES buffer, pH 7.5, 33 mM NaCl, 4 mM $MgCl_2$, 400 μM DTT. Add 476.6 mg HEPES, 96.4 mg NaCl, 40.7 mg $MgCl_2 \cdot 6H_2O$, and 20 μL DTT solution (prepared in Subheading 2.2.3) to a 50 mL Falcon tube and add 40 mL water. Adjust the pH to 7.5. Adjust the volume to 50 mL with water.

2.3.2 Generation ClbN Intermediate In Vitro

1. 100 mM L-asparagine (L-Asn) solution: Add 13.2 mg L-Asn to a 1.5 mL Eppendorf tube and add 1 mL water. Store at −20 °C.
2. 100 mM adenosine triphosphate (ATP) solution: Add 55.1 mg ATP to a 1.5 mL Eppendorf tube and add 1 mL water. Store at −20 °C.
3. 100 mM octanoyl-coenzyme A (octanoyl-CoA) solution: Add 8.9 mg octanoyl-coenzyme A to a 1.5 mL Eppendorf tube and suspend in 100 μL water. Store at −20 °C.

2.3.3 Generation of pSETAziA3Δ aziA6

1. Primers:

Primer	Sequence (5′-3′)
AziA6UF	TCAAGCTTCACACGAGCGAGAACGCG
AziA6UR	GTCTAGAGGAAGGACCTCTTCGTGA
AziA6DF	ATCTAGACGTGCTCAACGCGTCCCC
AziA6DR	CAGGATCCCGAGAACTATCCCGACCT
pSETAziA3F	TCAGATCTAGGAGGCTCTTCG TGACCACCGCCACCAGC
pSETAziA3R	TCTAGAAATTCAGTGGTGGTGGTGGTGG TGGCCGGCGATCCAGGCCGCCAG

2. LB media is prepared as detailed in Subheading 2.2.1.
3. 2× YT liquid medium: 1.6% tryptone, 1.0% yeast extract, and 0.5% NaCl in water. Add 1.6 g tryptone, 1.0 g yeast extract, and 0.5 g NaCl in 250 mL flasks, and add 90 mL water. Adjust pH to 7.0 and bring the volume to 100 mL with water. Autoclave and cool to room temperature.

4. 3% nalidixic acid stock solution: Add 1.5 g of nalidixic acid to a 50 mL Falcon tube and add 50 mL water. Store at 4 °C.
5. 10% apramycin stock solution: Add 5 g of apramycin to a 50 mL Falcon tube and add 50 mL water. Store at 4 °C.
6. 0.5 mg/mL nalidixic acid and 70 μg/mL apramycin solution: Dissolve 333 μL nalidixic acid stock solution and 14 μL of apramycin stock solution in 20 mL of water in a 50 mL Falcon tube. Store at 4 °C.

2.3.4 In Vitro Expression, Posttranslational Modification, and Purification of AziA3 [10, 11]

1. 2× YT liquid medium is prepared as detailed in Subheading 2.3.3.
2. GYM agar plate: 0.4% glucose, 0.4% yeast extract, 1.0% malt extract, 0.2% $CaCO_3$, and 1.2% bacto agar in water. Add 0.4 g glucose, 0.4 g yeast extract, 1.0 g malt extract, 0.2 g $CaCO_3$, and 90 mL water to a 250 mL flask. Adjust the pH to 7.5 and adjust the volume to 100 mL with water. Add 1.2 g of bacto agar just prior to autoclaving.
3. YEME liquid medium: 0.3% yeast extract, 0.3% malt extract, 0.5% peptone, 1.0% glucose, and 34% sucrose in water. Add 3 g yeast extract, 3 g malt extract, 5 g peptone, 10 g glucose, 340 g sucrose, and 1000 mL of water to 2 L flasks. Autoclave and cool to room temperature.
4. 100 mg/mL apramycin solution: Add 5 g apramycin sulfate and 50 mL of water to a 50 mL Falcon tube.
5. Lysis buffer solution: 50 mM Tris–HCl, pH 7.5, 500 mM NaCl, 1 mM β-mercaptoethanol (ß-ME), 10% glycerol, 5 mM imidazole, and 1 mM phenylmethylsulfonyl fluoride (PMSF) in water. Add 302.9 mg Tris base, 1.46 g NaCl, 17 mg imidazole, 5 mL glycerol, and 40 mL of water to a 50 mL Falcon tube. Add 50 μl ß-ME solution prepared in Subheading 2.2.1 and PMSF solution as detailed in Subheading 2.3.1. Adjust the pH to 7.5 and bring the volume to 50 mL with water.

3 Methods

3.1 Synthesis of the Biotin-Cys Thioester Capture Agent

1. 1 equivalent of N-(tert-Butoxycarbonyl)-S-trityl-L- cysteine is dissolved in DMF at 4 °C.
2. Add 2 equivalent of 4-dimethylaminopyridine (DMAP), 1.3 equivalent of N-(3-(dimethylamino)propyl)-N-ethylcarbodiimide (EDC), and 1.5 equivalent of N-(3-aminopropyl)biotinamide trifluoroacetate.
3. The reaction is stirred at 4 °C for 20 min, warmed to room temperature, and subsequently stirred for an additional 12 h.

4. DMF is removed by evaporation under a nitrogen stream to give the protected Biotin-Cys(Trt)-Boc.
5. The Biotin-Cys(Trt)-Boc (0.5 mmol scale) is dissolved in water with 5 mL of trifluoroacetic acid (TFA) and 100 mg of triethylsilane. The reaction is stirred for 24 h under nitrogen.
6. Evaporate the solvent and concentrate in vacuo.
7. The organics are resuspended in 15 mL of 1:1 water/dichloromethane.
8. Extract the mixture two times with dichloromethane and lyophilize the aqueous layer to dryness.
9. Dissolve the product mixture in 1 mL water and purify by HPLC (yield 70.8%, Phenomenex column Prodigy 5 μm C18 150 Å, 150 × 4.6 mm as example in this experiment).
10. HPLC gradient conditions are as follows: 0 min, 80% A, 20% B; 1 min, 80% A, 20% B; 23 min, 0% A, 100% B; 33 min, 0% A, 100% B; 35 min, 80% A, 20% B; and 40 min, 80% A, 20% B.
11. Biotin-Cys. ^{1}H-NMR (400 MHz, D_2O) δ 4.61 (dd, 1H, J = 7.9, 4.8), 4.43 (dd, 1H, J = 7.8, 4.8), 4.14 (t, 1H, J = 6.0), 3.38–3.20 (m, 6H), 3.06 (dd, 1H, J = 9.4, 4.4), 3.00 (dd, 1H, J = 12.9, 4.8), 2.78 (d, 1H, J = 13.1), 2.27 (t, 2H, J = 7.4), 1.80–1.53 (m, 6H), and 1.48–1.35 (m, 2H); ^{13}C-NMR δ 176.8, 167.8, 165.3, 62.0, 60.2, 55.3, 54.5, 39.6, 36.9, 36.4, 35.4, 27.9, 27.8, 27.6, 25.1, 24.8. HRESIMS m/z $[M + H]^+$ calculated for $C_{16}H_{30}N_5O_3S_2$, 404.1790; found, 404.1779.

3.2 Demonstration of the Capture Strategy with the PKS AziB (from the Azinomycin Biosynthetic Pathway)

3.2.1 In Vitro Expression, Posttranslational Modification, and Purification of AziB

1. Add 50 μL of the kanamycin solution (Subheading 2.2.1) to 100 mL of LB medium.
2. Pick a single colony from an LB agar plate containing *E. coli* BL21(DE)/pET24_aziB, and inoculate in 100 mL LB medium.
3. Incubate the culture at 37 °C, shaking overnight at 250 rpm.
4. Add 500 μL of the kanamycin solution to 1 L LB medium, prepare 5 × 1 L in 2 L Erlenmeyer flasks.
5. Transfer 10 mL of the overnight culture to each 5 × 1 L LB broth and culture at 37 °C, shaking at 250 rpm for about 3.5 h until an OD_{600} of 0.6 is reached.
6. Add 1 mL of IPTG solution to each of the 5 × 1 L flasks to initiate the induction process. Continue culturing for 24 h at 16 °C, shaking at 250 rpm.
7. Harvest the 5 L of cell culture by centrifugation, 7477 RCF for 15 min.
8. Add 50 μL of the ampicillin solution to 100 mL of LB medium.

9. Pick a single colony from an LB agar plate containing *E. coli* BL21(DE)/pET21_Svp, and inoculate in 100 mL LB medium.
10. Incubate the culture overnight at 37 °C, 250 rpm.
11. Add 500 μL of the ampicillin solution to 1 L LB liquid medium in 2 L Erlenmeyer flask, prepare 3 × 1 L.
12. Transfer 10 mL of the overnight culture to each 3 × 1 L LB broth and culture at 37 °C, shaking at 250 rpm for about 3.5 h until an OD_{600} of 0.6 is reached.
13. Add 1 mL of IPTG solution to each of the 3 × 1 L flasks to initiate the induction process. Continue culturing for 24 h at 16 °C, shaking at 250 rpm.
14. Harvest the 3 L of cell culture by centrifugation, 7477 RCF for 15 min.
15. Combine 5 L of harvested *E. coli* BL21(DE) cell containing pET24_aziB and 3 L of harvested *E. coli* BL21(DE) cell containing pET21_Svp, and resuspend in 50 mL cell harvest solution (*see* Subheading 2.2.1).
16. Lyse the resuspended cells by sonication (J = 50%) on ice, 6 cycles, 30 s on, and 2 min off for each interval.
17. Pellet the cell debris by centrifugation, 2 h at 10,785 RCF to give the AziB and Svp protein suspension.
18. Add 800 μL of the coenzyme A solution (*see* Subheading 2.2.3) to the supernatant.
19. Incubate the mixture at 28 °C for 1 h.
20. Centrifuge the mixture, 10,785 RCF for 1 h, to clarify the supernatant.
21. Filter the supernatant by passing it through a 0.22 mm membrane filter with a syringe prior to column purification.
22. Purify the supernatant using a HisTrap HP 5 mL column with 0 mM imidazole solution as medium A and 500 mM imidazole solution as medium B (*see* **step 1** in Subheading 2.2).
23. Load the AziB/Svp protein suspension onto the column with a peristaltic pump, 0.5 mL/min flow rate.
24. Wash the column with 8% medium B for 10 column volumes with a 0.5 mL/min flow rate.
25. Elute the fraction with 20% medium B for 10 column volumes with a 0.5 mL/min flow rate.
26. Analyze protein fractions with a 15%-sodium dodecyl sulfate polyacrylamide gel (*see* **step 2** in Subheading 3.2).
27. Concentrate the purified protein fractions using a 100 kDa centrifugal ultrafiltration unit (Amicon® Ultra centrifugal filter unit) to generate holo-AziB (20 mg/mL concentration).

3.2.2 Protein Analysis by 15% Sodium Dodecyl Sulfate Polyacrylamide (SDS-PAGE) Gel Electrophoresis

1. Mix 7.5 mL monomer solution, 3.75 mL 4× running buffer, 150 μL SDS solution, 3.5 mL water, 75 μL APS solution, and 10 μL of TEMED in a 50 mL Falcon tube. Stir to mix for 5 min.
2. Load 3 mL of the prepared solution into the sandwiched plates of the SDS-PAGE gel apparatus.
3. Apply water to the top and allow to sit for 1–2 h.
4. Mix 1.33 mL monomer solution, 2.5 mL 4× stacking buffer, 100 μL SDS solution, 6 mL water, 50 μL APS solution, and 10 μL of TEMED in a 50 mL Falcon tube. Stir for 5 min.
5. Pour the water out and add the prepared solution to the solidified gel, and cap with the gel comb.
6. Allow the apparatus to sit for an additional 1–2 h before the prepared SDS-PAGE gel is ready for use.
7. Add 20 μL of the protein fraction from 3.3 and 3.4 to a 1.5 mL Eppendorf tube with 20 μL of the 2× treatment buffer.
8. Incubate the sample at 65–80 °C for 10–15 min.
9. Load 10 μL of each sample to the prepared 15% SDS-PAGE gel.
10. Apply tank buffer to the apparatus and run the gel at 200 V for 60 min.
11. Remove the gel carefully from apparatus and glass plates.
12. Soak the gel in Coomassie blue staining solution and shake for 45 min.
13. Dispose off the stain solution, apply destain solution I, and shake for another 45 min until the protein band(s) are clearly shown.

3.2.3 Generation of AziB Intermediate In Vitro

1. Buffer-exchange 1 mL of purified holo-AziB into 1 mL exchange buffer by centrifugation at 3724 RCF for 1 h.
2. Add 1.75 μL of acetyl coenzyme A solution, 8.5 μL of malonyl coenzyme A solution, 1 μL of NADPH solution, and 1 μl DTT solution to holo-AziB.
3. Incubate the reaction at 28 °C for 12 h.
4. Buffer exchange the mixture into 1 mL of fresh exchange buffer by centrifugation at 3724 RCF for 1 h.

3.2.4 Reaction Between the AziB Intermediate and the Biotin-Cys Thioester Capture Agent

1. Add 0.76 g guanidine hydrochloride, and Biotin-Cys in 10:1 molar excess to AziB and 1 mM TCEP.
2. Incubate the reaction at 4 °C for 16 h.

3.2.5 Purification of Biotin-Cys-Captured AziB Intermediate (See **Note 4**)

1. Add 0.2 mL of Streptavidin mutein matrix to a 1 mL column.
2. After allowing the gel to settle, open the outlet to remove excess storage buffer.
3. Wash the gel bed with 10 bed volumes of solution 2 washing buffer.
4. Equilibrate the gel with 4 volumes of solution 3 equilibration buffer.
5. Mix two parts of the supernatant from Subheading 3.2.4 to one part of solution 4 equilibration buffer (3x).
6. Apply the sample onto the prepared column, open the outlet and allow the sample to slowly and completely penetrate into the gel. Collect the flow-through and subsequently close the column outlet.
7. Incubate at room temperature for 10 min.
8. Wash the gel with 10 gel bed volumes of solution 2 wash buffer in three steps: separately collect the first and second wash fractions, 1 gel bed volume each. Pool wash fraction 3–10 for another 8 bed volumes.
9. Close the column outlet.
10. Apply 1 gel bed of the solution 5 and collect as elution fraction 1.
11. Collect elution fraction 2 as **step 10**.
12. Repeat the steps to collect fraction 3 and 4.

3.2.6 Analysis of the AziB Intermediate-Capture Product by LCMS

1. Perform LCMS analysis with solvent A and solvent B (Phenomenex column Prodigy 5 μm C18 150 Å, 150 × 4.6 mm).
2. Solvent conditions were as follows: 0 min, A-80% B-20%; 1 min, A-80% B-20%; 23 min, A-0% B-100%; 33 min, A-0% B-100%; 35 min, A-80% B-20%; flow rate 0.75 μL/min.

3.3 Demonstration of the Capture Strategy with the NRPS ClbN (Colibactin Biosynthetic Pathway)

3.3.1 In Vitro Expression, Posttranslational Modification, and Purification for ClbN

1. Add 50 μL of the kanamycin solution to 100 mL LB medium.
2. Pick a single colony from *E. coli* BL21(DE)/pET28_ClbN on LB agar plate and inoculate into 100 mL of LB medium.
3. Incubate the 100 mL culture at 37 °C, shaking overnight at 250 rpm.
4. Add 500 μL of the kanamycin solution (Subheading 2.2.1) to 1 L LB medium, prepare 4 × 1 L medium in 2 L Erlenmeyer flasks.
5. Transfer 10 mL of the overnight culture to each flask containing 1 L LB medium, and continue culturing at 37 °C, 250 rpm for about 4.5 h until an OD_{600} of 0.6 is reached.
6. Add 1 mL of IPTG solution to each flask to initiate the induction process. Incubate for 16 h at 16 °C, shaking at 250 rpm.

7. Harvest the 4 L of culture by centrifugation at 7477 RCF for 15 min.
8. Resuspend the cells in 50 mL cell harvest solution. Add 50 μL of the PMSF solution.
9. Lyse the resuspended cell by sonication ($J = 50\%$) on ice, 30 s on, and 2 min off for each cycle, repeat six times.
10. Centrifuge to remove the cellular debris at 10,785 RCF for 2 h to give ClbN-containing protein suspension.
11. Filter the supernatant by syringe with a 0.2 mm membrane filter prior to HisTrap chromatography.
12. Purify ClbN using a 5 mL HisTrap HP column with 0 mM imidazole solution as medium A and 500 mM imidazole solution as medium B.
13. Load the ClbN-containing solution onto the column with a 0.5 mL/min flow rate.
14. Wash the column with 5% and 10% of medium B for 2 column volumes each, continuing with a 0.5 mL/min flow rate.
15. Elute fractions with 15%, 20%, 25%, 30%, and 50% medium B for 10 column volumes at a 0.5 mL/min flow rate.
16. Confirm the purified protein fraction by 15% SDS-PAGE gel analysis (*see* **step 2** in Subheading 2.2).
17. Desalt ClbN and buffer exchange by using spin desalting column into 1 mL exchange buffer (*see* **step 1** in Subheading 2.3).
18. Add 125 μM coenzyme A and 250 nM Sfp to the ClbN sample.
19. Incubate the reaction mixture at room temperature for 1 h.

3.3.2 Generation of the ClbN Intermediate In Vitro

1. Add 40 μL L-Asn, 50 μL ATP, 9 μL octanoyl-CoA, and 60 μL DMSO to 841 μL ClbN enzymatic solution.
2. Run the reaction for 3 h at room temperature.
3. Buffer exchange the mixture into 1 mL fresh exchange buffer by centrifugation at 3724 RCF for 1 h with an Amicon centrifugal unit.

3.3.3 Reaction Between the ClbN Intermediate and the Biotin-Cys Thioester Capture Agent

1. Add 0.76 g guanidine hydrochloride, and Biotin-Cys in 10:1 molar excess to ClbN and 1 mM TCEP.
2. Incubate the reaction at 4 °C for 16 h.

3.3.4 Purification of the Biotin-Cys-Captured ClbN Intermediate

Follow the same procedure as detailed in Subheading 3.2.5.

3.3.5 Analysis of the ClbN Intermediate-Capture Product by LCMS

1. Perform LCMS analysis with solvent A and solvent B. (Phenomenex column Prodigy 5 μm C18 150 Å, 150 × 4.6 mm).
2. Solvent conditions were as follows: 0 min, A-90% B-10%; 1 min, A-90% B-10%; 5 min, A-65% B-35%; 23 min, A-5% B-95%; 28 min, A-35% B-65%; 31 min, A-90% B-10%; 40 min, A-90 B-10%; flow rate 0.75 μL/min.

3.4 Demonstration of the Capture Strategy with the NRPS AziA3 (azinomycin Biosynthetic Pathway): Elucidation of an Unknown Intermediate

As substrates for the NRPS module is unknown, substrates are loaded onto the megasynthase in vivo.

3.4.1 Generation of the Genetic Knockout Plasmid pSETAziA3Δ*aziA6*

Construction of the Disruption Plasmid pKCAziA6

1. Amplify the upstream fragment and downstream fragment of *aziA6* from *Streptomyces sahachiroi* genomic DNA with Taq 2× Master mix and primers AziA6UF/AziA6UR and AziA6DF/AziA6DR.
2. Digest the upstream and downstream products with XbaI/HindIII and XbaI/BamHI.
3. Clone the sequence into the corresponding site within the pKC1139 plasmid to generate pKCAziA6.

Generation of the *S. sahachiroi* ΔaziA6 Disruption Mutant

1. Transform the pKCAziA6 plasmid into the chemically competent *E. coli* S-17 cells and grow on the LB agar plates overnight.
2. Pick one pKCAziA6/S17 colony from the agar plates and inoculate into 10 mL LB medium with 5 μL apramycin solution (*see* **step 2** in Subheading 2.3.3) and incubate at 37 °C overnight, shaking at 250 rpm.
3. Harvest the cells by centrifugation at 15,300 RCF for 1 min and wash the cell pellet twice with LB medium and resuspend in 600 μL of 2× YT medium.
4. Collect the *S. sahachiroi* spores from the 10 GYM agar plates using 2× YT medium.
5. Heat-shock the collected spores at 65 °C for 10 min and incubate at 37 °C for 3 h.
6. Collect the spores by centrifugation at 3724 RCF for 10 min and wash with 20 mL of 2× YT two times and subsequently resuspend in 4 mL of 2× YT.
7. Mix the recipient cells with 600 μL of pKCAziA6/S17 donor cells, and plate 800 μL of the mixture on ISP4 agar plates.
8. Incubate the agar plates at 28 °C for 24 h.

9. Overlay the plated cells with 1 mL 0.5 mg/mL nalidixic acid and 70 μg/mL apramycin solution to select for *S. sahachiroi* exconjugants.
10. Incubate the plates for another 10 days to allow for the appearance of apramycin-resistant exconjugants.
11. Exconjugants were screened for several generations until the double crossover allelic exchange was detected, as shown by the presence of apramycin sensitivity.

Construction of Plasmid pSET152_AziA3

1. Use the primers pSETAziA3F and pSETAziA3R to amplify *aziA3* from *S. sahachiroi* genomic DNA. The primers contained restriction sites (XbaI and BglII) for cloning as well as a polyhistidine tag.
2. Perform PCR with Phusion Master Mix polymerase.
3. Digest the PCR product and the pSET152 plasmid with XbaI and BglII and ligate using T4 DNA ligase.

Generation of *S. sahachiroi* pSETAziA3ΔaziA6

1. Transform the pSETAziA3 plasmid into the chemically competent *E. coli* S-17 cells and grow on the LB agar plates overnight at 37 °C.
2. Pick a pSETAziA3/S17 colony from the agar plates and inoculate into 10 mL LB medium, incubate the culture overnight at 37 °C, shaking at 250 rpm.
3. Collect the *S. sahachiroi ΔaziA6* spores from the 10 GYM agar plates using 2xYT medium.
4. Heat-shock the collected spores at 65 °C for 10 min and incubate at 37 °C for 3 h.
5. Collect the spores by centrifuge at 3724 RCF, wash with 20 mL of 2xYT two times, and resuspend in 4 mL of 2× YT.
6. Wash the S17 LB culture twice with LB medium and resuspend in 600 μL of 2× YT medium.
7. Mix the *S. sahachiroi ΔaziA6* recipient cells with 600 μL of pSETAziA3/S17 donor cells, and plate 800 μL of the mixture on ISP4 agar plates.
8. Incubate the agar plates at 28 °C for 24 h.
9. Overlay the plates with 1 mL 0.5 mg/mL nalidixic acid and 70 μg/mL apramycin solution to select for the *S. sahachiroi* pSETAziA3ΔaziA6 strains.
10. Incubate for another 10 days to allow for the appearance of apramycin resistant strains.

3.4.2 In Vitro Expression, Posttranslational Modification, and Purification of AziA3

1. Cut 1 × 1 cm agar from the *S. sahachiroi* pSETAziA3ΔaziA6 GYM plates, add to 100 mL 2× YT media in 250 mL baffled flasks.
2. Incubate the culture at 28 °C, shaking at 250 rpm for 24 h.
3. Add 500 μL of the apramycin solution to 1 L YEME medium, and prepare 4× 1 L medium in 2 L baffled flasks.
4. Transfer 10 mL of the overnight culture and inoculate into each of the four flasks containing 1 L YEME medium and continue to culture at 28 °C at 250 rpm for about 120 h.
5. Harvest 4 L of cells by centrifugation at 7477 RCF for 15 min.
6. Resuspend the cell pellet in 50 mL of lysis buffer.
7. Lyse the resuspended cells by sonication ($J = 50\%$) on ice, 30 sec on, 2 min off, and repeat for six cycles.
8. Centrifuge the suspension at 10,785 RCF for 2 h.
9. Filter the supernatant by passing it through a 0.22 mm membrane filter with a syringe prior to column purification.
10. Purify the protein with a HisTrap HP 5 mL column with 0 mM imidazole solution as medium A and 500 mM imidazole solution as medium B.
11. Load the protein sample onto the column with a peristaltic pump, 0.5 mL/min flow rate.
12. Wash the column with 92% medium A and 8% medium B for 10 column volume with 0.5 mL/min.
13. Elute the fraction with 20% and 50% medium B for 10 column volume with 0.5 mL/min.
14. Confirm the presence of the purified protein with a 15%-SDS-PAGE gel, as detailed in Subheading 2.2.2.
15. Concentrate the purified protein fraction using a 100 kDa centrifugal ultrafiltration unit (Amicon® Ultra centrifugal filter unit) to obtain AziA3, 20 mg/mL.

3.4.3 Reaction Between the AziA3 Intermediate and the Thioester Capture Agent Biotin-Cys

1. Add 0.76 g guanidine hydrochloride, and Biotin-Cys in 10:1 molar excess to AziA3 alongside 1 mM TCEP.
2. Incubate the reaction at 4 °C for 16 h.

3.4.4 Purification of Biotin-Cys-Captured AziA3 Intermediate

Follow the same procedure as detailed in Subheading 3.2.5.

3.4.5 Analysis of the AziA3 Intermediate-Capture Product by LCMS

1. Perform LCMS analysis with solvent A and solvent B (Phenomenex column Prodigy 5 μm C18 150 Å, 150 × 4.6 mm).
2. Solvent conditions were as follows: 0 min, A-90% B-10%; 1 min, A-90% B-10%; 5 min, A-65% B-35%; 23 min, A-5% B-95%; 28 min, A-35% B-65%; 31 min, A-90% B-10%; 40 min, A-90 B-10%; flow rate 0.75 μL/min.

4 Notes

1. LB medium should be made fresh prior to use.
2. Apramycin and nalidixic acid should be made fresh prior to use.
3. Ammonium persulfate solution and SDS solution should also be made fresh prior to use.
4. For additional information, refer to the streptavidin mutein matrix product information sheet.
5. ISP4 agar plates gives the highest conjugal transformation efficiency for the *S. sahachiroi* strain. Mannitol Soya flour agar plates can also be used for screening transformants.
6. Disruption mutants can be confirmed using PCR and sequencing or Southern Blot.
7. Products were initially identified in the LCMS by MS/MS fragmentation. The Biotin-cys backbone provided a characteristic fragmentation pattern to use as a filter for product identification.

Acknowledgments

This work was supported by the National Science Foundation (CHE-1904954) and the Welch Foundation (A-1828-20190330).

References

1. Mori S, Simkhada D, Zhang H et al (2016) Polyketide ring expansion mediated by a thioesterase, chain elongation and cyclization domain, in azinomycin biosynthesis: characterization of AziB and AziG. Biochemistry 55: 704–714
2. Brotherton CA, Balskus EP (2013) A prodrug resistance mechanism is involved in colibactin biosynthesis and cytotoxicity. J Am Chem Soc 135:3359–3362
3. Li Y, Yu HB, Zhang Y et al (2020) Pagoamide A, a cyclic depsipeptide isolated from a cultured marine chlorophyte, Derbesia sp., using MS/MS-based molecular networking. J Nat Prod 83:617–625
4. Paterson I, Lam NYS (2018) Challenges and discoveries in the total synthesis of complex polyketide natural products. J Antibiot (Tokyo) 71:215–233
5. Zhu HJ, Zhang B, Wang L et al (2021) Redox modifications in the biosynthesis of alchivemycin A enable the formation of its key pharmacophore. J Am Chem Soc 143:4751–4757
6. Wang G, Chen J, Zhu H et al (2017) One-pot enzymatic total synthesis of presteffimycinone, an early intermediate of the anthracycline

antibiotic steffimycin biosynthesis. Org Lett 19:540–543

7. Washburn LA, Nepal KK, Watanabe CMH (2021) A capture strategy for the identification of Thio-Templated metabolites. ACS Chem Biol 16:1737–1744
8. Du L, Lou L (2010) PKS and NRPS release mechanisms. Nat Prod Rep 27:255–278
9. Wu SC, Wang C, Hansen D, Wong SL (2017) Simple approach for preparation of affinity matrices: simultaneous purification and reversible immobilization of a streptavidin mutein to agarose matrix. Sci Rep 7:42849
10. Shepherd MD, Kharel MK, Bosserman MA et al (2010) Laboratory maintenance of Streptomyces species. Chapter 10, Unit–10E.1
11. Korn-Wendisch F, Kutzner HJ (1992) The family Streptomycetaceae. In: Balows A, Trüper HG, Dworkin M, Harder W, Schleifer K-H (eds) The prokaryotes. Springer-Verlag, New York, pp 921–995

Chapter 15

Chemical Labeling of Protein 4′-Phosphopantetheinylation in Surfactin-Producing Nonribosomal Peptide Synthetases

Fumihiro Ishikawa and Genzoh Tanabe

Abstract

4′-Phosphopantetheinylation is an essential posttranslational modification of the primary and secondary metabolic pathways in prokaryotes and eukaryotes. Several peptide-based natural products are biosynthesized by large, multifunctional enzymes known as nonribosomal peptide synthetases (NRPSs), responsible for producing virulence factors and many pharmaceuticals. The thiolation (T) domain serves as a covalent tether for substrates and intermediates in nonribosomal peptide biosynthesis and must be posttranslationally modified with a 4′-phosphopantetheinyl group. To detect 4′-phosphopantetheinylation of NRPS in bacterial proteomes, we developed a 5′-(vinylsulfonylaminodeoxy)adenosine scaffold with a clickable functionality, enabling effective chemical labeling of 4′-phosphopantethylated NRPSs. In this chapter, we describe the design and synthesis of an activity-based protein profiling probe and summarize our work toward developing a series of protocols for the labeling and visualization of 4′-phosphopantetheinylation of endogenous NRPSs in complex proteomes.

Key words Nonribosomal peptide synthetase (NRPS), Thiolation domain, 4′--phosphopantetheinylation, Adenylation domain, Activity-based protein profiling, *Bacillus subtilis* ATCC 21332, Surfactin-NRPSs

1 Introduction

Protein 4′-phosphopantetheinylation is an essential posttranslational modification (PTM) for cell viability in prokaryotes and eukaryotes [1]. PTM is mediated by a phosphopantetheinyl transferase (PPTase) with coenzyme A as a substrate [1]. In prokaryotes, PPTases posttranslationally modify modular synthases for chain elongation reactions in fatty acid synthetases, polyketide synthases, and nonribosomal peptide synthetases (NRPSs), which yield a large number of natural products with distinct structures and bioactivities [1].

NRPSs biosynthesize a wide ranging of bioactive peptide natural products, including the antibiotic daptomycin, antitumor bleomycin, and virulence factor pyoverdine. A typical NRPS module

Michael Burkart and Fumihiro Ishikawa (eds.), *Non-Ribosomal Peptide Biosynthesis and Engineering: Methods and Protocols*, Methods in Molecular Biology, vol. 2670, https://doi.org/10.1007/978-1-0716-3214-7_15,

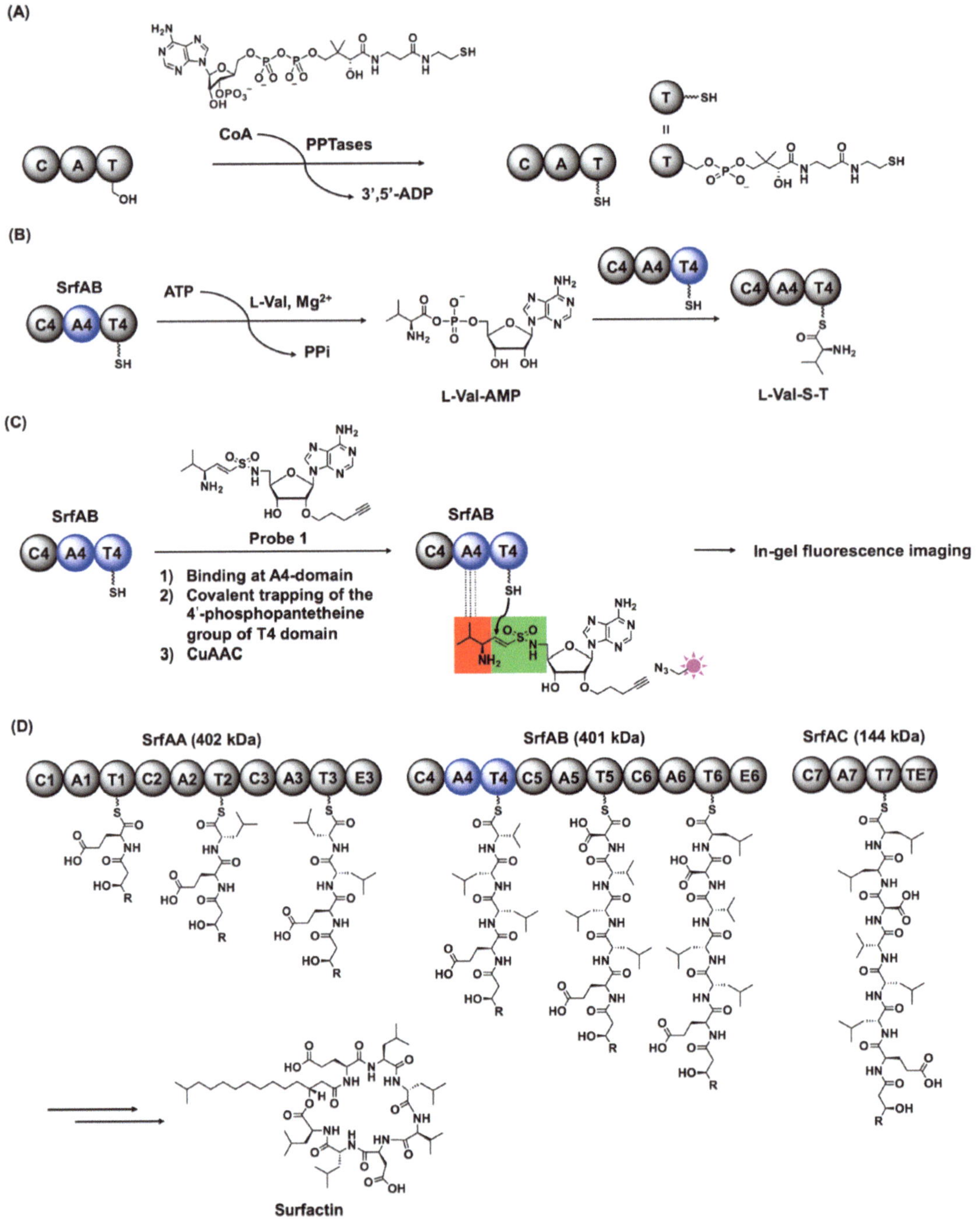

Fig. 1 (**a**) Posttranslational 4′-phosphopantetheinylation by a 4′-phosphopantetheinyltransferase (PPTase). Depicted by a typical NRPS module containing thiolation (T), adenylation (A), and condensation (C) domains. (**b**) Adenylation reaction and amino acid loading catalyzed by the A-domain. Depicted by module 4 of SrfAB-NRPS during surfactin biosynthesis. The NRPS module catalyzes the formation of L-Val-AMP by the A4-domain, which undergoes nucleophilic attack by the thiol group of the 4′-phosphopantetheine of the T4 domain, providing an L-Val-S-T species. (**c**) Structure of probe **1** and its labeling mechanism. Firstly, the A-domain of module 4 of SrfAB-NRPS recognizes probe **1**, which then accelerates the covalent trapping of the 4′--phosphopantetheine in the T4 domain of module 4 of SrfAB-NRPS via Michael addition of the

includes thiolation (T), adenylation (A), and condensation (C) domains [2]. The T domain functions as a covalent tether for amino acid substrates and the growing peptidyl intermediates. The T domain was posttranslationally modified with a 4′--phosphopantetheine functionality (Fig. 1a). This functionality is attached to a conserved Ser residue of the T domain by PPTases in an Mg^{2+}-dependent reaction involving coenzyme A (Fig. 1a) [1]. The A-domain functions as the gatekeeper domain of the NRPS machinery. This domain selects and activates an amino acid substrate for the corresponding aminoacyl adenylate intermediate at the expense of adenosine triphosphate (ATP) (Fig. 1b). The adenylated amino acid substrate is then transferred into the thiol group of the 4′-phosphopantetheine functionality of a downstream T domain, forming a thioester-bound aminoacyl-S-T domain (Fig. 1b). The C domain catalyzes peptide bond formation between two aminoacyl substrates bound to the T domains in the upstream and the downstream NRPS modules.

NRPSs are responsible for the biosynthesis of specific metabolites, which functions as virulence factors in pathogenic bacteria. Pyoverdine, the major siderophore produced by *Pseudomonas aeruginosa* is a good example. This is because the nonribosomal peptide pyoverdine is a key virulence determinant [3]. PPTases convert the inactive *apo*-synthases into the active *holo*-synthases. Accordingly, PPTases have received considerable interest as potential antibiotic targets [4]. Strategies for investigating 4′-phosphopantetheinylation of T domains in dynamic proteomic environments should facilitate insights into activity, transcriptional regulation, and PTM processes of NRPS as well as enable the assessment of PPTase activity inside bacterial cells.

We developed an activity-based protein profiling (ABPP) probe that can be used for labeling, visualizing, and tracking endogenous NRPSs in proteomes (Fig. 1c) [5]. The ABPP probe consists of a terminal alkyne for Cu(I)-catalyzed azide-alkyne [3 + 2] click chemistry (CuAAC) and a 5′-(vinylsulfonylaminodeoxy)adenosine amino acid for selective binding to the A-domains and capture the terminal thiol of the 4′-phosphopantetheine functionalities during the formation of thioester-bound aminoacyl-S-T domains (Fig. 1c) [6, 7]. Furthermore, we installed a clickable alkyne at the 2′-OH of the adenosine skeleton, which has less effect on the binding affinity of the A-domain of NRPSs (Fig. 1c) [8, 9]. To expand the

Fig. 1 (continued) vinylsulfonamide warhead. Finally, the samples are treated with 5/6-TAMRA-peg$_3$-azide under standard copper(I)-catalyzed azide-alkyne cycloaddition (CuAAC) conditions. (**d**) The nonribosomal peptide biosynthesis of surfactin. Modules comprise T (T1-T7), A (A1-A7) (A1, L-Glu; A2, L-Leu; A3, L-Leu; A4, L-Val; A5, L-Asp; A6, L-Leu, A7, L-Leu selective A-domains), C (C1-C7), E (E2 and E6), and TE (TE7) domains. Target A and T domains for probe **1** are colored blue. (Reproduced from Kasai et al. [5])

applicability of the ABPP probe, we have reported an advanced assay platform that enables visualization of the A-domain activities and protein–protein interactions of aryl acid adenylating enzymes [10]. We refer to excellent review articles that explain the basic technology of ABPP [11, 12].

In this chapter, we describe the method used for the synthesis of an ABPP probe, as well as a protocol for the ABPP of NRPSs for labeling, visualizing, and analyzing endogenous NRPSs from the lysate of the surfactin producer, *Bacillus subtilis* ATCC 21332, as a model bacterium (Fig. 1d).

2 Materials

2.1 Synthetic Procedure of 4′-Phosphopantetheinylation Labeling Reagent

All materials and reagents described below can be purchased from commercial suppliers and used without further purification:

1. Ethyl acetate.
2. Hexanes.
3. Chloroform.
4. Dichloromethane.
5. Tetrahydrofuran.
6. Acetonitrile.
7. *N*,*N*-Dimethylformamide.

2.2 Bacterial Culture

1. *B. subtilis* ATCC 21332.
2. Petri dishes.
3. Nutrient broth: Add 3.0 g of beef extract and 5.0 g of peptone to 1 L of Milli-Q water. The resulting mixture is sterilized at 121 °C for 20 min and stored at room temperature.
4. Nutrient agar: Add 3.0 g of beef extract, 5.0 g of peptone, and 15 g of agar to 1 L of Milli-Q water. The resulting mixture is sterilized at 121 °C for 20 min before being poured into Petri dishes and cooled to room temperature. The plates are stored at 4 °C.
5. 500 mM ammonium nitrate (NH_4NO_3) solution: Add 40 g of NH_4NO_3 to 1 L of Milli-Q water. The solution is sterilized at 121 °C for 20 min and stored at room temperature.
6. 300 mM disodium hydrogen phosphate (Na_2HPO_4) solution: Add 42.6 g of Na_2HPO_4 to 1 L of Milli-Q water. The solution is sterilized at 121 °C for 20 min and stored at room temperature.
7. 300 mM potassium dihydrogen phosphate (KH_2PO_4) solution: Add 40.8 g of KH_2PO_4 to 1 L of Milli-Q water. The solution is sterilized at 121 °C for 20 min and stored at room temperature.

8. 7 mM calcium chloride ($CaCl_2$) solution: Add 103 mg of calcium chloride dihydrate ($CaCl_2{\cdot}2H_2O$) to 100 mL of Milli-Q water. The solution is sterilized at 121 °C for 20 min and stored at room temperature.
9. 800 mM magnesium sulfate ($MgSO_4$) solution: Add 19.7 g of Magnesium sulfate heptahydrate ($MgSO_4{\cdot}7H_2O$) to 100 mL of Milli-Q water. The solution is sterilized at 121 °C for 20 min and stored at room temperature.
10. 8 mM ethylenediamine-*N,N,N′,N′*-tetraacetic acid (EDTA) and 8 mM iron(II) sulfate ($FeSO_4$) solution: Add 3.0 g of EDTA disodium salt dihydrate ($EDTA{\cdot}2Na_2{\cdot}2H_2O$) and 2.2 g of iron(II) sulfate heptahydrate ($FeSO_4{\cdot}7H_2O$) to 100 mL of Milli-Q water The solution is sterilized at 121 °C for 20 min and stored at 4 °C.
11. Glucose solution A: Add 40 g of D-(+)-glucose to 698 mL of Milli-Q water. The solution is sterilized at 121 °C for 20 min and stored at room temperature.
12. Glucose solution B: Add 10 g of D-(+)-glucose to 112 mL of Milli-Q water. The solution is sterilized at 121 °C for 20 min and stored at room temperature.
13. Iron-enriched minimal salt medium for seed culture: Mix 698 mL of glucose solution, 100 mL of 500 mM NH_4NO_3 solution, 100 mL of 300 mM Na_2HPO_4 solution, 100 mL of KH_2PO_4 solution, 1 mL of 7 mM $CaCl_2$ solution, 1 mL of 800 mM $MgSO_4$ solution, 500 μL of 8 mM EDTA, and 8 mM $FeSO_4$ solution on a clean bench. The medium is stored at room temperature.
14. Iron-enriched minimal salt medium for large-scale cultivation: Mix 112 mL of glucose solution, 25 mL of 500 mM NH_4NO_3 solution, 25 mL of 300 mM Na_2HPO_4 solution, 25 mL of KH_2PO_4 solution, 250 μL of 7 mM $CaCl_2$ solution, 250 μL of 800 mM $MgSO_4$ solution, and 62.5 mL of 8 mM EDTA and 8 mM $FeSO_4$ solution on a clean bench. The medium is stored at room temperature.
15. Phosphate-buffered saline (PBS) buffer: Add 9.6 g of Dulbecco's PBS to 1 L of Milli-Q water. The solution is sterilized at 121 °C for 20 min and stored at room temperature.
16. 80% (v/v) glycerol solution: Add 80 mL glycerol to 20 mL of Milli-Q water. The solution is sterilized at 121 °C for 20 min and stored it at room temperature.
17. Milli-Q water.
18. Screw cap tubes, 2 mL.
19. Culture tubes, glass.
20. Plastic tubes, 50 mL.

21. Eppendorf tubes, 1.5 mL.
22. Baffled flasks, 1 L.
23. Constant temperature incubator shaker.
24. Cell density meter.
25. High-speed refrigerated micro centrifuge.

2.3 Sample Preparation for the Labeling Studies

1. 20 mM Tris–HCl (pH 8.0): Add 3.15 g of tris (hydroxymethyl)aminomethane to 900 mL of Milli-Q water. Adjust pH to 8.0 with an appropriate aqueous hydrochloric acid (HCl) solution, and make up to 1 L with Milli-Q water. Store at room temperature.
2. Protease inhibitor cocktail, EDTA-free (100×): Dissolve the powdered product in a glass bottle with 1 mL of Milli-Q water. Aliquots are stored at −80 °C.
3. 1 M magnesium chloride ($MgCl_2$) solution: Add 10.2 g of $MgCl_2$ hexahydrate to 50 mL of Milli-Q water. The solution is stored at room temperature.
4. 1 M Tris(2-carboxyethyl)phosphine (TCEP) solution: Add 14.3 g of TCEP hydrochloride to 50 mL of Milli-Q water. Remove insoluble material using a syringe filter. Store aliquots at 4 °C.
5. 20 mM Tris–HCl (pH 8.0), 1 mM $MgCl_2$, 1 mM TCEP, and protease inhibitor cocktail: Add 10 μL of 1 M $MgCl_2$ and 10 μL of 1 M TCEP to 10 mL of 20 mM Tris–HCl (pH 8.0) and 100 μL of protease inhibitor cocktail (100×). To prevent the decomposition of the protease inhibitors, cool on ice prior to being used. Prepare the cocktail at the time of use.
6. 10 mg/mL lysozyme solution: Add 10 mg of lysozyme from egg white to 1 mL of 20 mM Tris–HCl (pH 8.0), 1 mM $MgCl_2$, and 1 mM TCEP. Prepare it at the time of use.
7. 2 mg/mL bovine serum albumin (BSA) solution for the Bradford protein assay: Add 2 mg of BSA to 1 mL of Milli-Q water. Store at 4 °C.
8. Coomassie Brilliant Blue (CBB) solution for the Bradford protein assay: Add 10 mL of protein assay CBB solution (5×) to 40 mL of Milli-Q water.
9. Milli-Q water.
10. Eppendorf tubes, 1.5 mL.
11. pH meter.
12. Constant temperature incubator.
13. High-speed refrigerated micro centrifuge.
14. Microplate reader.

2.4 Chemical Labeling of Protein 4′-Phosphopantetheinylation in SrfAB-NRPS

1. *B. subtilis* ATCC 21332 proteome.
2. 20 mM Tris–HCl (pH 8.0), 1 mM $MgCl_2$, and 1 mM TCEP: Add 50 μL of 1 M $MgCl_2$ and 50 μL of 1 M TCEP to 50 mL of 20 mM Tris–HCl (pH 8.0). Store at 4 °C.
3. Dimethyl sulfoxide (DMSO) for molecular biology research.
4. 10 mM DMSO solution of probe **1**: Add 4.9 mg of probe **1** to 1 mL of DMSO. Aliquots are stored at −30 °C.
5. 100 mM DMSO solution of L-Val-AMS (*see* Chapter 4).
6. 100 mM DMSO solution of L-Leu-AMS (*see* Chapter 4).
7. 100 mM DMSO solution of L-Asp-AMS (*see* Chapter 4).
8. 5 mM DMSO solution of 5/6-TAMRA-peg_3-azide: Add 1 mg of 5/6-TAMRA-peg_3-azide to 317 μL of DMSO. Aliquots are stored at −80 °C.
9. 5 mM DMSO solution of Tris[(1-benzyl-1H-1,2,3-triazol-4-yl)methyl]amine (TBTA): Add 2.7 mg of TBTA to 1 mL of DMSO. Aliquots are stored at −80 °C.
10. 50 mM TCEP solution: Add 14.3 mg of TCEP-HCl to 1 mL of Milli-Q water. Aliquots are stored at −80 °C.
11. 50 mM copper(II) sulfate ($CuSO_4$) solution: Add 12.5 mg of $CuSO_4{\cdot}5H_2O$ to 1 mL of Milli-Q water. Aliquots are stored it at −80 °C.
12. 0.25 M Tris–HCl (pH 6.8): Add 15.0 g of tris (hydroxymethyl)aminomethane to 900 mL of Milli-Q water. Adjust pH to 6.8 with an appropriate aqueous HCl solution and make up to 1 L with Milli-Q water. Store at room temperature.
13. 10%(w/v) sodium dodecyl sulfate (SDS) solution: Add 5 g of SDS to 50 mL of Milli-Q water. Store at room temperature.
14. 0.5 M dithiothreitol (DTT) solution: Add 771 mg of DTT to 10 mL of Milli-Q water. Aliquots are stored at −30 °C.
15. 0.1%(w/v) bromophenol blue (BPB) solution: Add 10 mg of BPB to 10 mL of Milli-Q water. Store at room temperature.
16. SDS-PAGE gel loading buffer (5×): 0.25 M Tris–HCl (pH 6.8) was mixed with 10% SDS, 0.5 M dithiothreitol (DTT), 50% glycerol, and 0.1% bromophenol blue (BPB). One aliquot of the resulting mixture can be stored at 4 °C until further use. The remaining aliquots are stored at −30 °C until required.
17. SDS-PAGE running buffer (10×): Add 30.2 g of tris (hydroxymethyl)aminomethane, 144 g of glycine, and 10 g of SDS to 1 L of Milli-Q water. Dilute the solution to 1× to run the gels. Store at room temperature.

18. CBB fixing solution: Add 400 mL of methanol and 150 mL of acetic acid to 1450 mL of Milli-Q water. Store at room temperature.
19. CBB staining solution: Add 100 g of ammonium sulfate, 1 g of Coomassie Brilliant Blue G-250, 30 mL phosphoric acid, and 200 mL of ethanol (99.5) to 1 L of Milli-Q water. Store it at room temperature.
20. 40%(w/v)-acrylamide/bis mixed solution (29:1).
21. Wide-range gel preparation buffer (4×) for PAGE.
22. 10% (w/v) ammonium persulfate solution.
23. SDS-PAGE gel (6%).
24. HiMark unstained protein standard.
25. ECL plex fluorescent rainbow markers.
26. Plastic container for SDS-PAGE.
27. Eppendorf tubes, 1.5 mL.
28. Tissue culture plate (96 well), flat bottom with low evaporation.
29. Typhoon laser-scanner platform.

3 Methods

3.1 Synthesis of Probe 1

Probe **1** was synthesized from adenosine, **2**, **6**, and 7 (Fig. 2). Compounds **2**, **6**, and 7 were synthesized as described in references [13–15], and, respectively.

3.1.1 Synthesis of 3

1. Dissolve adenosine in *N*,*N*-dimethylformamide under an atmosphere of nitrogen at room temperature.
2. Add 1.5 eq of sodium hydride (a 60%-suspension in mineral oil) and stir the solution at room temperature for 1 h.
3. Add 1.2 eq of compound **2** and stir the solution at room temperature for three days.
4. Quench the reaction by adding H_2O.
5. Evaporate the solution in vacuo.
6. Purify the residue by high-performance liquid chromatography (HPLC) (*see* **Note 1**).

3.1.2 Synthesis of 4

1. Dissolve compound **3** in dichloromethane under an atmosphere of nitrogen at room temperature.
2. Add 6.0 eq of *tert*-butyldimethylchlorosilane and 12 eq of imidazole, and stir the solution at room temperature for 19 h.

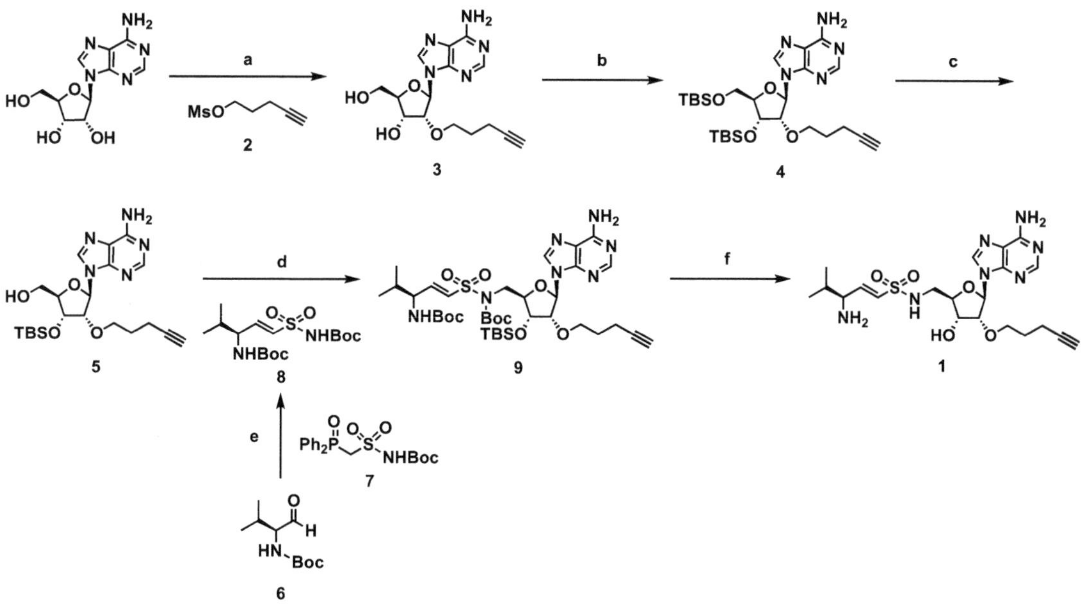

Fig. 2 Synthetic route to probe **1** [5]. *Reagents and conditions*: (**a**) NaH, **2**, DMF, 7.1%; (**b**) TBSCl, imidazole, CH_2Cl_2, room temperature, 81%; (**c**) trichloroacetic acid, THF, H_2O, room temperature, 75%; (**d**) **7**, DEAD, Ph_3P, THF, room temperature, 74%; (**e**) NaH, **7**, DMF, CH_2Cl_2, room temperature, 33%; (**f**) (1) TFA, CH_2Cl_2, 0 °C and (2) TBAF, THF, room temperature, 98%, over two steps. (Reproduced from Kasai et al. [5])

3. Dilute the solution with ethyl acetate and wash it with a 0.1 M aqueous hydrochloric acid, saturated aqueous sodium bicarbonate, and brine.
4. Dry the organic layer over anhydrous sodium sulfate.
5. Evaporate the solution in vacuo.
6. Purify the residue by flash chromatography (60:40 hexane/ ethyl acetate).

3.1.3 Synthesis of 5

1. Dissolve compound **4** in a 4:1 (v/v) mixture of tetrahydrofuran and H_2O at 0 °C.
2. Add 14 eq of trichloroacetic acid and stir the solution at 0 °C for 2 h.
3. Quench the reaction by adding a 1 M aqueous sodium hydroxide.
4. Dilute the solution with ethyl acetate and wash it with saturated aqueous sodium bicarbonate and brine.
5. Dry the organic layer over anhydrous sodium sulfate.
6. Evaporate the solution in vacuo.
7. Purify the residue by flash chromatography (17:83 hexane/ ethyl acetate).

3.1.4 Synthesis of 8

1. Dissolve 1.5 eq of compound 7 in *N,N*-dimethylformamide under an atmosphere of nitrogen at room temperature.
2. Add 2.5 eq of 1.3 M Lithium bis(trimethylsilyl)amide in tetrahydrofuran, and stir the solution at room temperature for 1 h.
3. Add compound **5**, and stir the solution at room temperature for 3 h.
4. Dilute the solution with ethyl acetate, and wash it with a 0.1 M aqueous hydrochloric acid and brine.
5. Dry the organic layer over anhydrous sodium sulfate.
6. Evaporate the solution in vacuo.
7. Purify the residue by flash chromatography (20:80 hexane/ethyl acetate).

3.1.5 Synthesis of 9

1. Dissolve compound **5** in *N,N*-dimethylformamide under an atmosphere of nitrogen at room temperature.
2. Add 1.0 eq of triphenylphosphine and 1.0 eq of a 2.2 M diethyl azodicarboxylate in toluene, and stir the solution at room temperature for 2 h.
4. Evaporate the solution in vacuo.
5. Purify the residue by flash chromatography (50:50 to 67:33 hexane/ethyl acetate and 99:1 to 97:3 chloroform/methanol).

3.1.6 Synthesis of Probe 1

1. Dissolve compound **9** in a mixture of trifluoroacetic acid and dichloromethane [50:50 (v/v)] at 0 °C.
2. Stir the solution at 0 °C for 12 h.
3. Evaporate the solution in vacuo.
4. Dissolve the residue in tetrahydrofuran under an atmosphere of nitrogen at room temperature.
5. Add 1.0 eq of a 1 M *tetra*-butylammonium fluoride in tetrahydrofuran, and stir the solution at room temperature for 10 h.
6. Evaporate the solution in vacuo.
7. Purify the residue by HPLC (*see* **Note 2**).

3.2 Bacterial Culture

This section describes the propagation and culture procedures used for the surfactin producer *B. subtilis* ATCC 21332.

3.2.1 Bacterial Propagation

1. Open vial containing culture lyophilizate of *B. subtilis* ATCC 21332.
2. Start with at least three culture tubes containing 5 mL of nutrient broth.
3. Remove 1 mL of nutrient broth using a 1 mL pipette and rehydrate the pellet.

4. Aseptically transfer this aliquot back into a culture tube and mix thoroughly.
5. Use several drops of this mixture to inoculate the remaining tubes of nutrient broth and nutrient agar plates.
6. Grow at 37 °C for 24 h.
7. Add 700 μL of liquid bacterial culture to each of 2 mL screw cap vials.
8. Add 100 μL of 80% steric glycerol to each vial, mix well, and store at −80 °C.

3.2.2 Bacterial Culture

1. Prepare nutrient agar plates (*see* Subheading 2.2).
2. Prepare an iron-enriched minimal salt medium for seed culture according to the literature procedure (*see* Subheading 2.2) [16, 17].
3. Streak *B. subtilis* ATCC 21332 for single colonies on nutrient agar plates and incubate the resulting plates at 37 °C for 24 h.
4. Pick a single colony.
5. Prepare a seed culture of 2–5 mL, iron-enriched minimal salt medium for seed culture in a culture tube and incubate the culture at 30 °C for 24 h on a reciprocating shaker at 200 rpm.
6. Inoculate 2 mL of the seed culture into 250 mL of an iron-enriched minimal salt medium for large-scale cultivation (*see* Subheading 2.2) in a 1 L baffled flask and grow at 30 °C for 19–22 h (optical density at 600 nm = 0.86–1.86) on a reciprocating shaker at 200 rpm.
7. Determine growth at specific time points by measuring the absorbance at 600 nm on a spectrophotometer.
8. Harvest the cells into 50 mL plastic tubes and centrifuge them for 15 min at 15,000 ×g at 4 °C.
9. Divide the cells into 1.5 mL Eppendorf tubes and centrifuge them for 10 min at 15,000 ×g at 4 °C.
10. Discard the supernatant and store the cell pellets at −80 °C until use.

3.3 Sample Preparation for Labeling Studies

This section describes the preparation of proteomes from *B. subtilis* ATCC 21332.

1. Resuspend the frozen cell pellets in 0.5−1 mL of 20 mM Tris–HCl (pH 8.0), 1 mM $MgCl_2$, 1 mM TCEP, and the protease inhibitor cocktail on ice.
2. Add 10−20 μL of 10 mg/mL lysozyme solution to the resuspended cells and gently shake the resulting mixture over ice for 30 min (*see* **Note 3**).

3. Transfer the mixture to a constant temperature incubator at 37 °C and incubate for 30 min with gentle shaking.
4. Centrifuge the mixture for 10 min at 15,000 ×g at 4 °C and collect the supernatant (*see* **Note 4**).
5. Determine protein concentration with Bradford assay in a 96-well plate using bovine serum albumin as a standard with an absorbance of 595 nm [18].

3.4 Chemical Labeling of Protein 4′-Phosphopantetheinylation in SrfAB-NRPS

This section describes the labeling of protein 4′--phosphopantetheinylation in endogenous SrfAB-NRPS in vitro (Fig. 1d and 3).

1. For labeling the endogenous SrfAB-NRPS, treat the resulting proteomes (45 μL, 2.0 mg/mL) with DMSO (0.5 μL) and probe **1** (0.5 μL of 10 mM stock in DMSO, final concentration: 100 μM) for 2–12 h at room temperature (*see* **Note 5**).
2. For inhibition studies for the labeling of an endogenous SrfAB-NRPS, L-Val-AMS (0.5 μL of 100 mM stock in DMSO, final

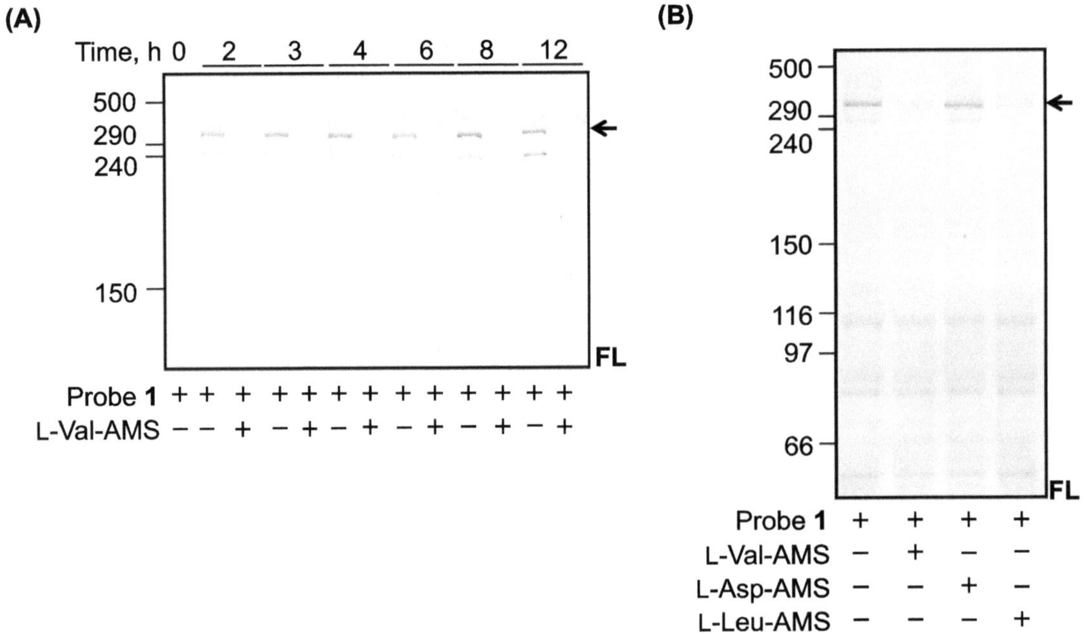

Fig. 3 Chemical labeling of protein 4′-phosphopantetheinylation using probe **1** [5]. (**a**) Chemical labeling of endogenous SrfAB-NRPS in a *Bacillus subtilis* ATCC 21332 proteome by probe **1**. The *B. subtilis* ATCC 21332 proteome (2.0 mg mL^{-1}) was treated with probe **1** (100 μM) for 2–12 h at 25 °C in either the absence or presence of L-Val-AMS (1 mM). (**b**) Selective labeling of the T4 domain of SrfAB-NRPS using a combination of probe **1** and inhibitors L-Val-AMS, L-Asp-AMS, and L-Leu-AMS. The *B. subtilis* ATCC 21332 proteome (2.0 mg mL^{-1}) was incubated with inhibitors L-Val-AMS, L-Asp-AMS, or L-Leu-AMS (10 μM) and reacted with probe **1** (100 μM) for 2 h at 25 °C. Allows point to the endogenous SrfAB-NRPS. The gel was visualized by in-gel fluorescence (FL). (Reproduced from Kasai et al. [5])

concentration: 1 mM) for 10 min at room temperature and incubate the mixture with probe **1** (0.5 μL of 10 mM stock in DMSO, final concentration: 100 μM) for 2–12 h at room temperature (*see* **Notes 5** and **6**).

3. For selective labeling of the T4 domain of an endogenous SrfAB-NRPS (Fig. 1d), L-Val-AMS, L-Asp-AMS, and L-Leu-AMS (0.5 μL of 1 mM stock in DMSO, final concentration: 10 μM) for 10 min at room temperature and incubate the mixture with probe **1** (0.5 μL of 10 mM stock in DMSO, final concentration: 100 μM) for 2 h at room temperature (*see* **Notes 5** and 7).
4. Take 40 μL of each sample for copper(I)-catalyzed alkyne-azide cycloaddition (CuAAC) click chemistry reaction.
5. Perform CuAAC click chemistry reaction by adding 4 μL of CuAAC click chemistry mix (1 μL of 50 mM $CuSO_4$, 1 μL of 5 mM TBTA, 1 μL of 50 mM TCEP, and 1 μL of 5 mM 5/6-TAMRA-peg$_3$-azide) to each sample (*see* **Note 8**).
6. Incubate each sample at room temperature in the dark for 1 h.
7. Take 40 μL of each sample for SDS-PAGE and in-gel fluorescence analysis.
8. Add 10 μL of reducing 5× SDS-loading buffer to each sample, mix thoroughly, and heat at 95 °C for 5 min.
9. Load 20 μL of each sample on an SDS-PAGE gel.
10. After electrophoresis, wash thoroughly with Milli-Q water before scanning with a Typhoon 9410 Gel and Blot Imager with 532 nm laser excitation and 580 nm emission using standard procedures.
11. Stain gels with CBB staining solution by standard procedures.

4 Notes

1. HPLC purification was performed under the following conditions: COSMOSIL 5C$_{18}$-ARII, φ 10 mm × 250 mm, methanol/aqueous trifluoroacetic acid (0.1%, 30:70), 8.0 mL/min, 210 nm, t_R: 34.0 min. Trifluoroacetic acid was removed in vacuo, and the water was removed by lyophilization to yield compound **3**.
2. HPLC purification was performed under the following conditions: Senshu Pak PEGASIL ODS SP 100 reverse-phase column, φ 20 mm × 250 mm, acetonitrile/aqueous trifluoroacetic acid (0.1%, 20:80), 8.0 mL/min, 210 nm, t_R: 14.5 min. Trifluoroacetic acid was removed in vacuo, and water was removed by lyophilization to give probe **1**.

3. Giant NRPS proteins are susceptible to mechanical cell disruption processes. Therefore, treat the bacterial cells gently to obtain the intracellular proteins [19].
4. If viscosity is observed in the bacterial cell lysates due to DNA content, add recombinant DNase I (RNase-free), incubate on ice for 5 min, centrifuge the mixture for 10 min 15,000 ×g at 4 °C, and collect the supernatant.
5. The final DMSO concentration is kept at 2.2% (v/v).
6. L-Val-AMS binds to the Val-activating domain (A4) of SrfAB-NRPS.
7. L-Asp-AMS and L-Lue-AMS bind to the Asp-activating domain (A5) and Leu-activating domain (A6) of SrfAB-NRPS, respectively.
8. $CuSO_4$, TBTA, TCEP, and 5/6-TAMRA-peg$_3$-azide were added at final concentrations of 1 mM, 100 μM, 1 mM, and 100 μM, respectively.

Acknowledgments

This work was partly supported by Grants-in-Aid for Research (C) (19 K05722 to F.I.) from JSPS and by grants from the Noda Institute for Scientific Research (F.I.), Research Foundation for Pharmaceutical Sciences (F.I.), Japan Foundation for Applied Enzymology (F.I.), and Institute Foundation for Science, Osaka (IFO) (F.I.). We are also thankful for the financial support provided by the Antiaging Project for Private Universities.

References

1. Beld J, Sonnenschein EC, Vickery CR et al (2014) The phosphopantetheinyl transferases: catalysis of a post-translational modification crucial for life. Nat Prod Rep 31:61–108
2. Hur GH, Vickery CR, Burkart MD (2012) Explorations of catalytic domains in non- ribosomal peptide synthetase enzymology. Nat Prod Rep 10:1074–1098
3. Minandri F, Imperi F, Frangipani E et al (2016) Role of iron systems in Pseudomonas aeruginosa virulenceand airway infection. Infect Immum 84:2324–2335
4. Leblanc C, Prudhomme T, Tabouret G et al (2012) 4′-Phosphopantetheunyl transferase PptT, a new drug target required for Mycobacterium tuberculosis growth and persistence in vivo. PLoS Pathog 8:e10003097
5. Kasai S, Ishikawa F, Suzuki T et al (2016) A chemical proteomic probe for detecting native carrier protein motifs in nonribosomal peptide synthetases. Chem Commun 52:14129–14132
6. Qiao C, Wilson DJ, Bennett EM et al (2007) A mechanism-based aryl carrier protein/thiolation domain affinity probe. J Am Chem Soc 129:6350–6351
7. Sundlov JA, Shi C, Wilson DJ et al (2012) Structural and functional investigation of the intermolecular interaction between NRPS adenylation and carrier protein domains. Chem Biol 19:188–198
8. Konno S, Ishikawa F, Suzuki T et al (2015) Active site-directed proteomic probes for adenylation domains in nonribosomal peptide synthetases. Chem Commun 51:2262–2265
9. Ishikawa F, Konno S, Suzuki T et al (2015) Profiling nonribosomal peptide synthetase activities using chemical proteomic probes for

adenylation domains. ACS Chem Biol 10: 1989–1997

10. Ishikawa F, Kasai S, Kakeya H et al (2017) Visualizing the adenylation activities and protein-protein interactions of aryl acid adenylating enzymes. Chembiochem 18:2199–2204
11. Cravatt BF, Wright AT, Kozarich JW (2008) Activity-based protein profiling: from enzyme chemistry to proteomic chemistry. Annu Rev Biochem 77:383–414
12. Sanman LE, Bogyo M (2014) Activity-based profiling of proteases. Annu Rev Biochem 83: 249–273
13. Rivero MR, Alonso I, Carretero JC (2004) Vinyl sulfoxides as stereochemical controllers in intermolecular Pauson-Khand reactions: applications to the enantioselective synthesis of natural cyclopentanoids. Chem Eur J 10: 5443–5459
14. Skiles JW, Miao C, Sorcek R et al (1992) Inhibition of human leukocytoelastase by *N*-substituted peptides containing α,α-difluorostatone residues at P_1. J Med Chem 35:4795–4808
15. Reuter DC, McIntosh JE, Guinn AC et al (2003) Synthesis of vinyl sulfonamides using the Horner reaction. Synthesis 15:2321–2324
16. Wei Y-H, Wang L-F et al (2004) Optimization iron supplement strategies for enhanced surfactin production with *Bacillus subtilis*. Biotechnol Prog 20:979–983
17. Yeh M-S, Wei Y-H, Chang JS (2005) Enhanced production of surfactin from *Bacillus subtilis* by addition of solid carriers. Biotechnol Prog 21: 1329–1334
18. Bradford MM (1976) A rapid and sensitive method for the quantitation of microgram quantities of protein utilizing the principle of protein-dye binding. Anal Biochem 72:248–254
19. Augenstein DC, Thrasher KD, Sinskey AJ et al (1974) Optimization in the recovery of a labile intracellular enzyme. Biotechnol Bioeng 16: 1433–1447

Part III

Bioinformatics Methods

Chapter 16

Norine: Bioinformatics Methods and Tools for the Characterization of Newly Discovered Nonribosomal Peptides

Areski Flissi, Matthieu Duban, Philippe Jacques, Valérie Leclère, and Maude Pupin

Abstract

In this chapter, we present Norine (https://norine.univ-lille.fr/norine), the unique resource dedicated to nonribosomal peptides. First, the content of the knowledgebase and the related tools are described. Then, a study case shows how to query Norine by annotations or structure and how to interpret the obtained results.

Key words Norine, Nonribosomal peptides, NRPS, Bioinformatics chemoinformatics

1 Introduction

Norine (https://norine.univ-lille.fr/norine) is the unique resource dedicated to computational biology analysis of nonribosomal peptides (NRPs) that represent a huge reservoir of microbial secondary metabolites with applications in human, animal, and environmental health. First released in 2006 [1], this resource is composed of a database together with tools useful for NRP analysis including visualization and structure comparison. A specificity of the database is that it can be queried according to both annotation and peptide structures. This project results from a University Lille (France) collaboration between computer science researchers from CRIStAL (*Centre de Recherche en Informatique et Automatique de Lille*) and microbiologists from ICV (*Charles Viollette Institute*) and BioEcoAgro crossborder joint research unit.

Michael Burkart and Fumihiro Ishikawa (eds.), *Non-Ribosomal Peptide Biosynthesis and Engineering: Methods and Protocols*, Methods in Molecular Biology, vol. 2670, https://doi.org/10.1007/978-1-0716-3214-7_16,

2 Materials

2.1 Norine

The name Norine was firstly attributed to the database part of the resource. Currently (May 2022), Norine contains about 1750 entries describing NRPs with various annotations classified in different tabs (Fig. 1):

- ***Peptide***: This tab gives general information about the NRP such as the Norine ID (e.g., NOR01984), the peptide name, family, synonyms, known biological activities or properties (antimicrobial, surfactant, antitumoral, siderophores, etc.), category (peptide, glycopeptide, lipopeptide, etc.), chemical formula, monoisotopic mass, and so on. Each validated NRP in Norine is associated with a DOI (Digital Object Identifier) like https://doi.org/10.26097/nor01984.
- ***Structure***: A specific notation, named monomeric structure, was designed to represent the complex structures of NRPs that can contain cycles and/or branches and that are composed of various monomers such as the 20 proteogenic amino acids, non-proteogenic amino acids like Ornithine (Orn), aminobutyric acid (Aib), and diaminobutyric acid (Dab) or more original building blocs as vancosamine (Van) and dihydroxybenzoic acid (diOh-Bz), chromophores and a variety of fatty acids. This notation includes the representation of the monomers composing an NRP and the chemical bonds between them. It allows to describe the structure of any NRP and determine its number of

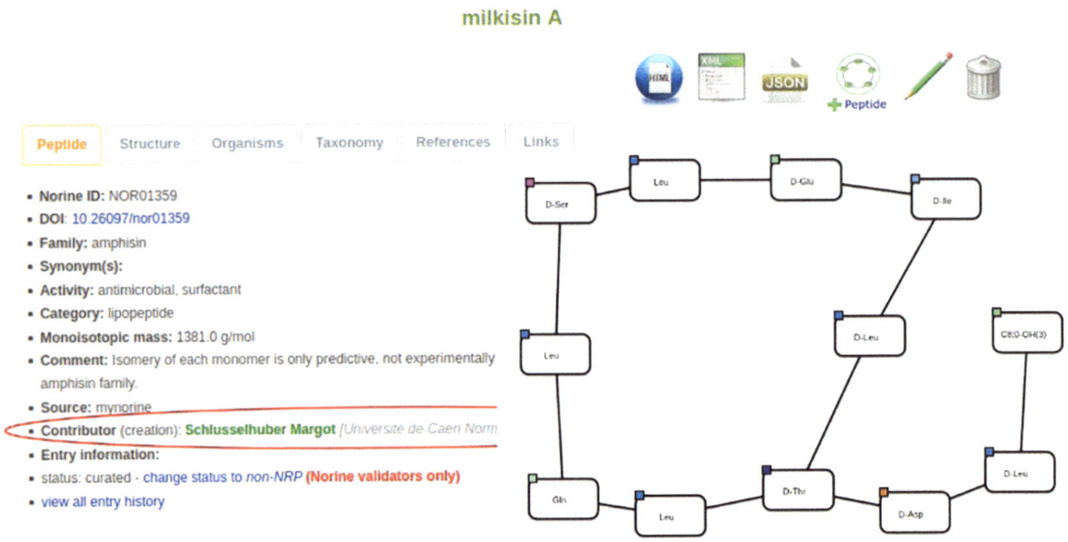

Fig. 1 An example of a peptide page corresponding to milkisin A
Contributors to the page are mentioned here highlighted surrounded in red

monomers and its structure type (linear, cyclic, branched, etc.). Three visualizations of NRPs are provided, the monomeric structure related to the biological mode of synthesis and two chemical (atomic) structures constructed by s2m and rBAN tools from the SMILES notation where the atoms of a monomer are in the same color to highlight the monomeric composition.

- ***Organisms*** that produce the NRP are represented in a taxonomic tree. The list comes from literature and is not exhaustive.
- ***References*** from which entry data were extracted (authors, title of the publication, DOI, and PubMed ID). Intentionally, not more than one or two references are proposed. They correspond to the articles describing for the first time the peptide with its structure and/or the synthetase, which proves it is indeed an NRP.
- ***Links*** to other resources in which the NRP is described (PubChem or ChEMBL for the atomic structure, UniProt or MiBIG for the synthetases, BIRD or PDB for 3D structure, etc.)

Norine database also contains annotations about all the monomers identified in at least one NRP including the full usual name, IUPAC names, or SMILES.

Since its creation, Norine's entries were manually extracted from scientific literature. Several articles are often exploited to collect comprehensive information for one peptide. To increase the amount of data while maintaining the quality, we have further developed a tool called MyNorine [2] to open the database to expert sourcing. Scientists all over the world can contribute to Norine by submitting new entries or suggesting modifications or additional annotations. All submissions are further manually validated by Norine curators. Currently, more than 50 scientists are registered in Norine for their contribution and they are cited on the corresponding peptide page as illustrated in the Fig. 1 (contributor item).

In 2019, we have developed a new pipeline [3] for Norine whose purpose is to semiautomatically import massive data (Fig. 2). External databases are mined to extract NRPs that are not yet in Norine or to complete the annotations. We ensure that high quality of the Norine data are maintained using filters to validate the new annotations, and the corresponding NRPs are tagged with an *unreviewed* status. This pipeline led to the introduction of about 540 new NRPs.

To increase interoperability with other databases, Norine provides a free web API to access the data in various formats (HTML, JSON, XML, and CSV). This is implemented with the RESTful architectural style [4]. The Norine REST services URIs use the following general syntax: https://.../norine/rest/<path>/

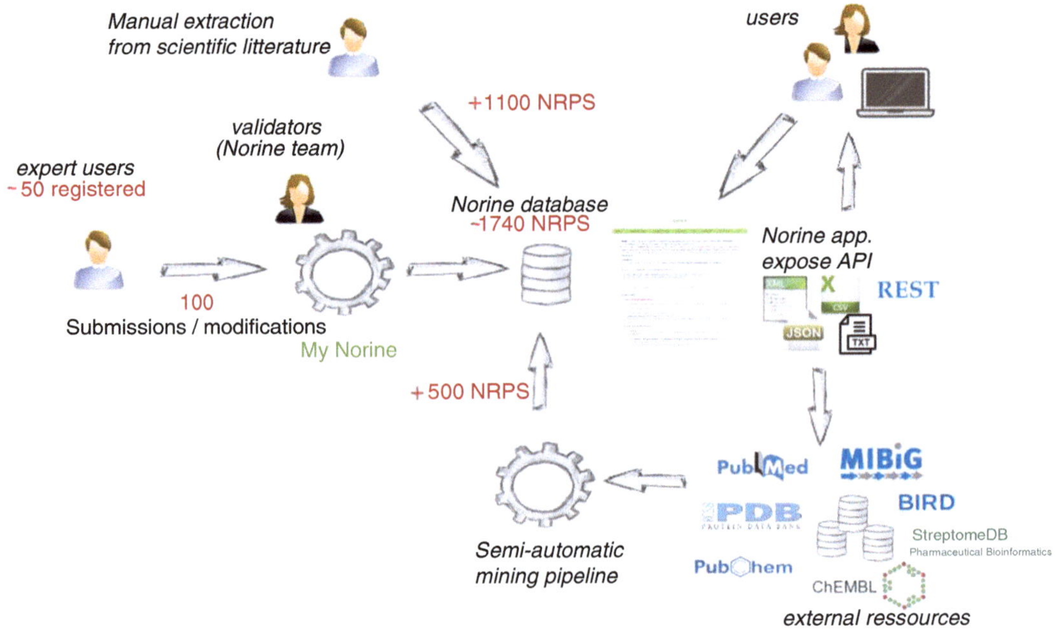

Fig. 2 Various processes to enrich Norine database

<format>/[parameter]. The *path* argument determines the type of service, format can be XML or JSON, and *parameter* corresponds to the user query.

For instance, anyone can access all the annotations of a unique NRP with an URL like https://norine.univ-lille.fr/norine/rest/id/json/NOR1984 (JSON) or get all the NRPs with "cyclosporin" in their name https://norine.univ-lille.fr/norine/rest/name/xml/cyclosporin (XML).

Thus, thanks to these results, Norine has been selected on October 2019 by the European bioinformatics ELIXIR consortium (https://elixir-europe.org) as an important resource for the scientific community in life science domain (*Service Delivery Plan*) and is referenced in bio.tools, the registry of software metadata and description (https://bio.tools/NORINE).

2.2 Smiles2-Monomers

Smiles2Monomers (s2m) [5] is a software to automatically infer monomeric structures from chemical structures represented by SMILES notation, mainly extracted from PubChem or PDB databases. s2m maps Norine monomers on the given atomic structure determine the best tiling to cover all the atoms of the NRP polymer. It outputs a graphical atomic structure, where atoms of a monomer are represented in the same color so that one color is attributed to each monomer. This tool has allowed to complete or correct about 570 SMILES or monomeric structure of NRPs in Norine.

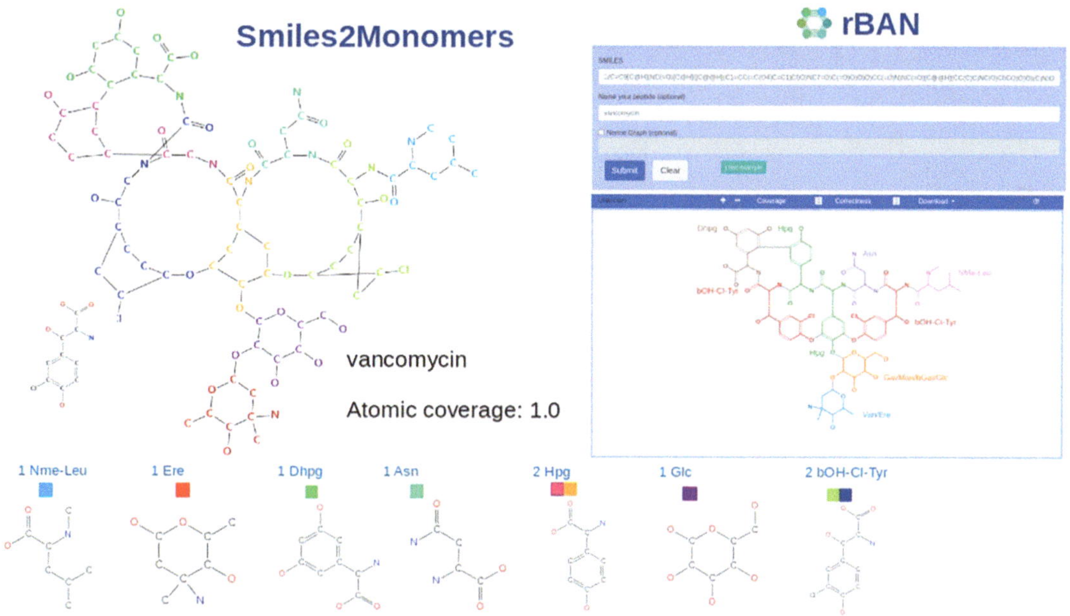

Fig. 3 Smiles2Monomers and rBAN results for Vancomycin

2.3 rBAN

rBAN (*retroBiosynthetic Analysis of Nonribosomal peptides*) software [6] has been integrated into Norine in 2019. As s2m, rBAN infers the monomeric structure of an NRP from a SMILES (Fig. 3). However, the strategy is different as, to simulate the retro-biosynthesis of NRPs, rBAN cuts polymers between two atoms identified as parts of bonds between two consecutive monomers such as a peptide bond or a disulfide bridge. Resulting fragments are then matched to Norine's monomers. When a monomer cannot be matched to Norine's monomers, missing substructures are searched in PubChem so as to suggest potential new ones. Figure 3 shows an example of execution of rBAN (it displays a directed graph format highlighting the bond types between monomers) and s2m for vancomycin antibiotic.

2.4 MyNorine

As stated above, the MyNorine [2] tool was developed to open the Norine database to experts in NRPs, and especially the biologists and biochemists involved in the discovery of new metabolites. Indeed, they are considered as the best experts to enrich the database and improve it in terms of quality and quantity. New data are carefully validated by Norine team before they are added in Norine. In other words, MyNorine distinguishes two roles: curators and validators. While curators contribute to the database by submitting new NRPs/annotations or suggesting modifications, validators are responsible of validation steps. Contributors firstly create an account and can use different sophisticated dynamic interfaces to enrich the database, as illustrated in Fig. 4.

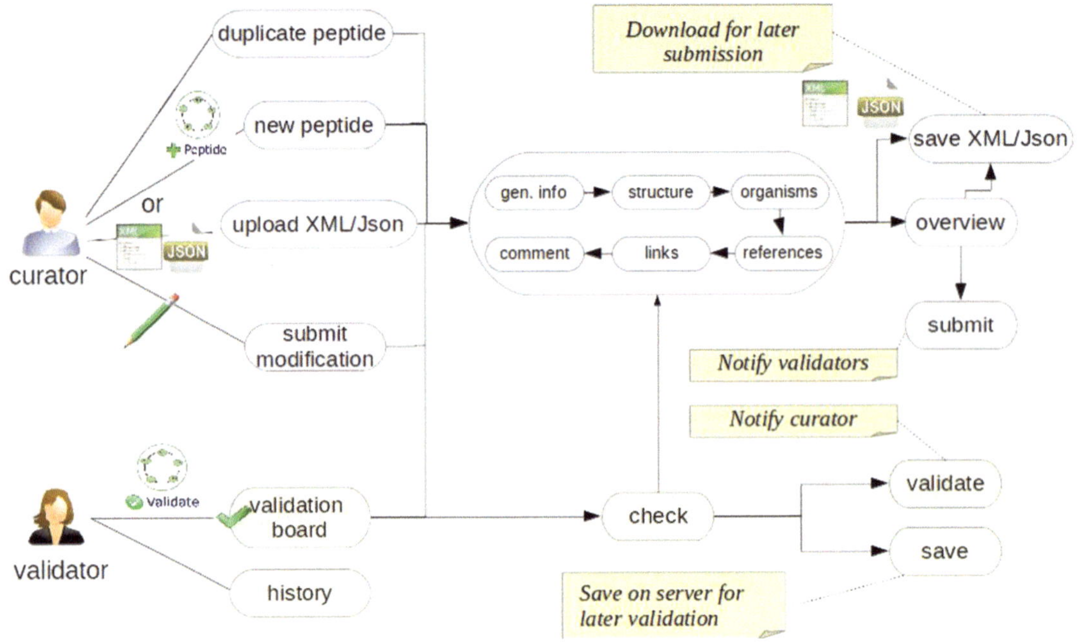

Fig. 4 Overview of MyNorine: roles and use cases

3 Methods

3.1 Querying the Norine Database

The Norine database can be queried based on two strategies: the first one is a classical querying using annotations through *annotation search* tab, while the second one is based on peptide structural features available through *structure search* tab.

3.1.1 How to Query the Database with Annotations?

The *annotation search* tab contains about 20 fields available to build the queries including the taxonomy related to the producing organisms, literature references, and the peptide itself such as its name, its category (peptide, lipopeptide, chromopeptide, etc.), the presence of monomers or their derivatives, the peptide structure type (linear, cyclic, partial cyclic, branched, etc.), or biological activities. Autocomplete helps to enter existing values. While a user starts filling a field, suggestions of existing values are dynamically proposed. Complex queries can be created by combining several fields with three operators "AND," "OR," and "AND NOT". As an example, if someone is interested in all curated peptides annotated with a SMILES, containing more than 10 monomers, introduced by the Norine team, excluding peptaibols and chromopeptides, the combined criteria will return a list of 68 peptides gathered into 25 families (Fig. 5).

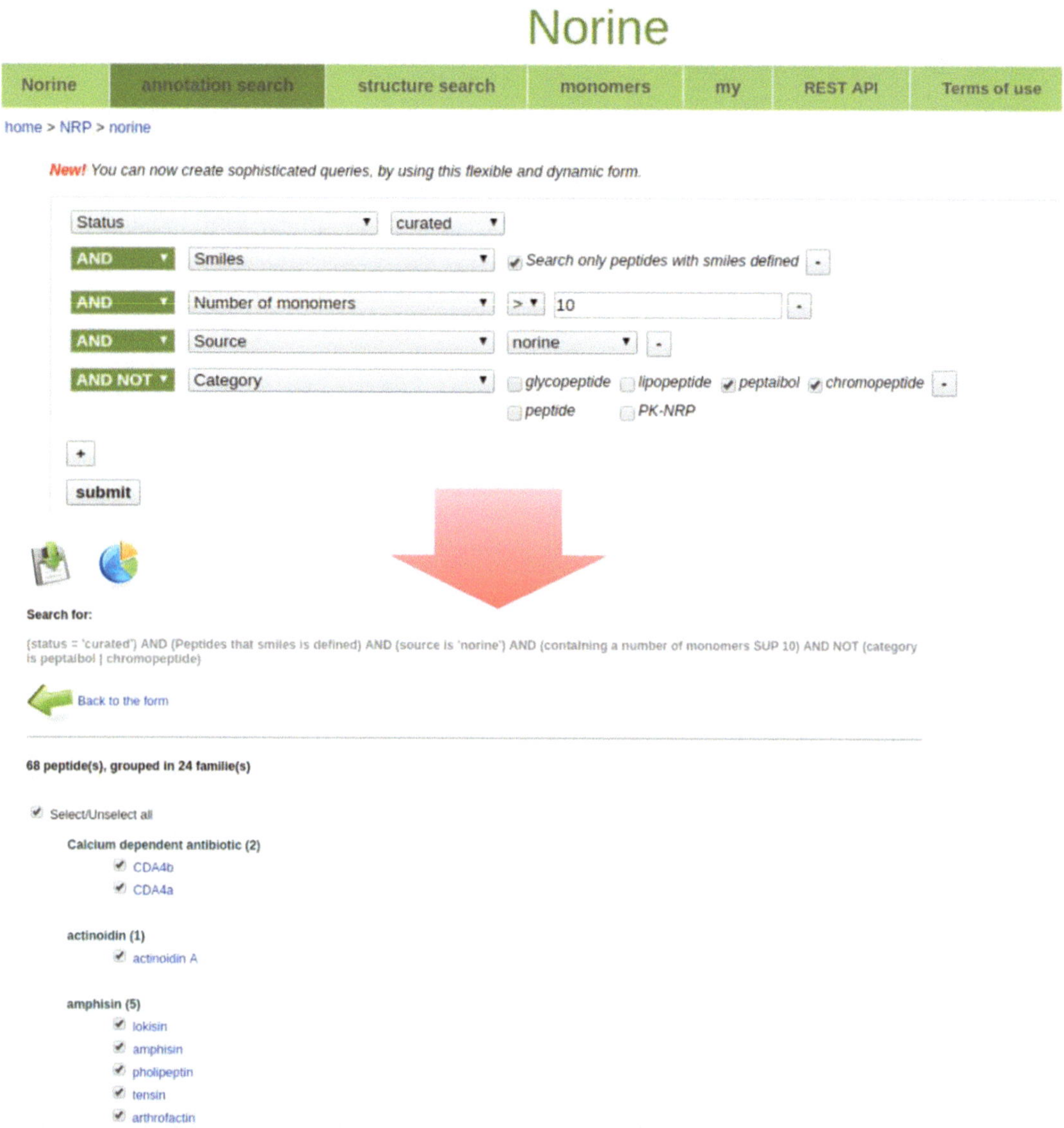

Fig. 5 Results returned after query using combined criteria

In the returned list, clicking on a peptide name opens the corresponding peptide page where all annotations are accessible and can be downloaded in XML or JSON format.

The annotations of all the peptides obtained by a query can be downloaded by clicking on the corresponding icon above the list. A form allows to select the exported fields, and the file format among XML, CSV, or HTML (Fig. 6).

Another icon that gives access to the distribution of the NRPs list are presented as pie charts or histograms to facilitate the overview of categories, structure type, or activities (Fig. 7). A subset of peptides corresponding to a specific criterion can be accessed by clicking on an area of interest within a graphic.

Fig. 6 Process to download and select annotations to export in various formats

3.1.2 Structure Search

The *structure search* tab has been created because peptides have to be considered as identical if they have the same structure (monomer composition and the same sequence including monomer isomery, branches, and cycles), whatever the peptide name or producing organisms. It is also very useful to identify if a newly characterized nonribosomal peptide is really new or is a new variant belonging to

Fig. 7 Example of graphical views obtained from results of the query

a known family. To facilitate the queries, an editor, that provides a column on the left side where the monomers are presented in a clustered list, has been developped. A structure can be easily drawn by clicking on a monomer and drag and drop it in the space available on the right side. Alternatively, a set of derivatives can also be chosen by clicking on a node of the tree.

3.2 Submission of a New Peptide

With the MyNorine tool, experts can submit new NRP and add or suggest modifications about existing peptides. The purpose of this module of Norine is to enhance the database by increasing the quantity of described NRPs and improve quality of annotations. Moreover, contributors are cited as the authors of the submission/modification. Users just have to create an account and can access to the tool to contribute by filling forms, as illustrated in Fig. 8.

3.3 Study Case

For this use case, we created a Fictive Lipopeptide containing 6 Monomers within the peptide moiety, so-called FL6M displaying the following structure: C12:0_D-OH-Asp_Dab_D-Ser_Leu_-Ser_Orn with a putative ring between D-Ser and Orn (where C12:0 is the fatty acid containing 12 carbons, Dab is a 2,4-diaminobutyric acid, OH-Asp is a hydroxylated aspartic acid, Orn is for ornithine, and D- specifies the D-isomery). The workflow presented below aims to identify if this peptide is original and to define its place among the NRP diversity. All the numerical results presented are those obtained after querying on May 2022.

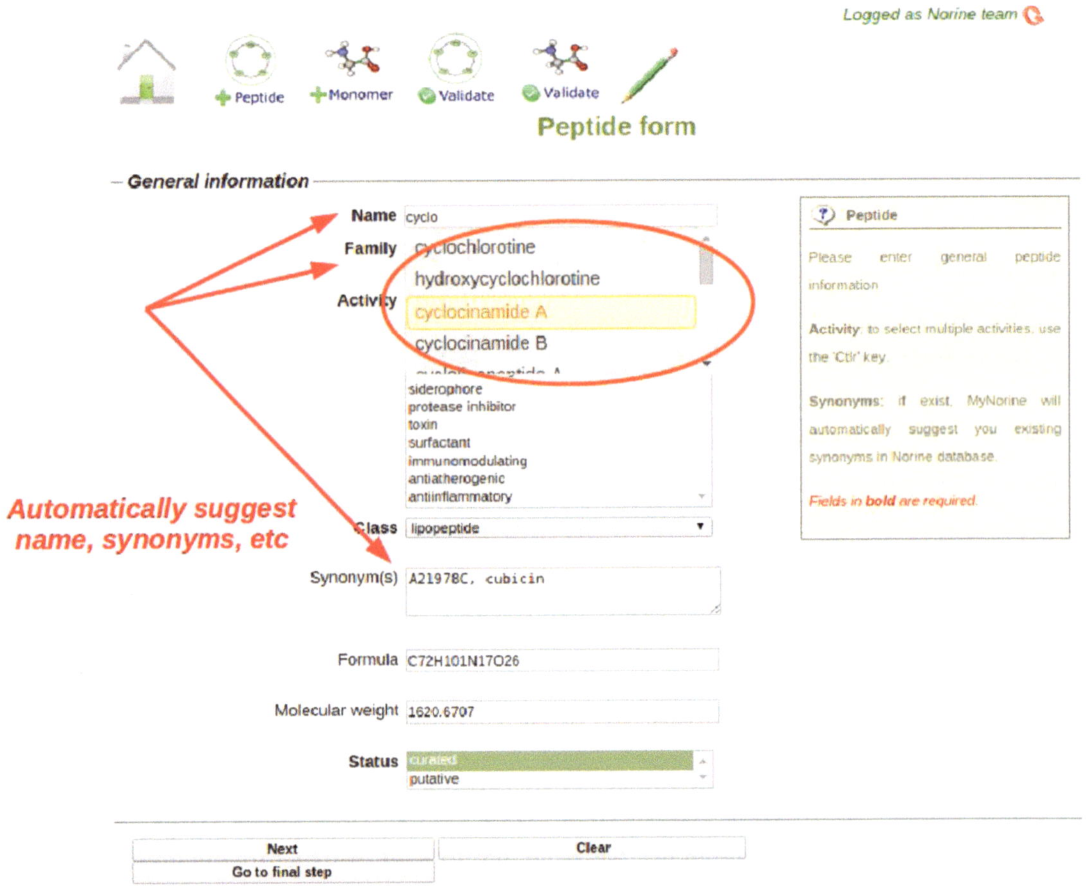

Fig. 8 Form filling to add a new peptide to Norine

3.3.1 Overview on NRPs Sharing Traits with FL6M

The first query is for the name and, as expected, no NRP named FL6M is already annotated in the Norine database. Then, the procedure described below aims to collect a maximum of information on Norine lipopeptides to check if lipopeptides with the same size (containing the same number of monomers) or sharing common specific monomers are described, in order to identify how similar with FL6M they are. For this purpose, it is possible to query the database following different pathways. They are finally leading to same results but differs in the intermediate information available. As illustrated in Fig. 9a, the query can start by searching for lipopeptides that are 314 gathered into 37 families. Clicking on graphical view icon directs toward various distributions. In the size distribution bar chart, it is easy to be informed that 20 peptides include 7 monomers (6 amino acids and 1 fatty acid) and by clicking on the corresponding bar, the list of these peptides is returned. Another way to query the database is to start from the number of monomers (Fig. 9b), 7 in this example. The database contains 176 peptides containing exactly 7 monomers, grouped

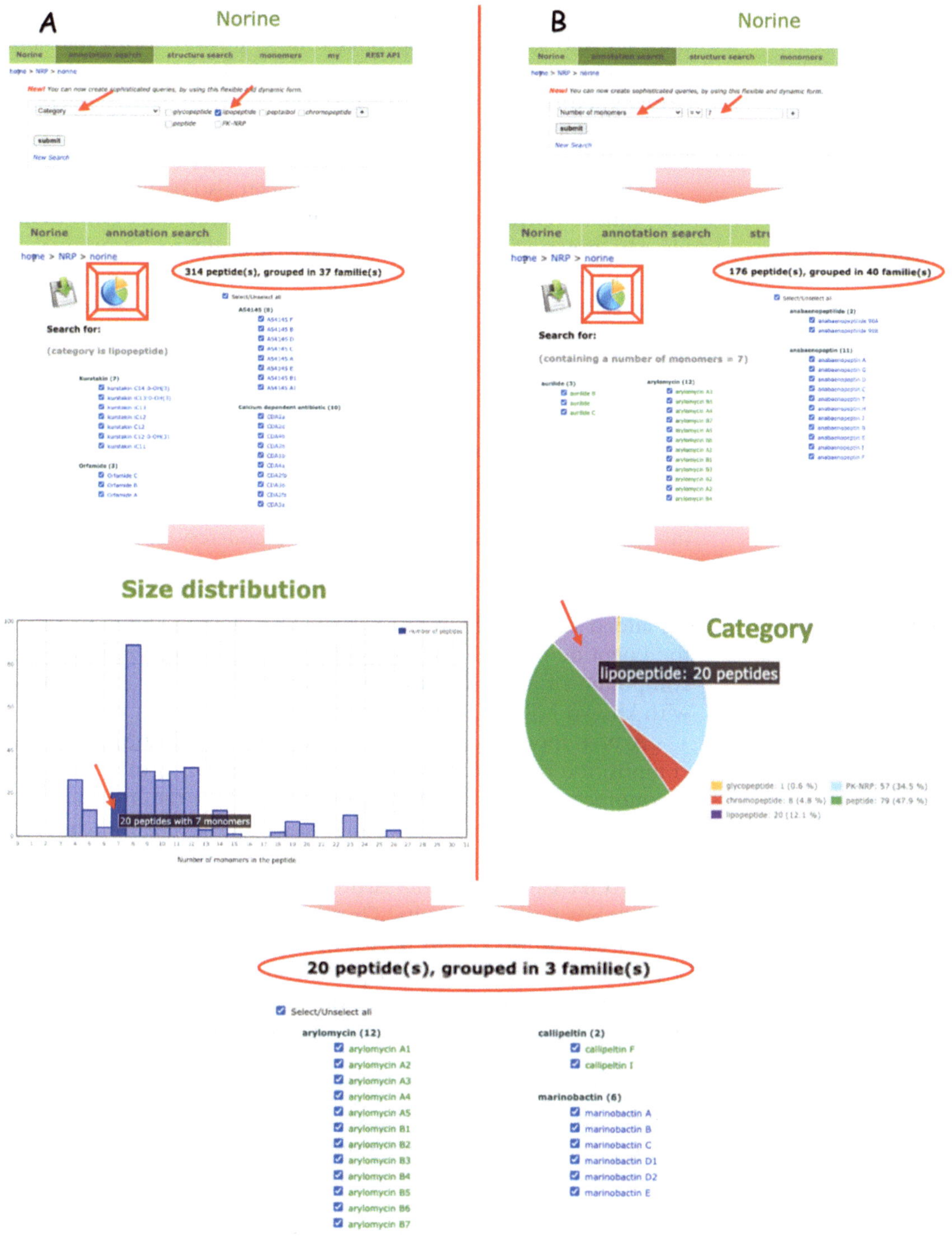

Fig. 9 Overview on NRPs sharing annotation with FL6M

into 40 families. The pie chart on categories indicates that among the 176 NRPs, 20 are lipopeptides. Both ways arrive to the same conclusion that 20 lipopeptides constituted of 7 monomers are stored in Norine. However, the first way allows to have an overview on lipopeptides, while the other way gives an overview on NRPs containing 7 monomers.

The presence of specific monomers can also be used as a starting point to retrieve information from Norine. In the FL6M example, the presence of a hydroxy-aspartate seems to be original. A search based on the presence of aspartate and its derivatives can be used as a query filter. Through graphical view tool, it is possible to see that among the 272 NRPs containing at least one aspartate derivative, 67 are constituted of 7 monomers. Among them, 6 are lipopeptides, all belonging to the same marinobactin family (Fig. 10a). As shown on Fig. 10b, the list of the 6 marinobactins can be directly obtained following a combined query with "asp derivatives" AND number of monomers =7 AND category lipopeptides. It is probably more direct but does not provide some intermediate information obtained as a courtesy of the graphical view tool.

3.3.2 Structure Comparison with NRPs Stored in Norine

An added value of the Norine database is the possibility to search for a peptide through its structure. This allows to determine if the compound of interest shares its structure, or part of its structure, with known NRPs.

A first approach can be the comparison of FL6M monomer composition to all Norine NRPs. This can be performed using the "monomer composition fingerprint search." In the graphical editor, drag and drop each of the 7 monomers from the clustered monomer list to the drawing panel. As shown in Fig. 11, the search for "D-OHAsp" instead of "Asp*" (meaning Asp derivatives) outputs slightly different results. This can be explained by the presence of OHAsp instead of D-OHAsp in the corrugatin, together with 3 Dab residues in this lipopeptide constituted of 9 monomers. Looking at the structures of both lipopeptides clearly indicate that they are not really similar (7 vs. 9 monomers, not the same sequence).

To further compare the structures, the "structure-based search" tries to map the given structural pattern to Norine NRPs. The graphical editor helps to design the pattern allowing to retrieve the monomers and bond them together. Figure 12 illustrates the different results obtained while searching for peptides containing the complete structural pattern of FL6M or pattern substructures with at least 2, 3, or 4 monomers.

To conclude if FL6M is a novel lipopeptide or belongs to a known family (new variant), it is necessary to compare carefully the structures. In the presented use case, while a score of 0.556 with

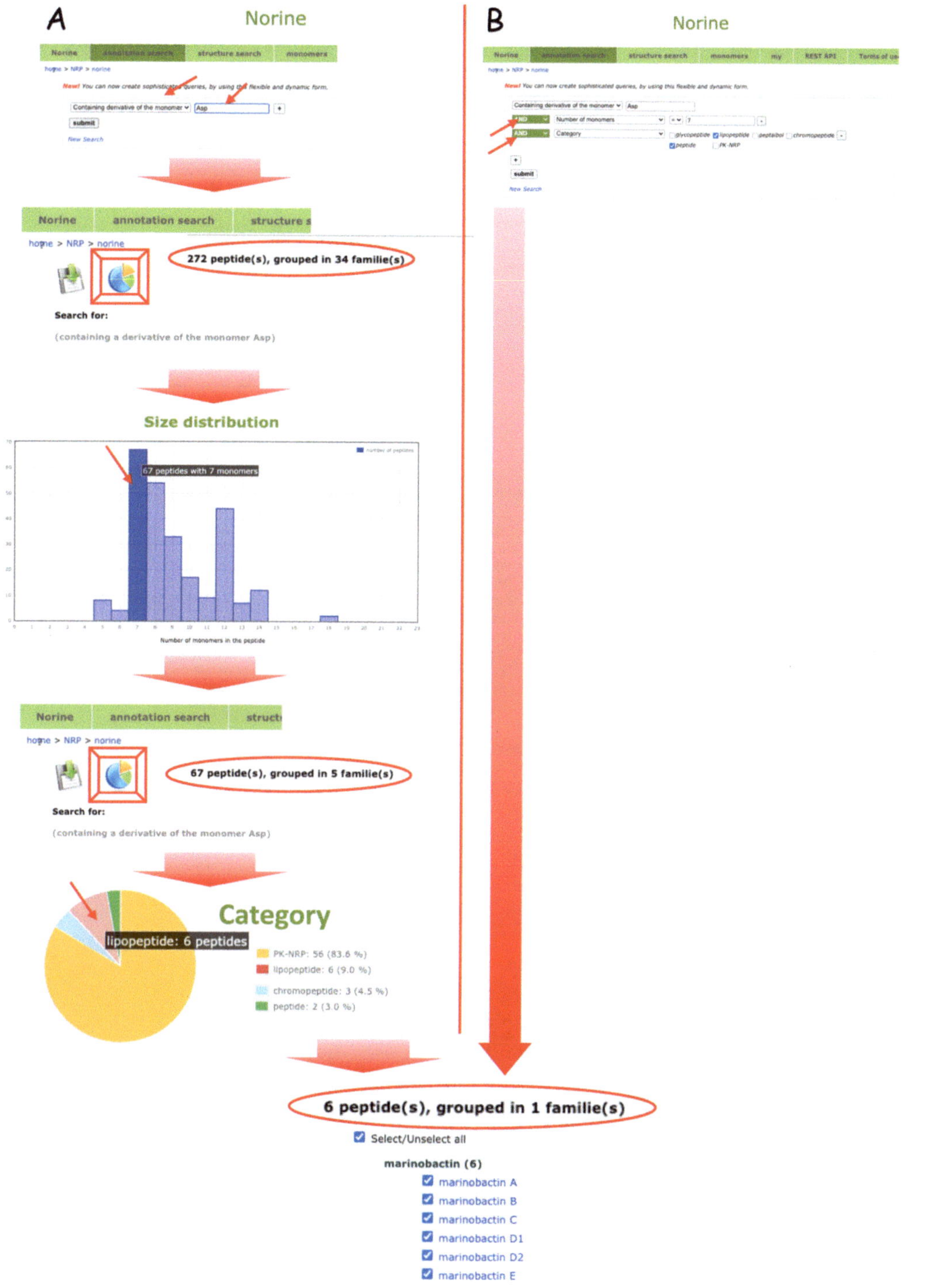

Fig. 10 Overview on NRPs sharing monomers with FL6M

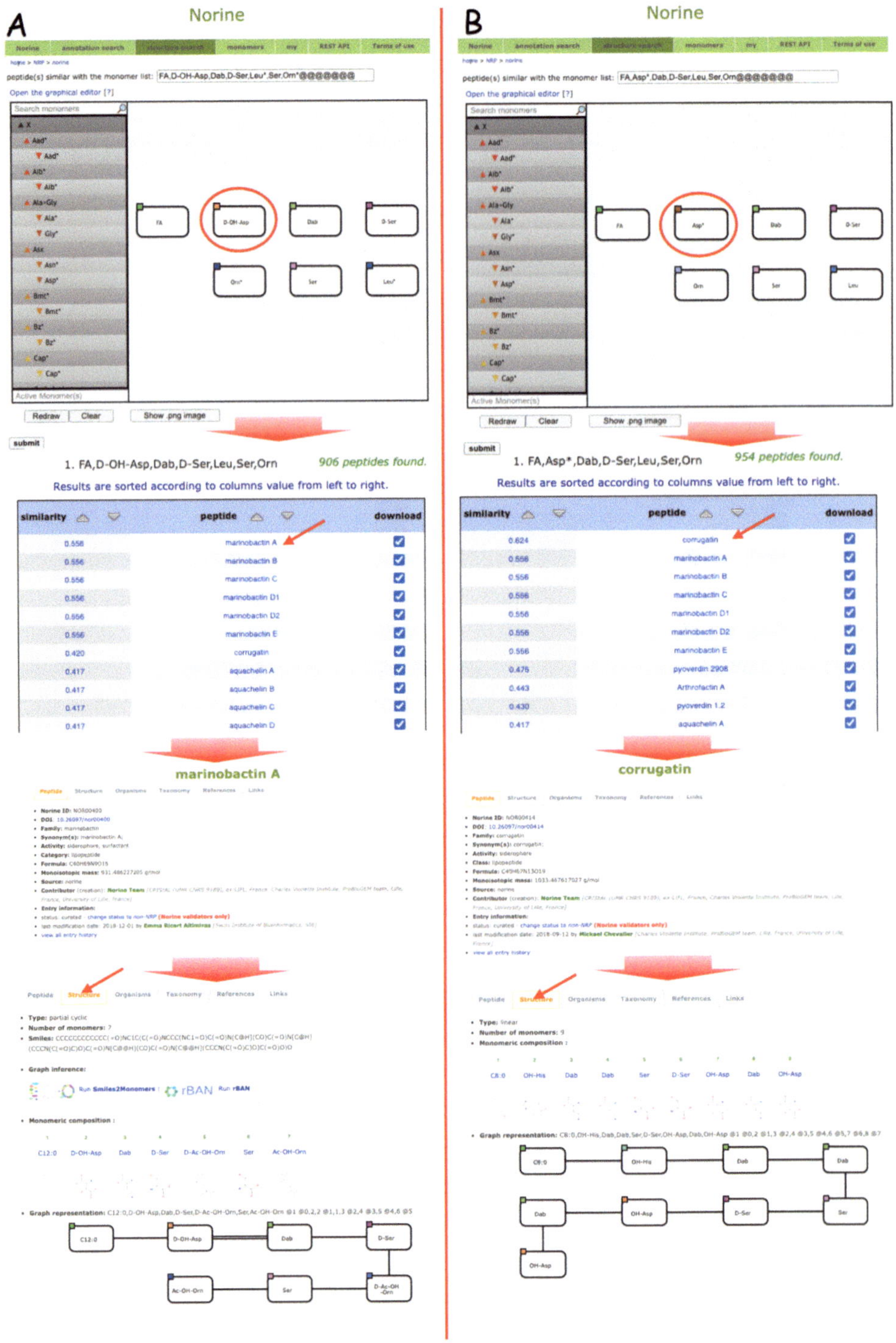

Fig. 11 Comparison of results obtain after D-OH Asp or Asp derivative query

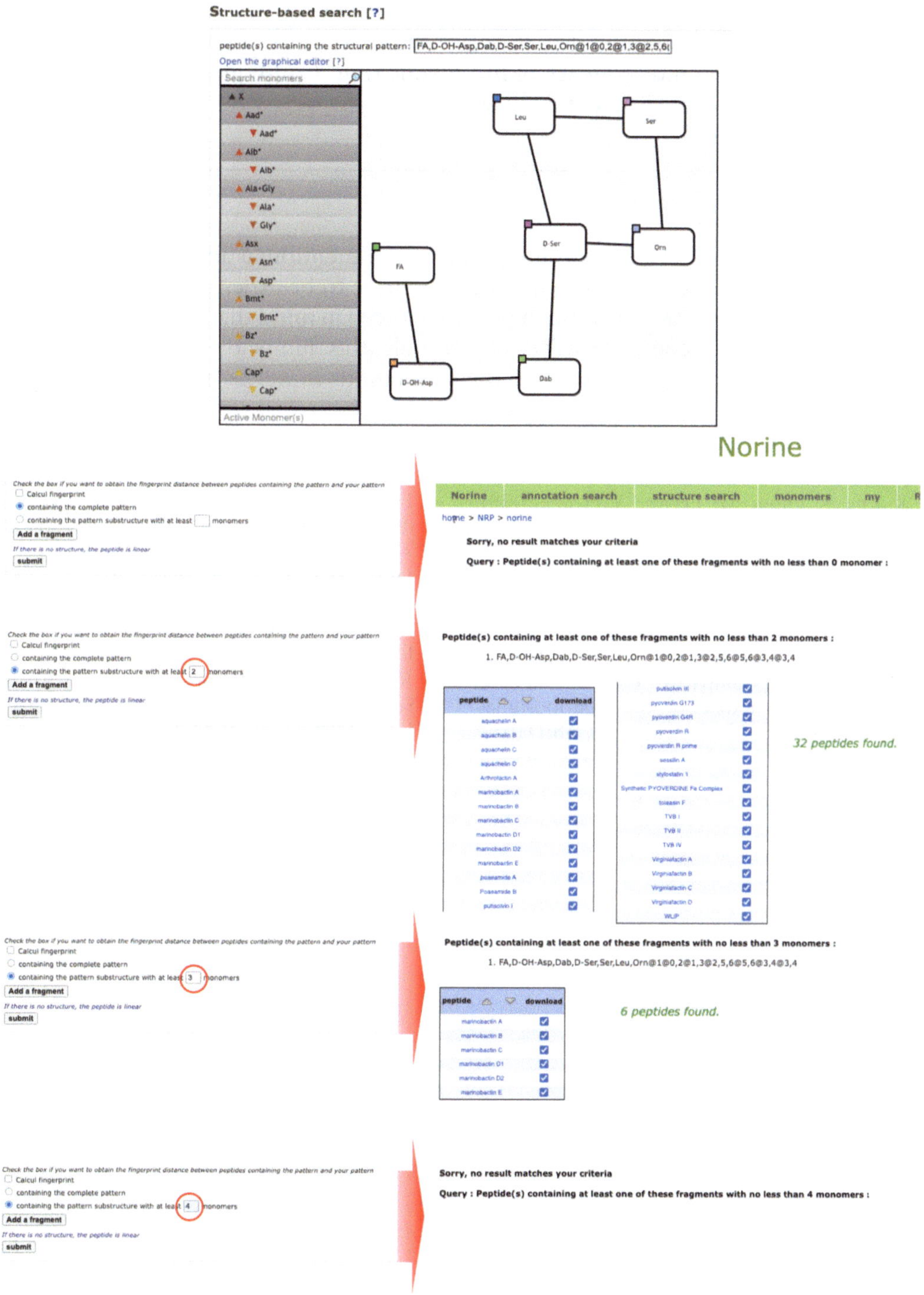

Fig. 12 Search for peptides containing complete structural pattern or pattern substructures

marinobactin A is obtained, a first glance at the structure clearly indicates that they are different, especially regarding the position and sequence of the lactone ring. Consequently, they likely belong to different families.

4 Notes

A same warning point can be mentioned about the assignment of putative activities that will have to be experimentally verified in any case. It is also important to keep in mind that Norine database is not comprehensive, accordingly this limits the completeness of the results related to the annotations as well as to the structure.

References

1. Caboche S, Pupin M, Leclère V et al (2008) NORINE: a database of nonribosomal peptides. Nucl Acids Res 36:D326–D331
2. Flissi A, Dufresne Y, Michalik J et al (2016) Norine, the knowledgebase dedicated to non-ribosomal peptides, is now open to crowd-sourcing. Nucl Acids Res 44:D1113–D1118
3. Flissi A, Ricart E, Campart C, Chevalier M et al (2020) Norine: update of the nonribosomal peptide resource. Nucl Acids Res 48:D465–D469
4. Fielding RT, Taylor RN (2002) Principled design of the modern web architecture. ACM Trans Internet Technol 2:115–150
5. Dufresne Y, Noé L, Leclère V et al (2015) Smiles2Monomers: a link between chemical and biological structures for polymers. J Cheminform 7:62
6. Ricart E, Leclère V, Flissi A et al (2019) rBAN: retro-biosynthetic analysis of nonribosomal peptides. J Cheminform 11:13

Index

Michael Burkart and Fumihiro Ishikawa (eds.), *Non-Ribosomal Peptide Biosynthesis and Engineering: Methods and Protocols*, Methods in Molecular Biology, vol. 2670, https://doi.org/10.1007/978-1-0716-3214-7,

F

G

H

J

L

M

N

O

P

R

S